Gold Usage

Gold Usage

W. S. Rapson

and (in part)

T. Groenewald

International Gold Corporation Ltd.
Johannesburg, South Africa.

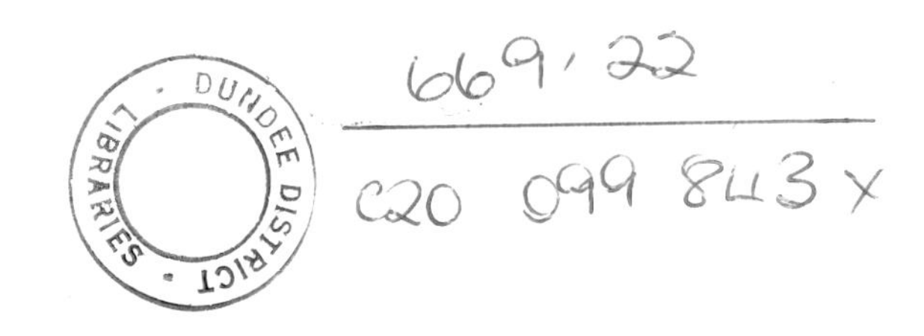

1978

ACADEMIC PRESS

London New York San Francisco

A Subsidiary of Harcourt Brace Jovanovich, Publishers

ACADEMIC PRESS INC. (LONDON) LTD
24/28 Oval Road,
London NW1

United States Edition published by
ACADEMIC PRESS INC.
111 Fifth Avenue
New York, New York 10003

Library of Congress Catalog Card Number: 77–80286
ISBN: 0–12–581250–7

Typeset in Photon Times by Kelmscott Press Ltd.
30 New Bridge Street, London E.C.4
Printed by Whitstable Litho Ltd, Whitstable, Kent

Authors' Preface

The number of papers, appearing in the scientific and technical literature, which have a bearing upon existing or potential applications of gold has been increasing steadily during the past two decades. Moreover, the researches which these papers describe embrace a very wide range of scientific and engineering disciplines. It is not a straightforward matter, therefore, for the individual engineer, scientist or craftsman involved, for example, in the fabrication of jewellery, to keep up to date with and take advantage of the advances which are being made in regard to the use of gold in other branches of industry.

In discussing gold usage, therefore, an attempt has been made to review the scientific and technical developments upon which all the various uses of gold are based. It is hoped that this will not only assist the specialist investigator or user of gold to acquaint himself with developments in areas other than his own, but that it will also provide for those who are concerned with the supply and demand for gold a broad picture of the science and technology upon which industrial consumption of gold is based.

It must be confessed that this task has strained the resources of the authors. They have been greatly assisted, however, by the appearance in the "Gold Bulletin" over the past few years of a number of reviews of various applications of gold, and by the willingness of so many friends and colleagues to assist them. Particular mention in this connection must be made of Dr. J. Chaston, formerly of Johnson, Matthey & Company, Limited, London; Mr. P. Gainsbury, Director of Research, Worshipful Company of Goldsmiths, London; Professor L. Graf of the Max-Planck-Institut für Metallforschung, Stuttgart; Dr. L. B. Hunt, Editor of the Gold Bulletin and Mr. F. H. Lancaster, Chief Information Officer of the Research Organisation of the Chamber of Mines of South Africa.

To Professor Dr.phil. Ernst Raub, former Director of the Forschungsingstitut für Edelmetalle und Metallchemie in Schwäbisch–Gmünd we are particularly indebted for the foreword which he has contributed. It is esteemed a great honour that this book should be introduced by this doyen of precious metal metallurgists who has himself contributed so extensively to the scientific and technical literature on gold.

W. S. RAPSON
T. GROENEWALD

October, 1977

Foreword

"Gold, however, and laughter, these he takes from the heart of the earth, for as you know full well, the heart of the earth is of gold." So writes the poet and philosopher Friedrich Nietzsche*.

Since the dawn of time, surely nothing has cast such a spell over man, or had such a lasting effect on his history as has gold.

For this, and for the high value placed upon it by man, the special and enduring physical and chemical properties of gold, as well as its rarity, have been responsible. Indeed gold is so firmly enthroned as King of the metals that the mixture of hydrochloric and nitric acids by which it is dissolved is known as *aqua regia* or royal water.

It is therefore not to be wondered at that gold has come to serve as a standard of value, to convey impressions of wealth and strength and to represent the essence of all that is good and beautiful. It is because of these facts that today, as in prehistoric times, gold finds its major application in the manufacture of such symbols of high living standards as jewellery and other precious articles or utensils.

Gold is also used in gold coins and although the number of countries in which such coins are used as legal tender may now be limited, gold coins are still minted throughout the world as collectors' pieces, and gold medals are awarded for outstanding or record performances in many spheres of life. Unworked gold bars are also purchased as fixed-value investments.

Gold is very resistant against chemical attack and because of this it is used in medicine, and especially in dentistry. The fact that it possesses also a number of distinctive physical properties has made it an indispensable material in modern technology.

As a general rule, gold is not used in the pure state, but in the form of its alloys. By use of appropriate alloying components its mechanical properties may be improved, and its colour modified almost at will. Both these attributes are important in the making of jewellery. Moreover, gold is more economically used in alloyed form. It must be borne in mind, however, that the dilution of gold is so great in 8 carat alloys that they can strictly speaking no longer be

* Friedrich Nietzsche's Werke, Vol. VII, Also sprach Zarathustra, Part 2, p. 195, C. G. Naumann Verlag, Leipzig, 1906.

considered as alloys of gold, but rather as gold-containing alloys whose properties are not to be compared with those of gold itself.

In the more functional applications of gold in industry, the use of gold alloys tends to be restricted, since it is essential that the special properties of gold which are being exploited are not altered significantly. This is critical since gold is used in industry only for applications of high functional importance.

The high value of gold dictates that it be used always in the most economical manner. Understandably, therefore, the use of gold in thin layers instead of in massive form evolved in early times, its excellent malleability being exploited in the beating of gold to produce gold leaf. More recently the need to produce for industrial purposes the very thinnest of gold layers, which nevertheless still retain all the properties of gold, has posed numerous problems for modern science and technology.

Although the history of gold and its usage is as old as the history of mankind, much new knowledge of alloying and fabrication processes has been acquired during the last few decades, which has proved to be of basic importance for both old and new applications of gold. Moreover, just as through the medium of their attempts to produce gold artificially the alchemists of earlier centuries often made valuable discoveries, so through studies of gold and its alloys the scientist of today is able to solve many problems of general and fundamental importance.

The aim of the present book by Rapson and Groenewald is to present accumulated knowledge of gold to the reader, and to facilitate its application by him. After careful statistics on production and consumption, the authors give a detailed account of such knowledge as it relates to the various uses of gold. In so doing Rapson and Groenewald do not set out to give a general description of all the voluminous scientific research which has been carried out on gold. They restrict themselves purposefully to research which is of interest to the user of gold, regardless of whether he is involved in modern industry or in one of the more traditional crafts in which gold is used. With admirable care all literature important for the usage of gold has been considered. Moreover, the scientific fundamentals are described in a manner which makes them understandable to the untrained reader also.

This book therefore fulfils a demand often made by our society today, namely, that the results of scientific work be presented to the engineer and the production man in a form which is understandable to him and which enables him to decide quickly on its usefulness and applicability in his own sphere of operations, thus speeding up the practical application of scientific findings.

The profound knowledge and experience of the authors and their analysis of over one thousand references provide a detailed insight into all that is worth knowing about the usage of gold.

It must be mentioned explicitly that this book does not provide stimuli for the engineer alone. It raises also many questions for the scientist which call for further study. Because of this it should certainly stimulate and accelerate both ongoing and future research on gold.

Practically everybody is interested in gold. The book by Rapson and Groenewald, which gives a comprehensive presentation of gold usage on the basis of present-day knowledge, will be found the world over. One can but thank the authors for their considerable effort and at the same time congratulate them on a task well done.

ERNST RAUB

Contents

Foreword by Dr. E. Raub v
Authors' Preface vii

Chapter 1
Demand, Supply and Distribution Patterns

1. Overall Patterns of Supply and Demand 1
2. Demands for Gold for Fabrication Purposes 9
   2.1. Carat gold jewellery and the arts 9
   2.2. Electronics 13
   2.3 Other industrial and decorative uses 17
   2.4. Dentistry 18
   2.5. Medals, medallions and fake coins 19
   2.6. Official coins 20
3. Demands for Hoarding and Investment 21
4. Distribution Patterns 24
   References 24

Chapter 2
Gold for Investment and Hoarding

1. Gold coins 25
2. Gold Bars and Gold in Other Forms 28
   References 29

Chapter 3
Carat Golds

1. Introduction 30
2. The Compositions of Commercially Available Carat Golds . 30
3. Yellow, Green and Red Carat Golds: The Au-Ag-Cu System . 33
   3.1. Metallurgical considerations 33
   3.2. 22, 20 and 18 carat Au-Ag-Cu alloys 37
   3.3. 16 and 14 carat Au-Ag-Cu alloys 39
   3.4. 10, 9 and 8 carat Au-Ag-Cu alloys 40
4. White Carat Golds 41
   4.1. Background 41

4.2. Nickel white golds 43
4.3. Noble metal white golds 47
4.4. White golds of other types 49
5. Violet Gold 49
6. Special Aspects of the Use of Carat Golds 50
6.1. Hardness and wearing properties: age hardening of gold alloys 50
6.2. Tarnish resistance, corrosion and stress corrosion cracking of gold alloys 55
6.3. Staining of the skin and of clothes by gold jewellery . . 64
6.4. Pickling and "colouring" or surface enrichment of carat golds 65
6.5. Electropolishing and electroetching 66
6.6. Chemical polishing and chemical etching 71
6.7. Investment casting of gold alloys 71
6.8. Continuous casting and hot working 74
6.9. Contamination of carat golds 76
6.10. Colour standardisation and matching 77
7. Carat Gold Solders 80
7.1. Caratage, colour and other requirements 80
7.2. Some observations on current practice 80
7.3. Compositions and melting ranges of some typical carat gold solders 81
7.4. Metallurgical aspects 83
References 85

Chapter 4
Dental Golds

1. Introduction 95
2. Gold Foil, Mat Gold and Gold Powder 95
3. Casting Alloys Based upon the Au-Ag-Cu System 96
3.1. Specifications 96
3.2. Metallurgical aspects 97
4. Casting Alloys for Coating with Enamel Veneers 99
5. Wrought Gold Alloys for Plates and Wires 104
6. Dental Gold Solders 104
References 106

Chapter 5
Gold and Gold Alloys in Instrumental Applications

1. Introduction 111
2. Gold Alloy Resistors for Standard Resistances and Potentiometer Wire 113

2.1. Gold-palladium alloys 113
2.2. Gold-iron and gold-palladium-iron alloys 115
2.3. Gold-chromium alloys 116
2.4. Gold-cobalt alloys 116
2.5. Gold-palladium-molybdenum and Gold-palladium-vanadium alloys 116
3. Gold Films in Resistance Thermometry 117
4. Gold Alloys for Low Temperature Resistance Thermometry . 118
4.1. Gold-manganese alloys 118
5. Gold Alloys for Thermocouples 118
5.1. Gold-palladium alloys (for higher temperatures) 118
5.2. Gold-cobalt and Gold-iron alloys (for low temperatures) . 119
6. Gold and Gold Alloys in Conductor Applications 120
6.1. Gold and gold alloys for electrical contacts 120
6.2. Gold wire and gold coatings in electronics circuitry . . 122
6.3. Gold coatings as current collectors in solar cells . . . 123
6.4. Gold in microwave tubes 123
6.5. Gold in vacuum tubes 123
6.6. Gold in semi-conductor devices 124
6.7. Super conductivity in gold alloys 125
7. Gold and Gold Alloys in Miscellaneous Instruments 125
7.1. Gold in an instrument for the detection of mercury vapour 125
7.2. Gold-chromium alloys for high pressure gauges 126
7.3. A gold-silver-copper alloy in moving coil meter suspensions 126
7.4. Gold as the basis of a high pressure calibration scale . 126
7.5. Gold containing membranes in ion-sensitive electrodes . 128
7.6. Gold coating of quartz crystals used for stabilising of oscillatory circuits 128
7.7. Gold coatings in a high sensitivity strain gauge 128
7.8. Gold grids in electronic humidistats 129
7.9. Magnetic alloys of gold in switching and other applications 129
7.10. Measurement of surface temperatures 130
References 130

Chapter 6

Gold and Gold Alloys in Engineering Applications

1. Industrial Gold Solders and Brazes 136
1.1. Introduction 136
1.2. Gold-copper alloys 136
1.3. Gold-silver and gold-silver-copper alloys 137
1.4. Gold-nickel alloys 140

1.5. Gold-nickel-copper, gold-nickel-chromium, gold-nickel-molybdenum and gold-nickel-tantalum alloys 140
1.6. Gold-palladium and gold-palladium-nickel alloys . . 141
1.7. Other high melting systems 142
1.8. Gold-silicon and gold-germanium alloys 142
1.9. Gold-tin-germanium alloys 145
1.10. Gold-tin alloys 145
1.11. Gold-indium and gold-indium-bismuth alloys . . . 145
1.12. Gold-antimony and gold-arsenic alloys 145
1.13. Gold-tantalum and gold-niobium alloys 146
2. High Strength, Corrosion Resistant Gold-Rhodium-Platinum Alloys for Spinnerets and Other Applications 147
3. Applications Based upon the Optical Properties of Gold and its Alloys 150
3.1. Introduction 150
3.3. Absorption, reflection and transmission of radiation by gold and its alloys 150
3.3. Gold and gold alloys for solar energy collectors and concentrators 156
3.4. Gold and gold alloys for heat reflectors in heating installations 161
3.5. Gold coated glasses for windows in buildings and transport vehicles 161
3.6. Electrically heated transparent gold coatings on glass . . 164
3.7. Gold and gold alloy coatings for protection from radiant energy 165
3.8. Gold blacks and their special properties 165
4. Miscellaneous Engineering Applications 166
4.1. Gold-palladium alloys in spark plug electrodes 166
4.2. Gold and gold plated condenser surfaces for dropwise condensation of steam 166
4.3. Gold-palladium alloys as thermal fuses for electric furnaces 167
4.4. Gold bursting discs and gold containers 167
4.5. Gold coatings as diffusion barriers 168
4.6. Gold in fuel cell construction 168
4.7. Gold coatings to assist heat dissipation 168
4.8. Gold plating in submarine telephone cable repeaters . . 169
4.9. Organic gold compounds as possible lubricating oil additives 169
4.10. Gold as a solid film lubricant 170
References 170

Chapter 7

The Bonding of Gold and Gold Alloys to One Another, to Other Metals and to Glass and Ceramics

1. Introduction 178

2. Solid Phase Bonding of Gold to Gold and to Other Metals and Alloys 178
2.1. Background 178
2.2. Diffusion bonding of gold 179
2.3. Cold bonding of gold 181
2.4. Hot bonding and ultrasonic bonding 182
2.5. Friction bonding and explosive bonding 183
3. Liquid Phase Bonding of Gold to Gold and to Other Metals and Alloys 184
3.1. Bonding with melting of one or both the joint components . 184
3.2. Bonding of gold with soft solders 185
4. Bonding of Gold to Glass 188
4.1. Areas of Application 188
4.2. Mechanism of bonding 188
4.3. Gold films deposited on glass by vacuum processes . . 190
5. Bonding of Gold to Ceramic Materials 191
5.1. Bonding mechanisms and techniques 191
References 192

Chapter 8

Electroplated Golds

1. Introduction 196
2. Chemical Considerations: Gold Complexes and Their Stabilities 197
3. Thermodynamic Aspects of the Deposition of Gold from Gold Complexes 201
4. Electrochemical Aspects of the Plating of Gold and Gold Alloys 204
4.1. Current efficiencies and the codeposition of hydrogen . . 204
4.2. Current density-potential relationships and limiting current densities 206
4.3. Current density-potential relationships in gold alloy deposition 208
4.4. The effects of current (or potential) modulation . . . 212
5. Kinetics and Mechanisms of Gold Plating 213
6. Electrocrystallisation Processes 214
6.1. The basic model 214
6.2. Parameters in gold plating which increase surface irregularities 215
6.3. Parameters in gold plating which decrease surface irregularities: macrothrowing power, microthrowing power, levelling and brightening 217
6.4 Modes of growth of pure gold deposits 219
6.5. The structure of gold alloy deposits 221
7. Electroplating of Pure Gold from Cyanide Baths 222
7.1. Introduction and historical development 222

7.2. Plating of pure gold from $Au(CN)_2^-$ in the presence of excess cyanide: alkaline baths 224
7.3. Plating of pure gold from $Au(CN)_2^-$ with no deliberate excess of cyanide: neutral and acid baths 225
8. Electroplating of Gold Alloys for Decorative Purposes From Cyanide Baths 230
8.1. Introduction 230
8.2. Plating processes 230
8.3. Pink, rose and red gold coatings 231
8.4. Green gold coatings 233
8.5. Yellow gold coatings 234
8.6. White gold coatings 235
9. Electroplating of Gold Alloys for Industrial Purposes from Cyanide Baths 236
9.1. Introduction 236
9.2. Electrodeposition of gold alloys containing cobalt and nickel from alkaline and neutral cyanide baths 237
9.3. Electrodeposition of bright gold alloys containing cobalt and nickel from acid cyanide baths 238
9.4. Electrodeposition of gold alloys containing iron . . . 240
9.5. Electrodeposition of gold alloys containing copper . . 241
9.6. Electrodeposition of gold alloys containing silver . . 241
9.7. Electrodeposition of gold alloys containing zinc or cadmium 242
9.8. Electrodeposition of gold alloys containing gallium, indium or thallium 243
9.9. Electrodeposition of gold alloys containing tin or lead . 244
9.10. Electrodeposition of gold alloys containing antimony or bismuth 245
9.11. Electrodeposition of gold alloys containing platinum metals 246
9.12. Electroplating of gold alloys containing manganese, chromium, molybdenum, tungsten or uranium 247
10. Electroplating of Gold and Gold Alloys From Other Types of Baths 248
10.1. The sulphite, amine-sulphite and other "mixed" sulphite complexes of gold 248
10.2. Plating of gold and gold alloys from sulphite and amine-sulphite baths 250
10.3. Electrodeposition from gold (III) halide complexes . . 252
10.4. Electrodeposition from gold(I) thiomalate compexes . . 252
10.5. Electrodeposition from gold(I) thiosulphate complexes . 252
11. Electroforming of Gold and Gold Alloys 254
References 255

Chapter 9

Gold and Gold Alloy Coatings of Other Types

1. Liquid Golds 271
2. Mechanically Clad Golds 273
   2.1. Introduction 273
   2.2. Manufacture 273
   2.3. Properties 273
3. Gold Coatings Formed by Thick Film Processes 276
   3.1. The nature of thick film pastes 276
   3.2. Gold conductors generated from thick film pastes . . 278
   3.3. Gold alloy conductors generated from thick film pastes . 278
   3.4. Gold epoxy conductors 279
4. Vacuum Coated Gold and Gold Alloys: Thin Film Processes . 279
   4.1.Evaporated and sputtered golds 279
   4.2. Sputtered gold alloys 281
5. Electroless Gold Coatings 281
   5.1. Introduction 281
   5.2. Systems for "electroless" plating of gold 283
   5.3. The use of gold as an activator in the plating of plastics and other non-conductors 284
   5.4. Electroless aerosol gold plating systems 285
6. Mercury Gilded Gold Coatings 286

References 286

Chapter 10

Gold and Gold Alloys in Miscellaneous Forms and Uses

1. Gold and Gold Alloy Wire 292
2. Gold Leaf 294
3. Colloidal Golds 295
4. Gold in Glass 297
5. Gold Powders 298
6. Gold Composites 300
7. Gold Skeletal Foams and "Expanded" Golds 302
8. Gold Alloys With Shape Memory 302
9. Gold in Photography 304
   9.1 Gold toning, image stabilisation and latensification . . 304
   9.2 Emulsion sensitisation by gold 305
   9.3 Photosensitivity of gold compounds 306
10. Radioactive Gold 306
    10.1. Introduction 306
    10.2. Production and properties of $^{198}_{79}Au$, $^{199}_{79}Au$ and of $^{195}_{79}Au$. 307

10.3. Therapeutic and diagnostic applications of $^{198}_{79}Au$ . . 307
10.4. Other applications of $^{198}_{79}Au$, $^{199}_{79}Au$ and of $^{195}_{79}Au$ . . . 309
11. The Use of Gold in Medicine 309
12. Gold as a Catalyst 310
12.1. Heterogeneous catalysis by gold 310
12.2. Heterogeneous catalysis by gold alloys 312
12.3. Homogeneous catalysis by gold 313
References 314
Author Index 323
Subject Index 343

1 Demand, Supply and Distribution Patterns

1. Overall Patterns of Supply and Demand

Gold from a variety of sources is offered for sale on the gold markets of the Western world. The greatest proportion is normally newly-mined gold, produced either as the principal product of gold mining companies or as a by-product of the electrolytic refining of other newly-mined metals, particulary copper. Significant quantities are also produced in the refining of gold-bearing scrap. Production from these three sources in the Western world can be assessed and even forecast with fair accuracy. Apart from gold from these sources, however, there is also gold bullion which is sold from official monetary holdings or from private holdings which have been built up for investment or hoarding purposes, as well as gold which may be offered for sale on Western gold markets from the communist bloc of countries. Sales of gold from these three sources have varied very considerably since 1964, and cannot easily be assessed or forecast.

Purchases of gold reveal a similar disconcerting dichotomy. On the one hand, there is gold which is purchased for fabrication purposes. This includes gold which is required for jewellery manufacture, for electronic and other industrial applications, for dentistry, for official gold coins and for medals, medallions and fake coins. While purchases for these purposes are susceptible to changes in the business cycle, and to changes in the price of gold like purchases of other commodities, they can be estimated with some measure of confidence. On the other hand, purchases of gold for official or monetary purposes or for private investment or hoarding are subject to unpredictable variations, which arise in considerable measure from political, economic or financial changes, and the changes in rates of inflation and interest which accompany them.

While studies of supply and demand in respect of gold in individual countries such as the U.S.A.[1] are available, the most comprehensive source of information in regard to gold supply and demand on a world basis is a series of annual

Table I[2]. Gold Bullion: Supply and demand, 1948–1975 (Metric tons)

	Free world mine production	Net trade with communist bloc	Total supplies	Official purchases or sales	Net private purchases
1948	702	—	702	369	333
49	733	—	733	369	337
1950	755	—	755	288	467
51	733	—	733	235	498
52	755	—	755	205	550
53	755	67	822	404	418
54	795	67	862	595	267
55	835	67	902	591	311
56	871	133	1004	435	569
57	906	231	1137	614	523
58	933	196	1129	605	524
59	1000	266	1266	671	595
1960	1049	177	1266	262	964
61	1080	266	1346	538	808
62	1155	178	1333	329	1004
63	1204	489	1693	729	964
64	1249	400	1649	631	1018
65	1280	355	1635	196	1439
66	1285	–67	1218	–40	1258
67	1250	–5	1245	–1404	2649
68	1245	–29	1216	–620	1836
69	1252	–15	1237	90	1147
1970	1237	–3	1270	236	1034
71	1236	54	1290	–96	1386
72	1182	213	1395	151	1244
73	1119	275	1394	–6	1400
74	1014	220	1234	–20	1254
75	951	149	1100	–25	1125

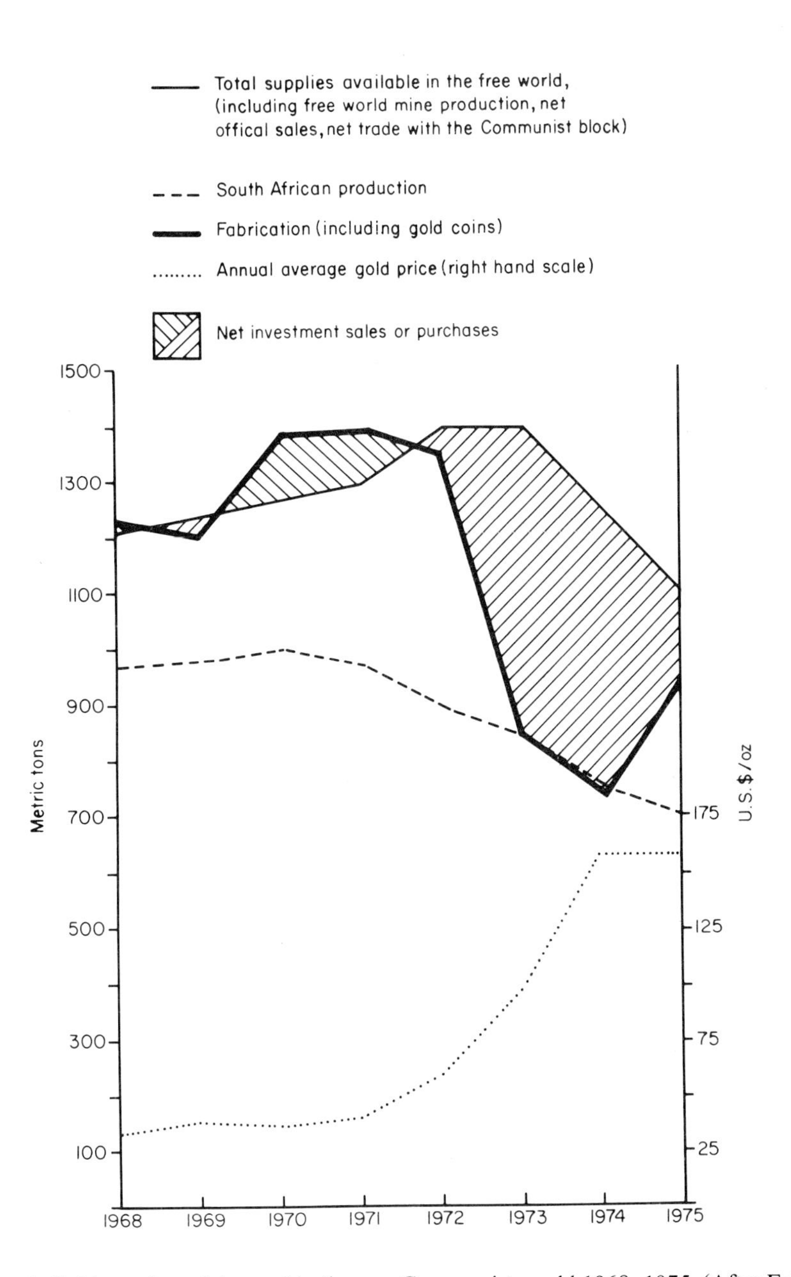

FIG. 1. Gold supply and demand in the non-Communist world 1968–1975. (After Fells and Glynn.[2])

reports from Consolidated Gold Fields Limited, which has appeared over the period 1969 – 1976. What follows is based upon these reports.

The supply and demand position is the non-communist (free) world over the years 1948 – 1975 is reflected in Table I which was taken from the latest of the above-mentioned reports.[2] It illustrates the fall in the level of official purchases of gold since 1964, and the manner in which this fall has since been compensated for, partly by private purchases and partly by very variable, but reduced, levels of net sales of gold from the communist bloc. It also illustrates how substantial official sales of gold such as occurred in 1967 and 1968, and like those of 25 million ounces (780 metric tons) of gold which are currently being made by the International Monetary Fund, can increase the amounts of gold which are available for private purchase. It seems almost certain, however, that the effects of such sales will be offset in some measure at least in the future as a result of action taken by central banking institutions. Some such institutions have emerged recently as buyers of gold, which they are using to augment their holdings which are being used as the need arises as collateral in the negotiation of loans. The activities of these institutions are currently tending to limit the amount of gold available for fabrication and private investment purposes and to set floor prices for gold.

This picture is complemented by the data in Figure 1. It gives an overall view of the world gold supply and demand position for the years 1968 – 1975, following the establishment of the two-tier market for gold in 1968, and the rapid increase in the free market as opposed to the monetary price of gold which followed. It shows not only the drop in sales of gold for fabrication needs which was to be expected and which occurred over this period, but also the manner in which available gold is excess of these needs was taken up by a substantial growth in net gold purchases for investment. It also reveals a rapid reaction in 1975 to a modest lowering in the average price of gold, resulting in a reversal of the downward trend in sales for fabrication purposes, and a decrease in sales for investment as the price of gold tended to level off, and as the prospects of escalating inflation rates diminished.

The changes in the demands for gold for different purposes over the period 1970 – 1975 are shown in Table II. This quantifies not only the very large falls in the demands for gold for carat jewellery, electronics and other industrial and decorative uses, and for medals, medallions and fake coins, but also the spectacular increase in the use of gold for fabrication of official gold coins which occurred during this period. These matters are analysed in greater detail later.

As regards production, South Africa and the U.S.S.R. (see Table III) are the major sources of supply of new gold. In South Africa, production is from the gold-bearing reefs of the Witwatersrand series (see Figure 2). These are exploited over a 400 km arc extending from Evander in the Eastern Transvaal

Table II[2]. Fabrication, speculation and investment (Metric tons)

	1970	1971	1972	1973	1974	1975
Purchases for fabrication by developed countries	768	842	1016	746	716	684
Purchased for fabrication by developing countries	611	549	332	111	19	263
TOTAL FABRICATION	1379	1391	1348	857	735	947
of which						
Carat jewellery	1063	1059	995	510	231	532
Electronics	92	88	108	128	95	65
Dentistry	63	69	71	73	61	65
Other industrial and decorative uses	62	69	70	71	59	48
Medals, medallions and fake coins	54	52	40	21	5	16
Official coins	46	54	63	54	285	221
Net private bullion purchases (sales)*	(345)	(5)	(104)	543	519	178
NET PRIVATE PURCHASES	1034	1386	1244	1400	1254	1125

*Excluding coins

to Virginia in the Orange Free State. Informaton relating to this production is very fully reported.[3] An interesting feature of South African production is that it normally decreases with increase in the price of gold. This is because on many mines an increase in the price of gold enables ores of lower average grade to be mined. Where hoisting and milling capacities are fully taken up, a decrease in the production of gold results, such as that which is portrayed in Figure 1.

In recent years the centre of gravity of South African production has shifted away from the West, Central and East Rand areas, and the bulk of current production is now from the Orange Free State, Klerksdorp and Far West Rand gold mining areas.

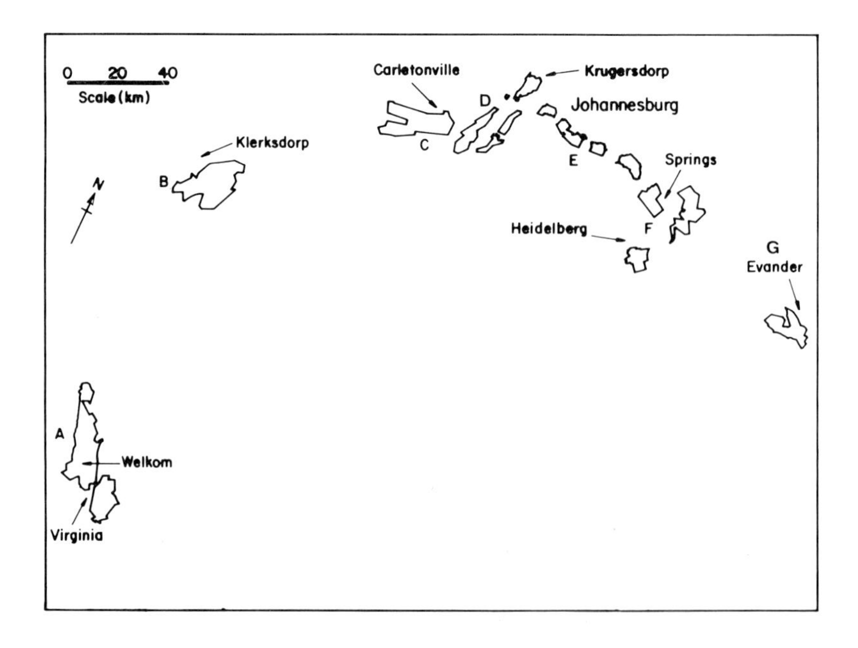

FIG. 2. Principal gold mining areas in South Africa and their gold production in metric tons in 1975.

A.	Orange Free State area (10 mines)	233.0
B.	Klerksdorp area (4 mines)	137.9
C.	Far West Witwatersrand area (9 mines)	228.8
D.	West Witwatersrand area (2 mines)	18.1
E.	Central Witwatersrand area (5 mines)	22.5
F.	East Witwatersrand area (6 mines)	20.4
G.	Evander area (4 mines)	30.1
	Total (40 mines)	690.8

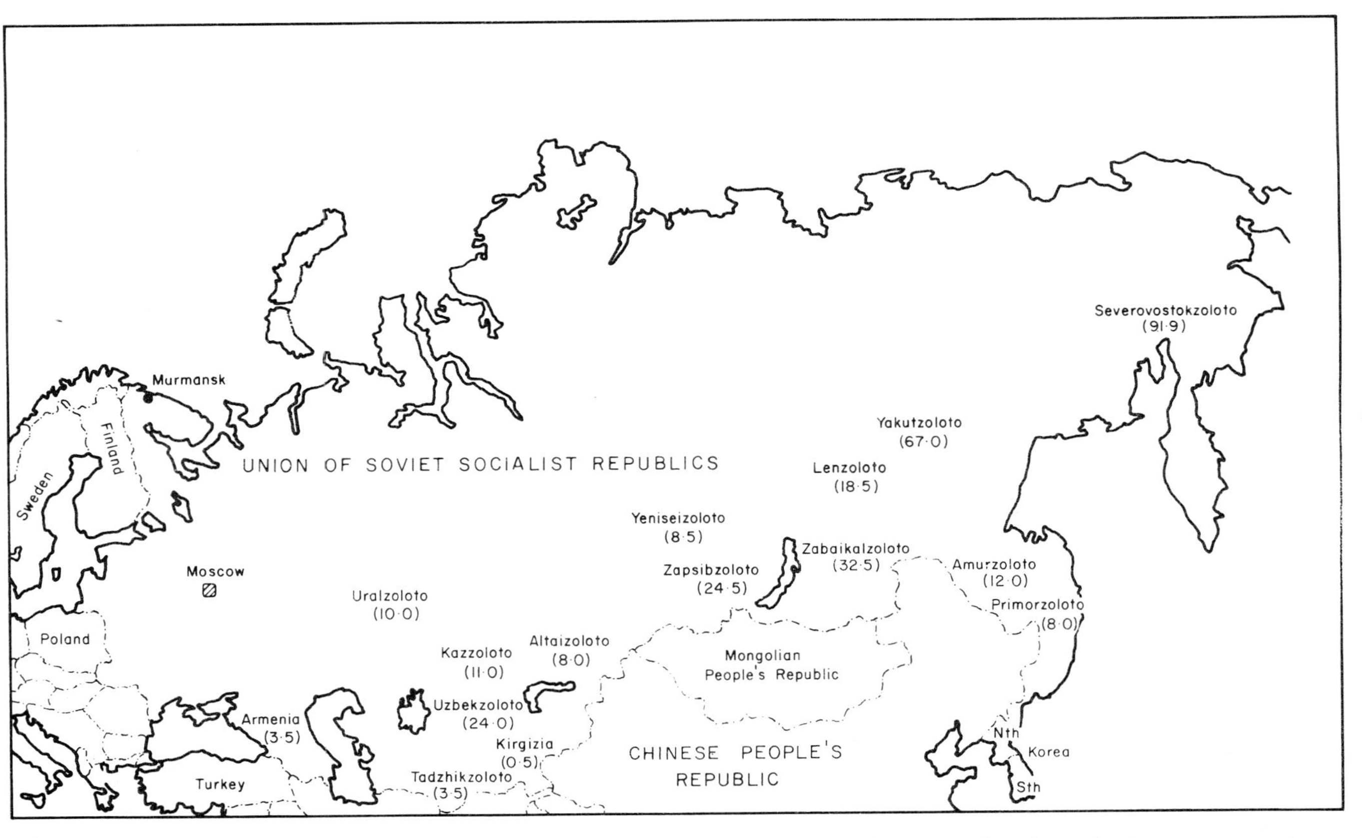

FIG. 3. Principal gold mining areas in the U.S.S.R. The figures in brackets after the name of each production centre are the production figures for that area in metric tons in 1975. (After Fells and Glynn.[2])

In the U.S.S.R. gold is produced from a number of centres and different geological formations, and information relating to Russian gold production is less easily accessible. Figure 3 and Table IV, however, which are reproduced from a report by Dowie,[4] provide a picture of the position in 1975. Whereas

Table III[2]. Gold Production 1970–1975 (Metric tons)

	1970	1971	1972	1973	1974	1975
SOUTH AFRICA	1000.0	976.6	908.7	852.3	758.5	708.1
CANADA	74.9	68.7	64.7	60.0	52.2	50.4
U.S.A.	54.2	46.4	45.1	36.2	35.1	32.4
OTHER AFRICA						
Ghana	21.9	21.7	22.5	25.0	21.4	21.4
Rhodesia	15.0	15.0	15.6	15.6	18.6	18.6
Zaire	5.5	5.4	2.5	2.5	4.4	3.6
Other	2.0	2.5	1.7	1.7	1.0	1.0
TOTAL OTHER AFRICA	44.4	44.6	42.3	44.8	45.4	44.6
LATIN AMERICA						
Brazil	9.0	9.0	9.5	11.0	13.8	12.5
Colombia	6.8	5.9	6.3	6.7	8.2	10.8
Mexico	6.2	4.7	4.6	4.2	3.9	4.7
Peru	3.2	3.0	2.6	2.6	4.0	4,5
Dominican Republic	—	—	—	—	—	3.0
Nicaragua	3.6	3.3	2.8	2.8	2.4	1.9
Other	6.6	8.2	9.0	7.9	5.9	6.0
TOTAL LATIN AMERICA	35.4	34.1	34.8	35.2	38.2	43.4
ASIA						
Philippines	18.7	19.7	18.9	18.1	17.3	16.1
Japan	8.4	7.7	9.6	10.4	5.5	4.7
India	3.2	3.7	3.3	3.3	3.2	3.0
Other	2.8	2.1	2.7	2.7	2.7	2.7
TOTAL ASIA	33.1	33.2	33.5	34.5	28.7	26.5
EUROPE	7.4	7.6	13.2	14.3	11.6	11.0
OCEANIA						
Papua/New Guinea	0.7	0.7	12.7	20.3	20.5	17.9
Australia	19.5	20.9	23.5	17.2	16.2	14.0
Other	3.6	3.1	3.2	3.2	3.2	3.2
TOTAL OCEANIA	23.8	24.7	39.4	40.7	39.9	35.1
FREE WORLD TOTAL	1273.2	1235.9	1181.8	1118.8	1009.6	951.5
U.S.S.R.	346.7	359.8	378.9	398.2	420.7	407.0
OTHER COMMUNIST COUNTRIES	18.4	18.4	18.4	19.4	20.4	20.0
TOTAL WORLD PRODUCTION	1638.3	1614.1	1579.1	1536.4	1450.7	1378.5

gold production in South Africa is unlikely to increase significantly in the future, the 10th Five-year Plan (1976 – 1980) for production in the U.S.S.R. indicates that annual production from this source, which has increased from 347 to 407 tons over the period 1970 – 1975, may increase possibly to as much as 500 tons per annum over the new five-year period. As mentioned above, however, the proportion of this gold which will reach free world markets cannot be foreseen.

2. Demand for Gold for Fabrication Purposes

An overall view of the position over the period 1970–1975 is provided in Table V

2.1 Carat Jewellery and the arts

Gold and its alloys are used in carat jewellery and the arts because of their beauty, their resistance to tarnish, their relative ease of working, and their intrinsic value. This does not mean that economic and technical considerations are not highly significant in dictating the manner, form and caratage in which gold jewellery are fabricated. Nevertheless at consumer level traditional social customs and fashion trends have strong influences on the market for gold jewellery.

Typical of fashion trends was the swing away from the use of gold in jewellery in favour of silver and "white" alloys which coincided with the rapid increase in the price of gold in 1972 and 1973. This fashion trend accentuated the reduction in purchases of gold for jewellery fabrication which was to be expected during this period. A return to favour of the use of gold in jewellery during 1975 is currently assisting sales, which are being further increased as a result of the levelling off and fall in the price of gold which has taken place in 1975 – 1976.

Table IV[2] Estimate of Soviet Gold Production 1975

Administrative unit	Metric tons	Administrative unit	Metric tons
Severovostok	91.9	Primorsky	8.0
Yakut	67.0	Altai	8.0
Zabaikal	32.5	Armenia	3.5
Zapsib	24.5	Tadzhik	3.5
Lena	18.5	Kirgiz	0.5
Uzbek	24.0	Other primary	1.5
Kazakh	11.0	Private producers	35.0
Ural	10.0		
Amur	12.0	By product	48.0
Yenisei	8.5		
		Total	407.9

Table V[2]. Fabrication of Gold for Sale in Non-communist Countries (Metric tons)

	1970	1971	1972	1973	1974	1975
EUROPE						
Italy	181.5	197.5	330.2*	111.9	59.4	79.8
Germany	82.1	79.1	84.4	80.0	60.0	56.5
Turkey	44.7	33.0	21.0	6.5	18.1	55.0
U.K. and Ireland	26.1	23.9	27.5	30.2	37.8	46.4
Spain	57.6	59.6	62.6	48.0	24.3	42.0
France	47.5	53.9	52.4	46.7	36.6	34.2
Greece	12.3	13.0	14.5	14.4	7.8	17.5
Switzerland	28.8	28.9	28.4	27.3	22.8	16.6
Austria	30.4	32.4	29.7	10.1	78.9	13.4
Yugoslavia	—	—	—	0.3	8.9	9.7
Belgium	7.0	7.8	8.6	8.3	7.8	6.7
Netherlands	5.3	6.0	6.5	5.4	4.3	5.0
Sweden	3.4	3.9	3.8	3.3	3.4	3.4
Portugal	11.1	14.0	12.4	9.0	4.5	3.0
Denmark	3.9	3.4	4.4	2.9	1.2	1.2
Finland	2.2	1.9	2.0	1.8	1.2	1.1
Norway	1.5	1.5	1.2	0.7	0.6	0.6
Cyprus and Malta	0.8	0.8	1.4	0.8	0.4	0.4
Hungary	—	—	—	—	17.0	—
Total EUROPE	546.2	560.8	691.0	407.6	395.0	392.5
N. AMERICA						
U.S.A.	185.4	215.3	226.6	209.3	150.2	118.8
Canada	7.2	8.2	11.2	13.0	11.5	11.5
Total N. AMERICA	192.7	223.5	237.8	222.3	161.7	130.3

LATIN AMERICA						
Mexico	25.5	22.2	28.1	22.2	72.9	19.2
Brazil	32.7	34.0	23.2	14.1	12.7	15.9
Chile	2.7	1.5	0.5	—	—	5.0
Colombia	3.0	3.1	1.7	3.4	0.8	1.2
Venezuela	5.9	5.8	5.0	0.8	–0.9	0.6
Peru	4.4	3.9	3.2	1.1	0.6	0.4
Argentina	17.2	12.5	0.5	—	–2.0	–10.0
Central America and Caribbean	3.5	4.3	2.2	—		
Other Latin America	2.2	2.1	2.0	1.0	–0.5	0.5
Total LATIN AMERICA	97.1	89.4	66.4	42.6	83.6	32.8
MIDDLE EAST						
Iran and Afghanistan	14.0	16.0	12.5	8.0	8.0	28.0
Arabian Gulf States & Kuwait	6.0	10.1	4.6	–12.0	–8.5	14.0
Saudi Arabia and Yemen	5.0	3.5	2.0	–2.0	–7.0	12.0
Iraq, Syria and Jordan	7.5	7.0	3.5	–18.0	–23.0	10.0
Israel	5.0	5.0	5.0	5.5	5.5	5.5
Lebanon	13.0	13.0	6.0	–0.5	2.0	2.0
Total MIDDLE EAST	50.5	54.6	33.6	–19.0	–23.0	71.5
INDIAN SUB/CONTINENT						
India	217.0	179.0	110.3	63.6	15.7	26.2
Pakistan	30.0	25.0	15.0	4.0	–0.5	5.5
Ceylon	4.0	6.0	3.0	2.0	–1.0	–0.5
Bangladesh and Nepal	—	—	–3.0	–3.0	–6.5	–3.5
Total INDIAN SUB-CONTINENT	251.0	210.0	125.3	66.6	7.7	27.7

Table V (contd)

	1970	1921	1972	1973	1974	1975
FAR EAST						
Indonesia	30.0	25.0	10.0	2.5	–15.0	15.0
Taiwan	4.0	4.5	2.0	5.0	—	6.0
Singapore	5.0	5.0	4.5	5.0	0.8	5.4
Thailand	12.0	8.0	–10.0	–24.0	–15.0	5.0
Malaysia	6.5	6.0	5.0	2.0	0.3	4.0
South Korea	6.0	6.0	4.4	3.7	1.5	4.0
China	—	—	—	0.5	1.0	1.0
Philippines	5.0	6.0	6.0	2.0	–3.0	0.8
Hong Kong	10.0	11.0	13.0	10.0	–1.0	0.5
Burma, Laos and Cambodia	7.0	6.0	–1.0	–3.0	–14.0	–2.0
South Vietnam	10.0	8.0	3.5	–6.0	–5.0	–5.0
Total EXCLUDING JAPAN	95.5	85.5	37.4	–2.3	–49.4	34.7
JAPAN	74.0	80.6	96.0	100.9	74.8	75.2
Total FAR EAST	169.5	166.1	133.4	98.6	25.4	109.9
AFRICA						
South Africa	10.4	21.3	25.3	33.3	102.7	151.0
Morocco/Algeria	26.0	30.0	12.0	1.5	—	10.0
Egypt	7.1	7.7	4.1	–6.0	–23.0	6.0
Libya	9.5	6.0	4.5	2.5	1.7	3.0
Tunisia	2.0	2.0	1.5	—	—	1.0
Other Africa	9.0	9.0	5.0	1.0	–2.0	6.0
Total AFRICA	64.0	76.0	52.4	32.3	79.4	177.0
AUSTRALIA	7.9	10.1	8.3	6.0	5.2	5.4
Total FREE WORLD	1378.9	1390.5	1348.2	857.0	735.0	947.1

These effects are illustrated in Figure 4, and the geographical distribution in the free world of the use of gold in jewellery is given in Table VI.

The figures in Table VI and Figure 4 are net figures and taken into account gold recovered from old jewellery.

2.2 Electronics

In industry, economic considerations dictate firstly, that gold be used only where its special properties cannot be matched by those of cheaper metals or

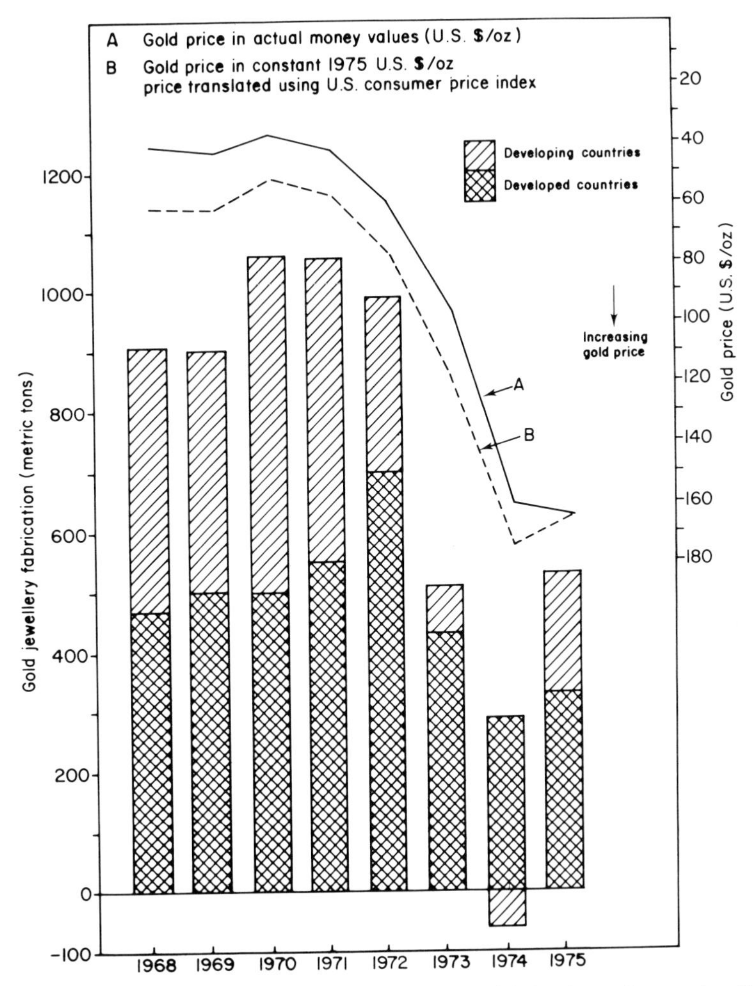

FIG. 4. Quantities of gold consumed annually in the fabrication of carat jewellery during 1968–1975 in relation to the variations in the price of gold over this period. (After Fells and Glynn.[2])

Table VI[2]. Fabrication of Gold in Carat Jewellery (net) (Metric tons)

	1970	1921	1972	1973	1974	1975
EUROPE						
Italy	160.5	174.3	313.0*	98.2	50.0	71.0
Turkey	32.0	23.0	13.1	3.0	12.0	41.4
Spain	53.0	55.0	58.0	45.2	21.5	38.7
Germany	47.6	47.1	45.4	38.0	25.0	27.0
France	25.7	30.8	29.4	22.4	16.1	19.0
U.K. and Ireland	14.7	15.5	18.0	20.3	15.8	18.3
Greece	11.0	12.0	13.5	13.5	7.0	16.5
Switzerland	23.0	23.0	22.0	16.0	11.6	8.2
Belgium	4.2	4.8	5.5	6.5	5.5	6.0
Yugoslavia	—	—	—	0.3	4.8	6.0
Portugal	11.0	13.9	12.3	8.9	4.4	2.9
Austria	6.1	6.8	6.6	4.1	2.5	2.5
Netherlands	2.9	3.4	3.4	2.4	1.4	2.1
Sweden	2.0	2.7	2.7	2.3	1.8	1.8
Finland	2.2	1.9	2.0	1.8	1.2	1.1
Denmark	3.6	3.1	4.1	2.6	0.9	0.9
Norway	1.3	1.3	1.0	0.5	0.5	0.5
Cyprus and Malta	0.8	0.8	0.6	0.6	0.2	0.2
Total EUROPE	401.6	419.4	550.6	286.6	182.2	264.1
N. AMERICA						
U.S.A.	97.8	125.8	127.3	102.0	72.7	64.2
Canada	5.9	6.7	9.0	9.7	9.4	9.2
Total NORTH AMERICA	103.7	132.5	136.3	111.7	82.1	73.4

LATIN AMERICA						
Brazil	29.0	30.0	19.0	12.0	10.0	13.0
Mexico	21.0	20.0	15.5	7.8	5.0	6.1
Colombia	2.8	2.9	1.7	1.9	0.8	1.2
Peru	4.2	3.7	3.0	1.0	0.5	0.3
Venezuela	5.0	4.9	4.2	0.3	−1.0	0.3
Chile	2.0	1.3	0.5	—	—	—
Argentina	16.5	12.0	0.5	—	−2.0	−10.0
Other Latin America	2.2	2.1	2.0	—		
Central America and Caribbean	3.5	4.3	2.2	1.0	−0.5	0.5
Total LATIN AMERICA	86.2	81.2	48.6	24.0	12.8	11.4
MIDDLE EAST						
Iran and Afghanistan	13.0	14.0	11.0	7.0	5.0	25.0
Saudi Arabia and Yemen	2.5	2.5	2.0	−2.0	−1.0	10.0
Iraq, Syria and Jordan	7.5	7.0	3.5	−18.0	−23.0	10.0
Arabian Gulf States, Kuwait	5.0	9.1	4.1	−12.0	−9.0	7.0
Israel	5.0	5.0	4.0	3.0	2.9	2.9
Lebanon	3.0	4.0	3.0	−2.0	1.0	2.0
Total MIDDLE EAST	36.0	41.6	27.6	−24.0	−24.1	56.9
INDIAN SUB-CONTINENT						
India	215.0	175.0	107.2	60.5	14.0	25.0
Pakistan	30.0	25.0	15.0	4.0	−0.5	5.5
Ceylon	4.0	6.0	3.0	2.0	−1.0	−0.5
Bangladesh and Nepal	—	—	−3.0	−3.0	−6.5	−3.5
Total INDIAN SUB-CONTINENT	249.0	206.0	122.2	63.5	6.0	26.5

Table VI (contd)

	1970	1921	1972	1973	1974	1975
FAR EAST						
Indonesia	30.0	25.0	10.0	2.5	−15.0	15.0
Taiwan	4.0	4.5	2.0	5.0	—	6.0
Thailand	12.0	8.0	−10.0	−24.0	−15.0	5.0
Malaysia	6.5	6.0	5.0	2.0	0.3	4.0
Singapore	5.0	5.0	4.2	4.2	0.5	3.0
South Korea	6.0	6.0	4.0	3.0	—	1.5
China	—	—	—	0.5	1.0	1.0
Philippines	5.0	6.0	6.0	2.0	−3.0	0.8
Hong Kong	10.0	11.0	13.0	10.0	−1.0	0.5
Burma, Laos and Cambodia	7.0	6.0	−1.0	−3.0	−14.0	−2.0
South Vietnam	10.0	8.0	3.5	−6.0	−5.0	−5.0
Total EXCLUDING JAPAN	95.5	85.5	36.7	−3.8	−51.2	29.8
JAPAN	34.5	34.5	40.0	45.0	39.9	38.6
Total FAR EAST	130.0	120.0	76.7	41.2	−11.3	68.4
AFRICA						
Morocco/Algeria	22.0	26.0	11.0	1.5	—	10.0
Egypt	6.5	7.0	3.5	−6.0	−23.0	6.0
Other Africa	9.0	9.0	5.0	1.0	−2.0	6.0
Libya	6.5	4.0	2.0	2.5	1.7	3.0
South Africa	3.2	3.5	3.2	3.0	2.8	1.4
Tunisia	2.0	2.0	1.5	—	—	1.0
Total AFRICA	49.2	51.5	26.2	2.0	−20.5	27.4
AUSTRALIA	6.9	7.1	7.1	4.7	4.1	4.2
Total FREE WORLD	1062.6	1059.3	995.3	509.7	231.3	532.3

* Includes inventories built up before introduction of V.A.T.

materials and, secondly, that it is used with the utmost economy. In industrial applications, therefore, the metal is used mostly in the form of fine wires or of thin coatings or films, and where feasible it is alloyed with a cheaper metal.

These effects are particularly evident in the use of gold in the fabrication of electronic equipment where the impact of rising gold prices (and a business recession) (see Table VII) has been to cause a rapid fall in the use of gold since 1973. Increased gold prices stimulated not only the use, where possible, of alternative materials, but also efforts to reduce the use of the metal to a minimum in situations where substitution of cheaper metals was not feasible. It seems probable that such economies in the use of gold are now reaching their limit and that in the future consumption of gold by the electronics industry will follow variations in its output more closely.

Table VII[2]. Gold use in Electronics (Metric tons)

Country	1970	1971	1972	1973	1974	1975
United States	44.8	42.8	51.2	61.6	44.0	20.4
Japan	18.1	20.1	22.5	26.0	15.0	17.1
W. Germany	10.0	8.0	11.0	15.0	12.0	8.0
France	4.5	4.0	4.5	5.0	5.5	4.5
United Kingdom	5.8	4.0	5.0	5.5	4.4	3.6
Switzerland	0.9	1.0	1.5	3.0	3.0	2.5
Italy	1.5	1.6	5.0	5.0	2.0	1.9
Yugoslavia	—	—	—	—	2.0	1.7
Netherlands	1.1	1.3	1.4	1.6	1.5	1.0
Canada	0.5	0.7	0.8	1.1	0.9	1.0
Korea	—	—	0.2	0.2	0.5	1.0
Spain	0.7	0.8	0.8	0.6	0.6	0.7
Australia	0.4	0.6	0.5	0.6	0.5	0.5
Austria	1.0	0.8	0.8	0.5	0.5	0.4
Brazil	1.5	1.8	1.5	0.3	0.3	0.4
Singapore	—	—	0.3	0.8	0.3	0.3
India	—	—	0.1	0.1	0.2	0.2
Belgium	0.7	0.8	0.9	1.0	0.9	—
Total	91.5	88.3	108.0	127.9	94.1	65.2

As might be expected, the demand for gold for the fabrication of electronic equipment is confined largely to developed countries, with the United States, Japan and West Germany being the largest consumers.

2.3 Other industrial and decorative uses

Included in this category are gold salts for electroplating, gold alloys used for the production of rolled gold or gold fill, as well as gold used in a variety of instrumental, engineering, medical and other applications.

Table VIII indicates that consumption of gold for end uses in this category has also fallen substantially since 1973, an effect which must be related partly to reduced production during the business recession and partly to efforts made to minimise the use of gold and to replace it by other metals as its price rose.

Table VIII[2]. Other Industrial and Decorative Uses (Metric tons)

Country	1970	1971	1972	1973	1974	1975
United States	22.5	23.4	24.8	24.6	17.0	13.0
Germany	10.0	10.0	11.0	12.0	12.0	8.0
Japan	5.8	8.0	9.7	7.4	5.4	6.1
France	9.6	10.6	10.0	11.7	8.6	5.2
Mexico	0.5	0.5	0.5	0.6	0.7	2.2
Switzerland	2.1	2.1	2.2	4.0	4.0	2.1
Italy	2.5	2.3	1.2	1.2	1.9	1.9
United Kingdom	3.0	3.1	3.2	4.0	2.3	1.8
Korea	—	—	0.2	0.5	1.0	1.5
Brazil	1.0	1.2	1.2	0.8	1.4	1.4
India	2.0	4.0	3.0	3.0	1.5	1.0
Yugoslavia	—	—	—	—	1.0	1.0
Spain	1.2	1.2	1.2	0.5	0.5	0.7
Canada	0.1	0.1	0.1	0.6	0.7	0.6
Belgium	0.5	0.5	0.5	0.3	0.3	0.3
Netherlands	0.4	0.4	0.4	0.3	0.2	0.2
Denmark	0.2	0.2	0.2	0.2	0.2	0.2
Sweden	—	0.1	0.1	0.1	0.1	0.1
Australia	0.1	0.1	0.1	0.1	0.1	0.1
Turkey	—	—	—	0.1	0.1	0.1
South Africa	—	—	0.2	0.1	0.1	0.1
Egypt	0.4	0.5	0.5	—	—	—
Argentina	0.2	0.2	0.2	—	—	—
Total	62.1	68.5	70.3	71.1	59.1	47.6

2.4 Dentistry

In dentistry, the use of gold and gold alloys is perhaps less affected by economic considerations and more by trends in dental practice, the traditional attitudes of both practitioners and patients, and the availability of alternative materials. Economic considerations appear, however, to favour the use of alternative materials, where dental services are available within the framework of state health insurance schemes. Despite the spread of such schemes, the overall use of gold in dentistry has remained steady during recent years and been relatively unaffected by variations in the gold price.

Data on the use of gold in dentistry are listed in Table IX.

Table IX[2]. Gold use in Dentistry (Metric tons)

Country	1970	1971	1972	1973	1974	1975
United States	20.4	23.3	23.3	21.1	16.5	18.5
Germany	8.5	9.0	11.0	12.0	10.0	12.0
Japan	9.6	10.0	10.3	12.5	10.0	10.0
France	6.0	6.5	7.0	7.0	6.0	5.0
Switzerland	1.3	1.3	1.4	4.0	4.0	3.8
Italy	4.0	4.5	5.0	5.5	4.0	3.5
Sweden	1.2	0.9	0.8	0.7	1.3	1.3
Israel	—	—	—	1.0	1.3	1.3
Spain	1.2	1.2	1.3	0.8	0.8	1.2
Brazil	1.0	1.0	1.0	1.0	1.0	1.1
Greece	1.3	1.0	1.0	0.9	0.8	1.0
Mexico	1.3	1.7	2.1	1.8	0.5	1.0
Yugoslavia	—	—	—	—	1.1	1.0
Austria	0.9	1.1	1.2	0.9	0.8	0.9
Netherlands	0.9	0.9	1.1	0.9	0.8	0.8
United Kingdom	1.0	0.8	0.8	0.7	0.6	0.8
Australia	0.5	0.6	0.6	0.6	0.5	0.5
Belgium	1.2	1.3	1.3	0.3	0.4	0.4
Canada	0.7	0.6	0.5	0.5	0.3	0.3
Denmark	0.1	0.1	0.1	0.1	0.1	0.1
Norway	0.2	0.2	0.2	0.2	0.1	0.1
Venezuela	0.4	0.4	0.3	0.2	0.1	0.1
Peru	0.2	0.2	0.2	0.1	0.1	0.1
South Africa	—	—	0.2	0.1	0.1	0.1
Portugal	0.1	0.1	0.1	0.1	0.1	0.1
Turkey	—	1.0	0.5	—	—	—
Egypt	0.1	0.1	0.1	—	—	—
Argentina	0.5	0.3	—	—	—	—
Chile	0.2	0.2	—	—	—	—
Columbia	0.2	0.2	—	—	—	—
Total	63.0	68.5	71.4	73.0	61.3	65.0

2.5 Medals, medallions and fake coins

The practice of fabricating fascimiles of well known and accepted official gold coins as a method of overcoming restrictions on the private ownership of gold in many parts of the world has diminished rapidly in the past few years, firstly as a result of the removal of restrictions on gold ownership in some countries and, secondly, as a result of increased production of official gold coins.

The changes in the demand for gold in this whole area are reflected in Table X.

Table X[2]. Gold in Medals, Medallions and Fake Coins (Metric tons)

Country	1970	1971	1972	1973	1974	1975
Kuwait	1.0	1.0	0.5	—	0.5	7.0
Japan	6.0	8.0	13.5	10.0	4.5	3.4
Saudi Arabia	2.5	1.0	—	—	– 6.0	2.0
Germany	6.0	5.0	6.0	3.0	1.0	1.5
Spain	1.5	1.4	1.3	0.9	0.9	0.7
Italy	13.0	15.0	6.0	2.0	0.5	0.5
Netherlands	—	—	0.1	0.2	0.4	0.3
Venezuela	0.5	0.5	0.5	0.3	—	0.2
Sweden	0.2	0.2	0.2	0.2	0.2	0.2
France	1.2	1.5	1.0	0.6	0.4	0.1
Canada	—	0.1	0.8	0.1	0.1	0.1
Austria	—	0.4	0.4	0.2	0.1	0.1
Lebanon	10.0	9.0	3.0	1.5	1.0	—
Belgium	0.4	0.4	0.4	0.2	0.7	—
Switzerland	1.0	1.3	1.2	0.3	0.2	—
Israel	—	—	1.0	1.5	—	—
Libya	3.0	2.0	2.5	—	—	—
Morocco	4.0	4.0	1.0	—	—	—
Iran	—	1.0	1.0	—	—	—
Egypt	0.1	0.1	—	—	—	—
Mexico	2.7	—	—	—	—	—
Chile	0.5	—	—	—	—	—
Brazil	0.2	—	—	—	—	—
Total	53.8	51.9	40.4	21.0	4.5	16.1

2.6 Official coins

There have been issues of official gold coins by an increasing number of countries in the past few years. The fabrication of these has usually been contracted out to established mints, and the quantities of gold employed for the purpose by such mints in different countries are listed in Table XI. The figures emphasise the extent to which the demand for official gold coins has increased in recent years. By far the biggest sales have been of the Kruger rand which contains 1 oz (31.10 g) of fine gold and has an advantage over small gold bars in that it can be bought and sold without assay. A recent (1975) similar introduction is the Russian Chervonetz which contains a $\frac{1}{4}$ oz (7.78 g) of fine gold.

Gold coins often sell at prices above the value of their gold content, partly because of their appeal to collectors and partly also because of their usefulness as portable and indestructible units of gold. The levels of the premiums at which coins are marketed vary, and they may be such as to limit popularity of the coins as an investment, particularly in countries where it is permissible to own gold bars which can be purchased at prices which include no premium over the value of the gold they contain. It is where there are restrictions upon the private ownership of gold, such as in the United Kingdom, that gold coins can most easily command a significant premium over the value of their contained gold, because they are the only legal method of investing in gold.

Table XI[2]. Gold in Official Coins (Metric tons)

Country	1970	1971	1972	1973	1974	1975
South Africa	7.2	17.8	21.7	30.1	99.7	149.4
United Kingdom	1.6	0.5	0.5	0.7	14.7	21.9
Turkey	12.7	9.0	7.4	3.4	6.0	13.5
Mexico	—	—	10.0	12.0	66.7	9.9
Austria	22.4	23.3	20.7	4.4	75.0	9.5
Chile	—	—	—	—	—	5.0
Iran	1.0	1.0	0.5	1.0	3.0	3.0
U.S.A.	—	—	—	—	—	2.7
Singapore	—	—	—	—	—	2.1
Israel	—	—	—	—	1.3	1.3
Italy	—	—	—	—	1.0	1.0
Netherlands	—	—	0.1	—	—	0.6
France	0.5	0.5	0.5	—	—	0.4
Canada	—	—	—	1.0	0.1	0.3
Malta	—	—	0.8	0.2	0.2	0.2
Australia	—	1.7	—	—	—	0.1
Hungary	—	—	—	—	17.0	—
Brazil	—	—	0.5	—	—	—
Switzerland	0.5	0.2	0.1	—	—	—
Columbia	—	—	—	1.5	—	—
Total	45.9	54.0	62.8	54.3	284.7	220.9

3. Demands for Hoarding and Investment

The identification of gold purchases as being for hoarding or investment is fraught with considerable difficulty, particularly where gold is purchased on one or other of the main markets. Moreover, a large proportion of the gold fabricated into official gold coins must be regarded as satisfying essentially an investment demand for the metal. Nevertheless Fells and Glynn[2] have recorded data (Table XII) relating to gold bars identified as purchased for hoarding during 1970–1975. These data do not take into account gold which may have been sold in bar form in major gold markets for investment or hoarding.

The overall figures for investment gold purchases are almost certainly best assessed as in Table I and Figure 1 as the difference between total gold becoming available for sale and sales identified as being for fabrication purposes.

Table XII[2]. Identified Hoarding of Bars (Metric tons)

Country	1970	1971	1972	1973	1974	1975
Asia						
Indonesia	45.0	17.0	– 15.0	14.0	– 13.0	24.0
Taiwan	—	17.0	2.0	10.0	—	14.0
South Korea	—	7.0	—	2.0	—	6.0
Singapore	—	—	—	1.0	5.0	2.0
Okinawa	—	3.5	– 0.5	—	—	—
Laos	—	—	—	– 0.5	—	– 5.0
Hong Kong	—	5.0	– 5.5	10.0	– 14.0	– 7.5
India	14.0	13.5	22.0	12.0	—	– 15.0
South Vietnam	28.0	17.0	– 1.0	– 12.0	– 15.0	– 20.0
Total	87.0	80.0	2.0	36.5	– 37.0	– 1.5
Middle East						
Israel	—	—	—	2.7	1.6	2.0
Iran	—	—	—	—	2.0	1.0
Saudi Arabia	—	—	– 4.0	—	—	1.0
Jordan	—	—	– 0.5	—	—	1.0
Kuwait	—	—	– 1.5	– 0.5	1.5	—
Lebanon	—	—	—	—	1.0	—
Iraq	—	—	– 1.0	—	—	—
Syria	—	—	– 5.5	—	—	—
Dubai	—	—	—	7.0	– 5.0	—
Total	—	—	12.5	9.2	1.1	5.0
America						
Brazil	—	—	—	—	2.1	—
Venezuela	—	—	—	—	0.5	—
Central America	—	– 0.3	0.1	—	—	—
Netherlands Antilles	—	—	—	– 0.1	—	—
Peru	0.5	0.3	—	—	—	—
Argentina	—	—	– 1.0	– 0.5	—	– 5.0
Total	0.5	—	– 0.9	– 0.6	2.6	– 5.0
Europe (Greece)	—	—	2.5	4.1	5.0	2.0
Africa (Egypt/Morocco)	—	0.2	0.5	—	—	—
Total Hoarding	87.5	80.2	16.6	49.2	– 28.3	0.5

As indicative of the effects of political or economic restraints in the purchase or sale of gold either in bars or in jewellery form, Fells and Glynn have estimated that with the end of the American involvement in South Vietnam, South Vietnamese refugees brought out a total of at least 20 tons of gold in bar form, in addition to 5 tons in the form of jewellery. Part of this gold apparently

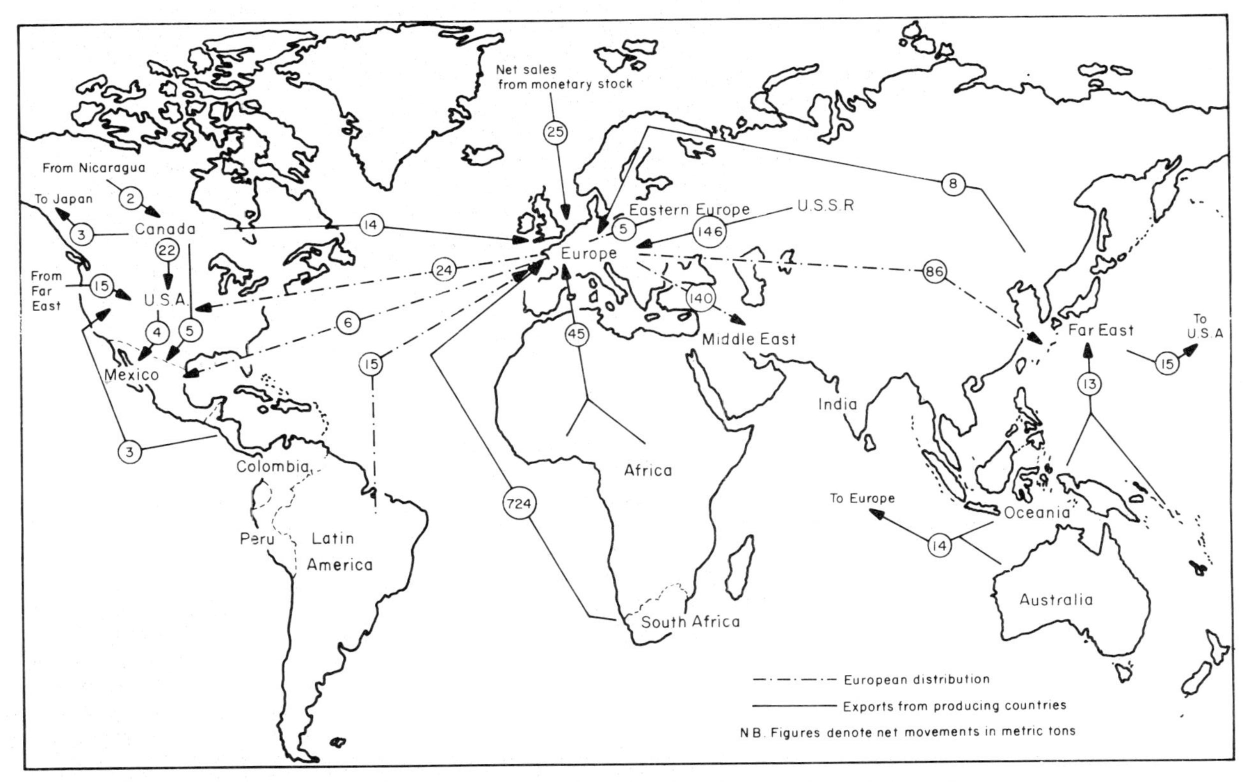

FIG. 5. Major gold bullion flows in 1976. (After Fells and Glynn.[2])

went to Hong Kong, part to Thailand, but the bulk was sold, officially and unofficially, in the United States.

The effects of price trends on investment in gold is clearly seen (Figure 1) in the growth of such investments during the period of rising gold prices from 1972–1974, and the fall off or liquidation of such investments when the price began to fall in mid-1975.

4. Distribution Patterns

Gold produced both in South Africa and in the U.S.S.R. is currently marketed almost exclusively in Europe, while most of that produced in other countries is also shipped to Europe. Europe is therefore the principal centre for the onward shipment of gold to the Far East, India and the Middle East.

Figure 5, reproduced from Fells and Glynn,[2] portrays most effectively the main bullion flows to Europe in 1975, as well as the flows of gold from Europe to supply the Middle East and the Far East over this period. In the Middle East there was a considerable shift of trade from Beirut to Damascus, caused by unrest in Lebanon in 1975. There was also considerable growth of Jeddah as a distribution centre in this area. The distribution centres in the Far East are Hong Kong and Singapore.

Books by Green[5-7] provide information concerning not only the gold markets in Europe, but also the distribution and smuggling of gold in the Middle and Far East.

REFERENCES

1. Michalopoulos, C., "The Gold Market in the United States: 1975". Appendix 1 to Ref. 2 (cf. van Tassel, R. C. and Michalopoulos, C., "The Commercial Demand for Gold in the United States". Clark University, 1973).
2. Fells, P. D. and Glynn, C., "Gold 1976". Consolidated Gold Fields Limited, London, 1976 (cf. Wolfe, T. W., "Report on the World Production, Marketing and Use of Gold". Gold Market Report to the United States Department of the Treasury, 1976).
3. Annual Reports, Chamber of Mines of South Africa, Johannesburg.
4. Dowie, D., "Soviet Gold in 1975: The 9th and 10th Five-year Plans", Appendix 2 to Ref. 2.
5. Green, T., "The World of Gold". Michael Joseph, London, 1968.
6. Green, T., "The World of Gold Today". Arrow Books Limited, London, 1973.
7. Green, T., "How to Buy Gold". Walker and Company, New York, 1975.

2 Gold for Investment and Hoarding

1. Gold Coins

Gold has a long history of use in alloyed form in coinage.[1-6] Friedberg's standard work[1] contains not only a complete catalogue of all known gold coins since 600 B.C., but also a comprehensive bibliography of specialist publications dealing with the coins of single countries or periods.

While gold coins issued prior to 1800, and especially those of ancient times, are no longer easily accessible and therefore of specialist interest only, there are numerous gold coins issued since 1800 which can still be acquired fairly readily. These have been described by Hoppe[7] and listed in a catalogue of the world's most important gold coins of the 19th and 20th centuries by Federal Coin and Currency Inc.[8] This catalogue gives not only the years of minting of various gold coins issued by 69 countries during this period, but also their dimensions, weights, finenesses, and other details concerning them, including the number of each type issued.

Most of these coins were minted during the period of the gold standard which began in 1823 when England became the first country to change to a gold-based currency in place of the silver currency in general use before that date. During this era, which virtually ended with the outbreak of World War I, large numbers of gold coins were issued by countries such as Austro-Hungary, England, France, Germany, Russia and the United States, and it has been estimated[8] that a total of 19,000 tons of gold was used for coinage over this period. England alone used 6,500 tons (over 90 million half sovereigns and over 600 million sovereigns were issued), the United States 5,000 tons, and France 3,500 tons.

There is a certain uniformity in respect of the weights, finenesses and dimensions of many of these coins which has its origin in an agreement, known as the Latin Monetary Union (L.M.U.) which was reached between France, Italy, Belgium and Switzerland in 1862. This agreement established the permissible finenesses, weights, sizes and denominations of the gold (and silver) coins of

Table XIII. Characteristics of Gold Coins as Agreed Upon by the Latin Monetary Union

Denomination*	Fineness in thousandths	Gross weight (g)	Fine weight (g)	Diameter (mm)
Fr. 100 – gold	900	32.25806	29.03225	35
Fr. 50 – gold	900	16.12903	14.51612	28
Fr. 20 – gold	900	6.45161	5.80644	21
Fr. 10 – gold	900	3.22580	2.90322	19
Fr. 5 – gold	900	1.61290	1.45161	17

*The names given to the coins in different countries vary, e.g. in Greece, drachma; in Serbia, dinars; in Spain, pesetas; in Rumania, lei.

these countries. The effects of this agreement were far-reaching, since a large number (over 20) of other countries also began minting gold coins according to the L. M. U. standard. This standard allowed for the minting of gold coins with characteristics as set out in Table XIII.

Within the British sphere of influence (Canada, Australia, India and South Africa) a similar uniformity developed in the characteristics of the £½, £1, £2 and £5 value coins which were issued from 1820 to 1937. Table XIV lists characteristics of these coins, which are all 916.66 fine or 22 carat.

While gold coins issued by other countries may have different gross weights and dimensions, they are today almost all of either 900 or 916.66 fineness. Notable exceptions are the Austrian 1 Ducat and 4 Ducat pieces of 986.6 fineness. No mention is normally made of the nature of the other metals in coinage alloys, but this is almost without exception copper. At finenesses of 900 or 916.66, the alloys of gold with copper are single-phase solid solutions and admirably suited to minting processes. Moreover, the amount of copper is sufficient to enrich the pure gold colour without distorting it.

The procedure followed in the production of Kruger rands is probably typical of that used in the production of other gold coins. In this,[9] electrolytic gold is melted in an induction furnace and cathode copper added in requisite

Table XIV. Characteristics of British Gold Coins 1820–1937

Denomination	Fineness	Gross weight (g)	Fine weight (g)	Diameter (mm)
£½	916.66	3.9940	3.6612	19
£1	916.66	7.9881	7.3224	22
£2	916.66	15.9761	14.6448	27
£5	916.66	39.9403	36.6120	35

amount afterwards, to produce an Au-Cu alloy containing 91.63–91.67% Au. This alloy is poured into vertically-closed moulds to produce billets 12 mm thick and the billets are then rolled to the requisite thickness, being reduced initially by 1.5 mm per pass through the rolls. When the billet is almost to size, it is passed several times at the same setting to ensure maximum uniformity and is then punched to produce coin blanks which are slightly overweight. These are tumbled in soap solution to remove sharp edges and to clean their surfaces. The blanks are then sorted by weight and those still overweight are ground down to the correct weight by allowing them to roll diagonally across a fast moving endless belt of aluminium oxide paper. The finished blanks are then ready for the minting process.

Table XV. Finenesses and Gold Contents of Some Recently Issued and Some More Freely Available Gold Coins[8]

Coin	Fineness	Weight (g)	Gold content (g)
Queen Elizabeth II Sovereign	916.66	7.9881	7.3224
Mexican 50 Peso	900	41.6666	37.4999
South African 2 Rand	916.66	7.9881	7.3224
South African Kruger Rand	916.66	33.9335	31.1035
Russian Chervonetz	900	8.6399	7.7759
Austrian 1 Ducat 1915 (restrike)	986.6	3.4909	3.4448
Austrian 20 Crown 1915 (restrike)	900	6.7751	6.0976
Austrian 4 Ducat 1915 (restrike)	986.6	13.9636	13.7864
Austrian (or Hungarian) 100 Crown 1915 (restrike)	900	33.8753	30.4878
U.S. $20 Double Eagle	900	33.4370	30.0933
U.S. $10 Eagle	900	16.7185	15.0466
U.S. $5 Half Eagle	900	8.3592	7.5233

Of the gold coins issued more recently the most important are the Queen Elizabeth II sovereign, the South African Kruger rand and two-rand pieces, the Mexican 50-peso piece or centenario, restrikes by the Austrian Mint in Vienna on the pattern of the crowns, florins and ducats of the former Austro-Hungarian Empire, and the Russian chervonetz. The United States ceased the issue of gold coins when it went off the gold standard in 1934, but up to that date 75 million half-eagle ($5), 56 million eagle ($10) and 88 million double-eagle ($20) pieces were issued.

The finenesses and gold contents of some of these are listed in Table XV.

The growing popularity of these and other gold coins, and particularly of the Kruger rand both as a medium for investment in gold and as collectors' pieces has already been referred to in Chapter 1. Faking, or the making of facsimiles of official gold coins has until recently been a profitable business, with the producers of such facsimiles taking their profit normally from the premium which the genuine coins command over and above the value of the gold which they contain. The reduced levels of such premiums in the case of recent coin issues, however, as well as increasing liberalisation in regard to the private ownership of gold, in forms on which no premium is paid, has reduced production of fake coins very greatly.

2. Gold Bars and Gold in Other Forms

Gold in two grades of purity is cast into bars with a wide range of weights and dimensions for investment purposes.

The first grade of purity is that which is characteristic of bars conforming to the specifications of the London Gold Market for "good delivery" bars. These must contain a minimum of 99.5% Au (995 parts per 1000 fine gold), conform to certain physical specifications, and be marked by a serial number and the stamp of an "acceptable melter or assayer". If not marked with the fineness and stamp of an acceptable assayer each bar must be accompanied by a certificate issued by an acceptable assayer stating the serial number of the bar and its fineness. Lists of the names of acceptable melters and assayers are issued by the London Gold Market.

"Good delivery" bars are produced, for example, from South African mine bullion by the Miller process.[9] The crude mine bullion contains metals such as iron, zinc, lead, copper and silver which are removed by injection of chlorine into the molten bullion. First the iron, zinc and lead react with the chlorine after which the copper and silver start to form their chlorides which also separate into the slag. Injection of chlorine is continued, however, only until the requisite degree of fineness has been reached, and the normal "good delivery" bars contain silver and copper as the main alloying metals. They are marked with their actual gold assays. Platinum metals which are present in some Witwatersrand ores[10] and crude mine bullions are not removed in the refining process and may therefore be present in the final product. "Good delivery" bars from the Rand Refinery are not routinely assayed for all elements, but the ranges of concentration in which certain elements may be expected to be present are as follows:

Au	99.50 – 99.80%
Ag	0.17 – 0.50%
Cu	150 – 600 ppm
Pd	0 – 200 ppm
Pt	0 – 500 ppm
Zn	< 20 ppm
Fe	< 20 ppm
Pb	below detection limit by method used

The second grade of purity is that which is achieved by electrolytic refining,[9] from which bars containing up to 99.99% Au can be produced according to requirements. Because mainly of the "lock-up" of gold in the process, electrolytic refining is more expensive than refining by the Miller process. In South Africa it is used[9] therefore only to the extent necessary to meet the demands for the higher purity gold which it yields.

REFERENCES

1. Friedberg, R., "Gold Coins of the World". The Coin and Currency Institute, New York, 3rd Ed., 1971.
2. Harris, R. P., "Gold Coins of the Americas". House of Collectibles Inc., Box D, Florence AL 35630.
3. Kenyon, R. L., "Gold Coins of England". Bernard Quaritch, 1884, Firecrest, 1970.
4. Schlumberger, H., "Gold Coins of Europe Since 1800". Sterling Publishing Co., New York, 1968.
5. Taxay, D., "The U.S. Mint and Coinage". Arco Publishing Co. Inc., New York, 1966.
6. Roberts-Austen, W. C., Alloys used for coinage, *J. Soc. Arts*, **32** (1884), 835 ff.
7. Hoppe, D. J., "How to Invest in Gold Coins". Arlington House, New Rochelle, N.Y., 1970.
8. "Gold Coins". Federal Coin and Currency Inc., 25 Broad Street, New York.
9. Adamson, R. J. (Ed.), "Gold Metallurgy in South Africa". Chamber of Mines of South Africa, Johannesburg, 1972.
10. Cousins, C. A., Platinoids in the Witwatersrand system, *J. S.A. Inst. Min. Metal.* **73** (1973), 184–199.

3 Carat Golds

1. Introduction

In jewellery applications the gold content of gold alloys is expressed either in terms of their caratage, which is the weight fraction of gold expressed in 24ths, or alternatively in terms of fineness which is the weight fraction of gold expressed in 1000ths. An 18 carat gold alloy therefore contains 18/24ths or 75% by weight of gold, and might be described as of 750 fineness. Hallmarking of gold jewellery,[1,2] although compulsory in Great Britain, is not compulsory in most countries.

The use of gold in alloyed rather than pure form in jewellery is based on the softness and poor wear characteristics of pure gold, on the beauty and range of colours that can be achieved in gold alloys, and on cost considerations. The most common caratages employed are 22, 18, 14 and 10, 9 or 8. In the lower caratages the tarnish resistance and rich colour tones so characteristic of the high caratage alloys are understandably decreased.

A number of publications is available in which formulations, properties and applications of carat golds in jewellery fabrication are described.[3-10] Fundamental aspects of the alloying behaviour of gold have recently been reviewed by Raynor.[11]

2. The Compositions of Commercially Available Carat Golds

Standards for carat golds have not so far been introduced, and data concerning the compositions of the various carat golds which they market are made available only by certain manufacturers. Those presented in Table XVI, for example, have been derived from the catalogue of one American manufacturer; and those in Table XVII are for a range of alloys which is available from the South African Mint.

Except for the white golds, it will be noted that the carat golds listed are essentially Au-Ag-Cu alloys, which may be modified to a greater or lesser extent by additions of zinc, particularly at caratages less than 18. Although not

Table XVI. The Composition of Carat Gold Alloys Marketed by Manufacturer in the U.S.A.

Caratage	Colour*	Composition (wt. %)				
		Au	Ag	Cu	Zn	Ni
18	y	75.00	15.00	10.00	—	—
18		75.00	13.00	12.00	—	—
18	w	75.00	—	2.23	5.47	17.80
18	g	75.00	22.50	2.50	—	—
18	r	75.00	5.00	20.00	—	—†
13.5		56.25	5.30	31.50	6.95	—†
14		58.33	5.00	30.00	6.63	—†
13.5	y	56.25	8.75	30.65	4.35	—
14		58.33	8.31	29.19	4.17	—
13.5		56.25	4.20	32.80	6.15	—
14		58.33	4.00	31.24	6.43	—
13.5		56.25	9.62	28.45	5.63	—
14		58.33	16.50	24.97	0.20	—
13.5		56.25	22.26	21.18	0.31	—
14		58.33	21.20	20.17	0.30	—
14		58.33	24.78	16.75	0.14	—
13.5		56.25	10.50	31.15	2.10	—
14		58.33	10.00	29.67	2.00	—
13.5		56.25	7.83	27.92	6.95	1.05
14		58.33	7.46	26.58	6.63	1.00
13.5	w	56.25	—	23.13	9.21	11.36
14		58.33	—	22.10	8.77	10.80
13.5		56.25	—	29.00	5.75	9.00
14		58.33	—	28.32	4.80	8.55
13.5		56.25	3.18	26.62	8.08	5.87
13.5		56.25	—	24.63	6.30	12.82
14		58.33	—	23.47	5.99	12.21
13.5	g	56.25	36.70	6.80	0.25	—
14		58.33	32.50	8.97	0.20	—
14		58.33	35.00	6.47	0.20	—
13.5	r	56.25	2.20	41.55	—	—
14		58.33	2.08	39.59	—	—
9.5	y	39.60	12.08	42.28	6.04	—†
10		41.70	11.66	40.81	5.83	—†
9.5		39.60	5.80	45.30	9.30	—
10		41.70	5.50	43.80	9.00	—
9.5		39.60	6.82	49.65	3.93	—
10		41.70	6.60	48.00	3.70	—
9.5	w	39.60	4.40	36.75	11.15	8.10
9.5		39.60	—	34.00	8.70	17.70
10		41.70	—	32.82	8.40	17.08
10		41.66	—	29.15	12.12	15.05‡
9.5	g	39.60	50.65	9.40	0.35	—
10		41.70	48.90	9.05	0.35	—
9.5	r	39.60	3.03	57.37	—	—
10		41.70	2.82	35.48	—	—

* y = yellow, w = white, g = green, r = red.
† 0.02% boron/Cu and 0.03% silicon/Cu used as deoxidiser in casting alloys of these types.
‡ 1.97% Co is also added to this alloy.

Table XVII. Composition of Gold Jewellery Alloys Supplied by the South African Mint

Caratage	Colour	Composition (wt. %)			
		Au	Ag	Cu	Zn
22	reddish	91.66	—	8.34	—
22	yellow	91.66	6.20	2.14	—
22	pale yellow	91.66	8.34	—	—
22	deep yellow	91.66	1.23	7.11	—
18	rich yellow	75.00	9.00	16.00	—
18	yellow	75.00	20.00	5.00	—
9	reddish yellow	37.50	12.48	46.58	3.50
9	yellowish red	37.50	5.50	53.50	3.50

exemplified in Table XVI, cadmium may be added also, particularly in the formulation of green golds, even to the exclusion of copper altogether.

In the white golds as listed, on the other hand, the silver in such alloys is replaced either wholly or in part by nickel and zinc.

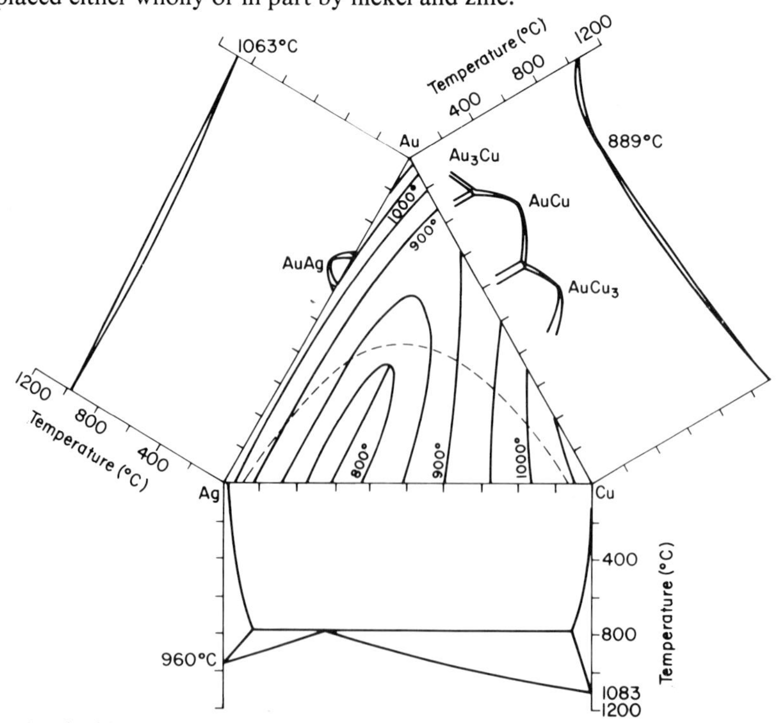

FIG. 6. Liquidus isotherms of the Au-Ag-Cu system in relation to the phase diagrams of the binary Au-Ag, Au-Cu and Ag-Cu edge systems. (Compositions in weight per cent). (After Gmelin[12] and Hansen.[13])

Although binary alloys — for example 18 carat alloys with silver and copper respectively — may be used in jewellery, the alloys in most general use can therefore conveniently be discussed in relation to the ternary Au-Ag-Cu and Au-Cu-Ni systems. Certain white gold alloys which are of other types will be discussed separately.

3. Yellow, Green and Red Carat Golds: The Au-Ag-Cu System

3.1 Metallurgical considerations

The liquidus surface[12] and appropriate binary edges[13] of the Au-Ag-Cu ternary system are illustrated in Figure 6. Isothermal sections[14] definining the two-phased regions are depicted in Figure 7, and vertical sections showing the liquidus, solidus and solid-solid transformations of 18, 14 and 9 carat Au-Ag-Cu alloys are given in Figure 8. These vertical sections are based on limited data from a number of sources,[3,4,12,14-17] and interrupted lines have been used where the data have been found inadequate or ill-defined.

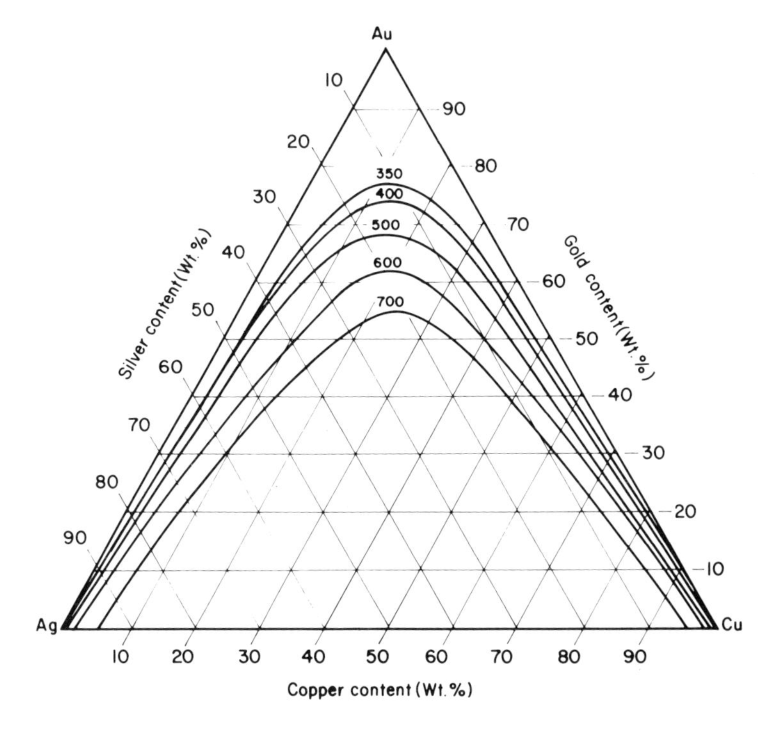

FIG. 7. Isothermal sections defining the two phased regions in the Au-Ag-Cu system at various temperatures. (After McMullin and Norton.[16])

In regard to the binary edge systems, both Au-Ag and Au-Cu alloys form simple solid solutions over their full composition ranges, with little separation between solidus and liquidus temperatures. In the case of Au-Cu alloys, however, the liquidus falls to a congruent minimum at 889°C and 80% Au. In contrast silver and copper form a simple eutectic melting at 779°C, and containing 72% Ag. The liquidus surface of the ternary system therefore shows a shallow minimum starting at the Au-Cu edge at 889°C, which transforms into a eutectic gutter which outcrops on the Ag-Cu binary edge at 779°C.

At lower temperatures neither the Au-Ag or Au-Cu binary alloys form duplex systems, but gold and copper form three ordered intermetallic phases, namely $CuAu_3$, CuAu and $CuAu_3$, and gold and silver form one possible ordered Au-Ag phase. The ternary Au-Ag-Cu system does, however, give rise to duplex systems at lower temperatures over wide composition ranges (Figure 7). The fact that these are generated in rapidly cooled or quenched alloys only on subsequent heat treatment or ageing is very important in practice.

Thus a significant feature to be noted in relation to Figure 8 is that rapid cooling or quenching from above 400°C for the 18 carat alloys, or from 650°C for the 14 carat alloys, yields workable alloys. The subsequent heating of these to 300°C, however, permits partial decomposition of the solid solution into two solid solutions α_1 and α_2, the one containing essentially gold and silver, and the other essentially gold and copper. Age hardening (see later) of these alloys therefore occurs,[18-21] and the extent of the hardening which can be achieved varies with alloy composition at each caratage,[3] with Brinell (BHN) hardnesses of over 200 being realised with some alloys.

Ordering[22] in the essentially Au-Cu phase can contribute materially to this age hardening process,[17,23] and Au-Cu ordering has been studied not only in the Au-Ag-Cu and Au-Cu-Ni systems but also in a number of other ternary systems containing gold and copper. It occurs at order-disorder transition temperatures and rates which are dependent to varying extents on the nature of the third metal in the system.[24-27] In some Au-Ag-Cu alloys it has been found that the onset of ordering causes a change from discontinuous to continuous segregation of phases.[23] Ordering may be accompanied by volume changes which can give rise to buckling or cracking when certain copper-rich annealed jewellery alloys, such as the 75 Au-25 Cu alloy known as French red gold, are re-heated during soldering or other fabricating processes. In such cases the rate of the ordering process and therefore the tendency to buckling or cracking can be controlled by addition of small amounts of other metals such as iron, cobalt or nickel.[27]

Since the solubilities of silver in the copper-rich phase and of copper in the silver-rich phase decrease with temperature some degree of precipitation hardening in these phases is also likely to occur on cooling.

During ageing segregation of more easily corrodable phases can affect the corrosion resistance of alloys of some compositions, whilst annealing can lead to extensive grain growth in passing back to single phase structures. Such grain growth can be inhibited by addition of nickel or cobalt[29] and nickel used in yellow carat golds acts as a grain refiner. It is also inhibited by work hardening (see later) and by the presence of small percentages of certain other metals such as iridium, ruthenium or rhenium.[28] These latter metals were first used as grain refiners in dental gold alloys, under which their application, which is being increasingly extended to other gold alloys, is discussed more fully.

A practical consideration is that zinc may be incorporated in carat gold alloys either in small quantities of 0.5% or less as a deoxidiser, in which case it affects the physical properties of the alloys to a negligible extent, or in larger amounts of up to 10% or more in the cases of copper-rich 10 or 14 carat alloys, where its main function in some cases is to offset the change of colour of the alloys to a reddish hue which is induced by the increase in the Cu:Au ratio, and in other cases is to modify mechanical and other properties. Cadmium may be used together with zinc for this latter purpose. Cadmium has also been incorporated in 18 carat alloys of this type to the exclusion of silver altogether.[29] When used in larger quantities zinc and cadmium not only lower the liquidus and solidus temperatures, but also restrict the range of conditions under which duplex solid phases co-exist, and therefore the possibilities of age hardening. They also tend to suppress ordering reactions in the copper-rich phases, probably because of the space lattice misfits to which their solution in these phases gives rise. Boron and silicon may be added as deoxidisers to certain casting alloys. Silicon additions are said to reduce losses of zinc during melting of such alloys.

3.2 22, 20 and 18 carat Au-Ag-Cu alloys

Of these the 22 and 20 carat alloys are less suited for use in jewellery because of their lower hardness and strength, though 22 carat alloys are traditionally used for wedding rings in Great Britain. They tend to be used only where more massive construction is possible. The 18 carat alloys, however, enjoy a high reputation because of their tarnish resistance and outstanding mechanical properties. The latter are such that the addition of zinc or other metals to modify working properties is not necessary.

The appearance and properties of these 18 carat alloys are therefore determined by the relative proportions of silver and copper which they contain. Age hardening is more pronounced with the redder copper-rich alloys, segregation of solid phases (see Figure 8) being accompanied by Au-Cu ordering. It is scarcely to be observed in 22 carat alloys, however, and becomes significant only with alloys of a caratage of 18 or less. The magnitude of the age hardening effects has been discussed by Wise[3] and is illustrated in Figure 10 using the

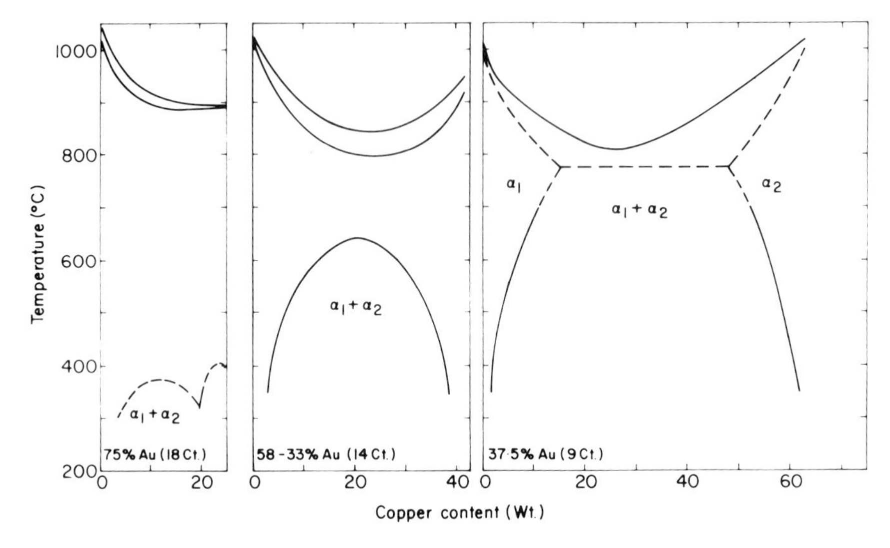

FIG. 8. Liquidus, solidus and solid-solid transformations of 18, 14 and 9 carat Au-Ag-Cu alloys
(a) 18 carat alloys (from references 3, 4, 12, 13 and 14. Data from a trade publication of Handy and Harman, New York, have also been used).
(b) 14 carat alloys (from references 3, 4, 13, 14 and 35).
(c) 9 carat alloys (from references 3, 4, 13 and 14).

data of Sterner-Rainer[4] for the quenched alloys and the data of Spanner and Leuser[18] and Leuser and Wagner[29a] for age hardened alloys.

Work hardening is also pronounced in the 18 carat alloys, and the data in Table XVIII illustrating this effect have been taken from Steiner-Rainer.[4,6]

Because of work-hardening the extent of working of gold alloys as measured by the resulting reduction in cross sectional area must be controlled. The

Table XVIII. Work Hardening Effects in 18 carat Au-Ag-Cu Alloys

Colour of alloy	Cu %	Brinell hardness for various deformations*		
		15%	30%	60%
Deep red	25.0	143	178	202
Orange red	21.4	156	177	205
Reddish yellow	16.7	166	176	197
Bright yellow	12.5	148	160	182
Greenish yellow	8.3	141	149	176
Yellowish green	3.6	114	127	138
Pale yellow green	0.0	69	78	93

*i.e. % reduction in cross section.

percentage reduction which different alloys can tolerate between annealing treatments varies, but is usually less than 70%. On the other hand, annealing treatments should not be carried out at too frequent intervals during working, since if the percentage reduction is decreased to much less than 30%, then as a rule grain growth during annealing is excessive, and weak coarse grained structures develop.[30] These form surface patterns ("orange peel" effect) on further working, and with severe working the grain boundaries which delineate these patterns may deepen with disastrous effects. High carat red alloys in particular must not be annealed after they have had only small deformation.

The green shades that can be achieved in 18 carat Au-Ag-Cu alloys (see Table XVIII) are pale and either yellowish or whitish. More definite green colours can be obtained[31] by the incorporation of cadmium as in the 75Au-11.5Ag-9.5Cu-4Cd quaternary alloy, or by the use of ternary alloys of such compositions as 75Au-16.5Ag-8.5Cd and 75Au-12.5Ag-12.5Cd from which copper is omitted altogether.

Although traditionally acceptable, the hardnesses of ca. 200 which can be achieved with 18 carat golds, are in fact insufficient to make them adequately

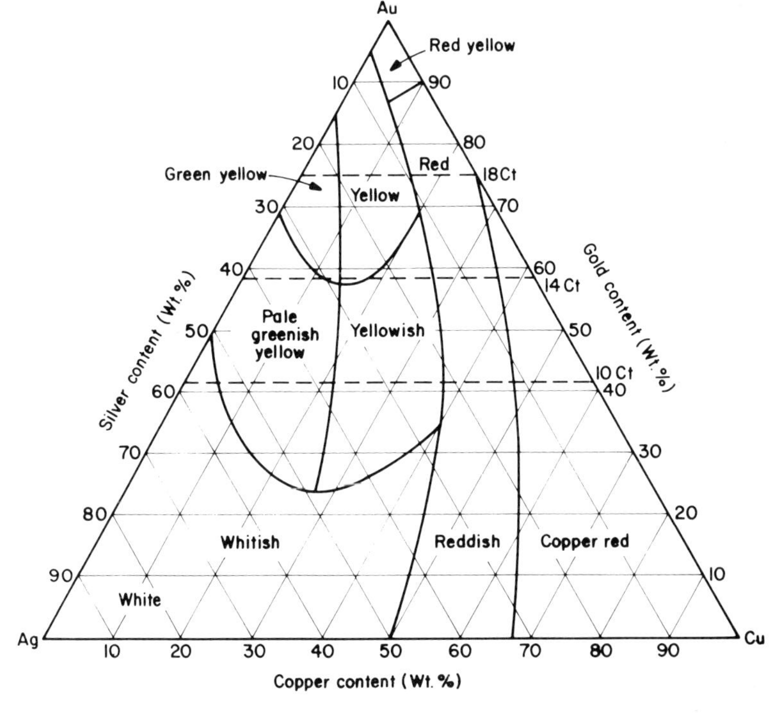

FIG. 9. Weight per cent composition ranges for various colours of Au/Ag-Cu alloys. (After Gmelin.[12])

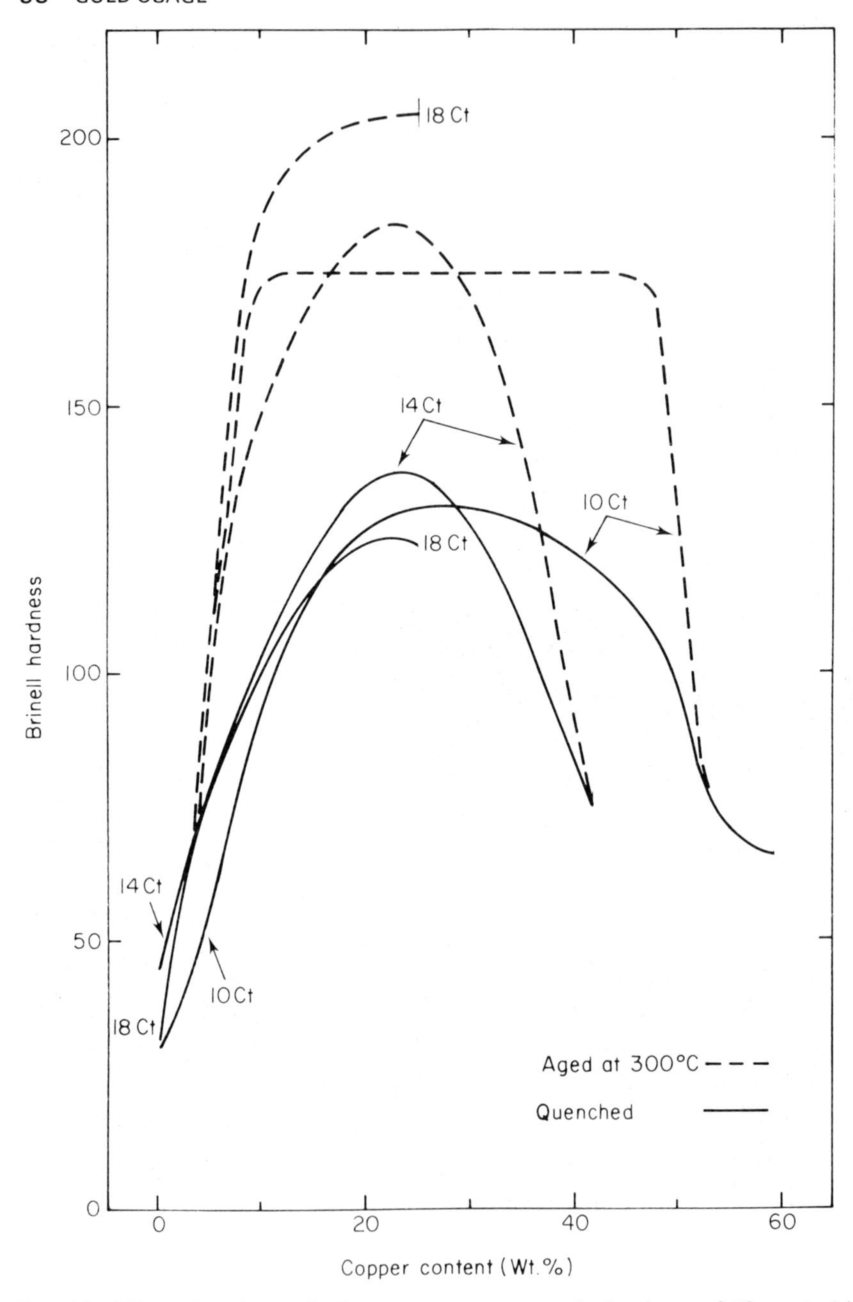

FIG. 10. Effect of ageing and of copper content upon the hardness of 18 carat, 14 carat and 10 carat Au-Ag-Cu alloys.

resistant to wear and abrasion, particularly in watchcase construction. There is therefore interest in the development of high caratage gold alloys which can be age hardened so that they are more wear resistant. Reference will be made to this later.

The incorporation of small quantities of other metals such as indium, palladium and cobalt in 18 carat alloys of this type has been claimed by Andryushchenko *et al.*[32-38] and others as one means of improving them in respect of such properties as hardness, strength, ductility and melting range. The effects of platinum and nickel have also been studied.[18]

3.3 16 and 14 carat Au-Ag-Cu alloys

In comparison with the 14 carat gold alloys, which are widely used in jewellery, 16 carat alloys are little used and they merit mention only in relation to their age hardening characteristics.

As will be noted from Figure 10, 14 carat alloys are susceptible to age hardening and ordering like their higher caratage counterparts. The extent of age hardening which can be achieved, however, passes through a maximum at about 15% Cu for 16 carat alloys[3] and at about 22% Cu for 14 carat alloys. The effects of additions of aluminium, nickel and zinc on such age hardening of 14 carat alloys have been studied by Spanner and Leuser.[18]

In practice age hardening of these and other carat gold alloys has been conveniently carried out[3] by immersing them in oil baths held at 300 – 325°C for periods which depend on the alloy composition, but which may be as short as half an hour.

Work hardening effects, illustrated for 14 carat alloys in Table XIX,[4,6] are also apparent.

Table XIX. Work Hardening Effects in 14 Carat Au-Ag-Cu Alloys

Colour of alloy	Cu %	Brinell hardness for Various deformations		
		15%	30%	60%
Deep red	41.5	132	152	185
Orange red	35.6	142	167	204
Reddish yellow	31.1	162	183	207
Orange yellow	27.7	147	159	173
Dark yellow	24.9	154	166	181
Yellow	20.7	156	171	186
Light yellow	16.6	176	189	215
Greenish light yellow	13.8	169	182	206
Light green	5.9	139	157	182
Pale green	0.0	105	118	139

The susceptibility to tarnish and corrosion of these alloys is dependent in large measure upon the nature of the phases into which they separate in ageing. In general, alloys which have been quenched but not aged, and are homogeneous, are more resistant to superficial tarnish and corrosion than aged alloys, though not necessarily to stress corrosion.[30,35]

Many alloys of these types as used in practice (see Table XVI) contain zinc which is added to modify colour and/or working properties.[21] In particular, zinc additions are effective in preventing cracking during rolling or drawing of alloys containing about equal percentages of copper and silver. Alloys containing zinc in high enough concentrations do not separate so easily into two phases on heat treatment, and are therefore usually more resistant to tarnish and superficial corrosion than alloys which do not contain zinc.

3.4 10, 9 and 8 carat alloys

These are very inferior to the widely used 14 carat and 18 carat alloys in respect of tarnish and superficial corrosion resistance, as might be anticipated from the very low atomic percentages of gold which they contain. Thus an 8 carat Au-Ag-Cu alloy with equal percentages of silver and copper contains only 17 atomic per cent of gold. They may also (see later) be particularly susceptible to stress corrosion cracking.

They are also inferior to 14 carat golds, containing the alloying elements copper and silver in the same ratios, in hardness, strength, and wearing properties, for although they can be age hardened, the degree of hardness which can be achieved is limited (see Figure 10).

Gold alloys of these low caratages which are employed in practice, however, are rarely simple Au-Ag-Cu alloys. They contain significant quantities of zinc, to the point where they can be considered as being essentially substitution brasses, in which some of the copper is replaced by gold and silver. This approach is a particularly valuable one in any consideration of ways and means of producing gold alloys of these types with modified or improved properties.

The addition of zinc in sufficiently high concentrations prevents separation of the alloys into more than one phase on heat treatment, and thereby increases their resistance to superficial tarnish and corrosion. It may (see later), however, decrease their resistance to stress corrosion cracking. Addition of zinc also lowers the melting points of these alloys, and tends to reduce their melting ranges, to the point where problems can arise in soldering.[39] In order to maintain an adequate differential between the melting ranges of alloys of low caratage for jewellery, and those of solders to be used with them, it is therefore necessary to restrict the zinc content of the alloys. Zinc addition also counteracts the reddening effect of the high percentage of copper in these alloys, and is important where yellow alloys are required.

The range of alloy compositions — and therefore the range of properties — which can be achieved with these low caratage gold alloys is considerable. The hardest, strongest and least ductile 9 carat Au-Ag-Cu alloys are those in which ca.15–45% Cu is present. Alloys for stamping, drawing or spinning therefore tend to have compositions outside this range, whilst where hardness and high elasticity are required, as in springs or clips, compositions may be selected within this range. The harder and stronger alloys therefore tend to be reddish. Generalisations are difficult, however, because as indicated above modifications of properties is usually achieved not only by changes in the Cu:Au ratio in the alloy, but also by additions of other metals such as zinc, nickel, etc. which affect the constitutions of the alloys and the extent to which their properties can be modified by heat treatment and other processing procedures.

Table XX illustrates the range of 9 carat alloys offered by one manufacturer, together with the information made available concerning their properties and their suitability for various types of applications.

The compositions of alloys such as these[40] fall mostly in the range 37.5Au/9-12Ag/46.5-39.5Cu/6-11Zn. The popular yellow DF alloy approximates to the composition 37.5Au-10Ag-45Cu-7.5Zn. The red alloys contain large amounts of copper and lesser amounts of silver and zinc, whilst the reverse is true for the green and pale yellow alloys. Both silicon and boron are added in small amounts to certain yellow 9-carat casting alloys, but although they are mentioned (see Table XVI) as deoxidisers, their full role is obscure.

The complexities involved in any attempt to develop improved low caratage gold alloys for specific applications are well illustrated by a recent report [40] describing attempts to produce an improved 9-carat yellow gold casting alloy.

4. White Carat Golds

4.1 Background

White gold alloys were originally developed as substitutes for platinum. They had therefore not only to look like platinum, but also to resemble it in respect of tarnish resistance and workability. As will be noted from Figure 9, white or whitish gold alloys are found in the Au-Ag-Cu system only with high silver contents and low contents of gold and copper. White carat golds of high caratage or with high tarnish resistance cannot be obtained within this system.

The white golds in use today are therefore based on other alloy systems and are of two main types, namely, the nickel white golds and the palladium or noble metal white golds. Although metals other than nickel and palladium, such as manganese, cobalt, iron and indium are also capable of bleaching the colour of gold to yield alloys which are platinum-like in appearance, no white

Table XX. Characteristics of 9 Carat Golds offered by one Manufacturer

Alloy	Colour	Melting range C Solidus Liquidus	Maximum hardness of annealed sheet Hv	Minimum elongation of annealed sheet 50 mm gauge per cent	Annealing Temp. C.	Annealing Method of cooling see note below	A – General purpose	B – Spinning	C – High relief stamping	D – Deep drawing	E – Enamelling	F – Chain making	G – Pen nib manufacture	H – Investment casting	I – Snaps, pins, springs
9 carat P	Pale Yellow	905–960	170	30	650	1								S	R
9 carat BY	Pale Yellow	890–920	145	20	650	3					R			S	
9 carat DF	Yellow	880–900	120	40	650	3	R			R		R		R	S
9 carat C	Yellow	885–895	120	45	650	3	S	S				R			
9 carat SC	Yellow	875–890	110	60	650	3	S	S				R		S	
9 carat G	Green	800–820	100	65	650	3	S	R							
9 carat BR	Red	890–915	140	30	650	1					R			S	
9 carat MR	Deep Red	900–920	110	35	650	3	S		R			S		S	
9 carat HW	White	975–1025	160	35	700	2	R			R		S		S	R
9 carat MW	White	910–940	90	40	650	3	R		S	S	R			R	
9 carat SW	White	990–1005	45	40	650	3			R						

R = Recommended
S = Suitable

Method of cooling:
1. *Must* be quenched from above 500°C
2. *Must not* be quenched
3. *May* be quenched once the metal has cooled to black heat (450°–500°C)
Alloys marked 1 are age hardenable

alloys of gold with these metals have so far been accepted for general use in jewellery manufacture.

It is difficult with the nickel white golds to achieve both good working properties and good colour. Most operators therefore select compositions which give good working properties, and achieve uniformity of colour in production by applying an electroplated coating of rhodium to finished articles.

The literature on white golds was summarised in 1972 by Atkinson and considerable use of his report[41] has been made in what follows. The effects of alloying elements on the colour of gold alloys have also been reviewed by Starchenko and Lifshits.[42]

4.2 Nickel white golds

One of the earliest patents taken out for a nickel white gold was one by Belais[43] for simple Au-Ni-Zn alloys containing 75-85Au/10-18Ni/2.9Zn. These were difficult to work, and were liable to crack on rolling or annealing. Later it was claimed that these alloys were improved by small additions of platinum [44] or of manganese.[45,46] They were later found to be improved even further by addition of Cu[20] and as will be noted from Table XVI, nickel white golds today are essentially Au-Ni-Cu-Zn alloys.

The effects of composition on the colour and other properties of Au-Ni-Cu-Zn alloys have been reported upon anonymously,[47] by Jarrett[48] and by Wise.[49] Wise states that increasing the copper content at the expense of the nickel and zinc contents results in alloys of less satisfactory colour.

The low-zinc high-copper content alloys tend to be pink, and the high-zinc moderate-nickel alloys tend to be slightly greenish.

The effects of composition upon working properties have been reported upon by Belousova *et al.*[50] who have studies difficulties arising from the brittleness of a 75Au-10.5Cu-10Ni-4.5Zn alloy, and by Wise[49] who has related the properties of these gold alloys to those of the ternary alloys of the Au-Ni-Cu system.

The liquidus diagram for this system (the solidus is close to the liquidus over most compositions), together with the diagrams for the three binary edge systems, is shown in Figure 11. This reveals a valley so far as liquidus temperatures are concerned which extends from a minimum in the Ni-Au system at 17.5% Ni to a more broadly based and shallower minimum in the Cu-Au system at 20% Cu. In this valley differences between liquidus and solidus temperatures vanish. Just below the solidus all the Au-Cu-Ni alloys are single phase, but over a considerable area of compositions[51] they separate out on slow cooling or heat treatment into two solid solutions, one of which is gold-rich and one nickel-rich. This has important consequences.

Thus the nickel-rich phase may contain little gold and be more susceptible to corrosion than the single phase alloy from which it is formed. For example, an

Table XXI. The Compositions of some Nickel White Gold Alloys

No.	Au	Ni	Cu	Zn	Other metals	References
1	58.5	14.5	20.0	7.0	—	4
2	58.5	15.3	25.8	0.4	—	4
3	75.0	13.5	8.5	3.0	—	4
4	75.0	13.5	11.5	—	—	4
5	75.0	7.0	12.0	2.0	—	4
6	58.5	15.0	18.0	5.5	—	4
7	75.0–85.0	10.0–18.0	—	2.0–9.0	—	43
8	75.0–91.7	24.7–3.3	—	—	5.0–0.25 (Mn)	45
9	83.3	16.7	—	—	—	54
10	75.0–85.0	10.0–18.0	—	2.0–9.0	0.15 (Pt)	43
11	80.0	13.9	1.0	5.0	0.1 (Pd)	55
12	58.35	17.6	16.95	7.0	0.15 (Mn)	49
13	75.0–85.0	10.0–18.0	—	2.0–9.0	0.5–2.0 (Mn)	46
14	75.0	16.23	3.75	5.04	0.02 (Re)	28
15	75.0	17.0	3.0	5.0	—	56
16	58.4	41.6	—	—	—	57
17	65.0	25.0	—	—	10.00 (Cr)	58
18	80.0–99.0	1.0–20.0	—	—	—	59
19	20.0–50.0	20.0–50.0	—	—	15.0–45.0 (Mo)	60
20	25.0–70.0	15.0–40.0	—	—	4.0–12.0 (Cr)	61
21	39.0–80.0	8.0–20.0	0.0–45.0	0.0–15.0	5.0–30.0 (Pd) 0.01–2.0 (Y) 0.0–2.5 (Co) 0.005–0.05 (Re)	28
22	73.0–77.0	2.5–7.5	10.0–15.0	1.5–4.0	2.5–7.5 (Cd) and (optionally) small amounts of Ag, Mg, Mn, Fe, Co, Al, Pd, In and Sn	62

18 carat Au-Ni alloy, after slow cooling to induce separation of the solid solution is readily attacked by nitric acid. The corrosion resistance of this alloy, and of Au-Cu-Ni alloys, may be greatly affected by phase separation. Moreover the phase separation, together with the subsequent Au-Cu ordering which occurs in the copper-rich phase[24] leads to age hardening. This is undesirable, since these alloys are sufficiently hard when single phase, and phase separation leads to loss of ductility and working properties. The effects on phase separation of the addition of copper to binary Au-Ni alloys are complex[51] and when zinc is present in addition as is the case with the quaternary nickel white golds in use today they are unknown.

Wise[3] comments that although the solid state transformations in the high carat alloys of this type are generally troublesome, they can be used to secure

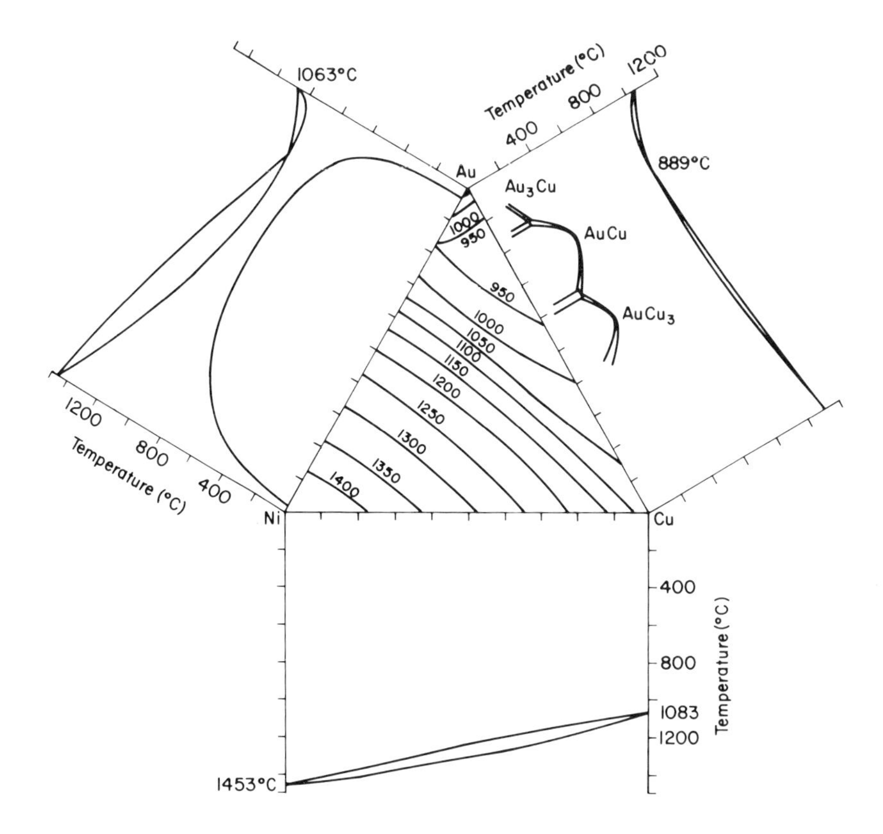

FIG. 11. Liquidus isotherms of the Au-Cu-Ni system in relation to the phase diagrams of the binary Au-Cu, Au-Ni and Cu-Ni edge systems. (Compositions in weight per cent.)

great improvement in machinability. Thus an 18 carat alloy which has been slowly cooled to promote phase separation can be machined with precision, whilst a subsequent quench from the annealed state homogenises the alloy and restores its corrosion resistance.

The compositions and some properties of some nickel white golds taken mainly from Atkinson[41] are reflected in Table XXI. Of these only numbers 1–8 have been claimed for jewellery applications. Numbers 9–11 have been developed for electrical uses, 12 as a spinneret alloy, and 13–14 as brazing alloys. They illustrate the wide range of compositions over which nickel, in conjunction with other metals, is an effective bleaching agent for gold, and in the case of alloys Nos. 14, 20 and 21 the use[28] of grain refining agents such as rhenium and yttrium to improve the physical characteristics of the alloys.

In connection with the casting and heat treatment of white golds, Stagno and Pinasco[52] and Stagno and Ienco[53] have reported upon metallographic studies of a 75Au-8.5Ni-13.1Cu-3.4Zn white gold alloy and of related Au-Cu, Au-Zn, Au-Ni and Au-Ni-Zn alloys. They were examined not only as cast under a number of conditions, and in different positions in the castings, but also in cast samples subjected to four different heat treatments.

Although the nickel white golds are less costly than palladium white golds and therefore preferred by many manufacturers, a number of limitations to their use must be borne in mind. Thus their content of zinc makes them unsuitable for vacuum casting, since they lose zinc by volatilisation and become brittle. Brittleness may also develop in nickel white golds as a result of melting and re-melting for the same reason. Also, the whiter alloy compositions of higher nickel content are unsuitable for stamping, doublé or chain-making because of their hardness. For such applications the more easily workable compositons of lower nickel content and lesser whiteness may be used and subsequently plated with rhodium.

Rhodium plating is not to be recommended, however, in the case of jewellery items such as wedding rings which are liable to mechanical damage. Moreover, rhodium-plated low-carat white gold jewellery may exhibit localised corrosion. The application of a nickel undercoating does not improve the resistance to corrosion in such cases, since it is itself corroded through pores in the thin rhodium deposit. Rhodium plating is, however, more successful when applied to low-carat white gold rings which are high in silver content and which contain little or no nickel. Silver-rich low-carat white gold alloys are much less sensitive to local corrosion by perspiration after rhodium plating than are low-carat nickel white golds after similar treatment.

A further feature of nickel white golds is their susceptibility to attack by sulphur. Contamination with sulphur even in very small concentrations results in the formation of low melting sulphides at grain boundaries.[30] A nickel white gold — or indeed any gold alloy containing nickel — which has been contaminated with sulphur therefore loses its strength on heating. Sulphur contamination can arise not only from the use of sulphur containing gases for heating, but also as a result of reducing reactions during heating from small residues of gypsum (calcium sulphate) used in casting, or of sulphuric acid used in pickling. It can also arise from the use of graphite crucibles, since any sulphur which may be present in the graphite will be taken up by a nickel white gold which is melted in contact with it. For this reason silicon carbide crucibles are often used in the melting of nickel white golds. Apparently — though no authority can be quoted for this — addition of a little manganese to a nickel white gold increases its resistance to attack by sulphur, which tends then to be taken up as manganese sulphide. This does not concentrate at grain boundaries like nickel sulphide, and so weaken the alloy on heating. Contamination with silicon, which may be generated if the alloys are melted in reducing atmospheres in contact with silica, produces effects similar to those of sulphur.

The special measures necessary to remove the dark oxide films which form on nickel white golds when heated in air, and which are referred to later, also create difficulty both in normal working and in soldering. To keep the formation of such oxide films to the minimum the use of boric acid or borax-sodium phosphate fluxes during heating of nickel white golds is desirable. If too much nickel in the surface layers is oxidised during heat treatment and subsequently removed, then the surface of the alloy may be enriched in gold and take on an undesired yellowish shade.

As with the yellow, green and red carat golds grain growth during heating can give rise to difficulties when nickel white golds are annealed.

The most important aspect of the annealing of nickel white golds, however, is that depending on their composition quenching from the solution treated condition may not yield ductile products, as is the case with yellow, green and red carat golds. Instead, hard and highly stressed products may result, which even if they are not cracked initially, rapidly crack when at attempt is made to work them. Since as mentioned above, slow cooling from the solution treated condition tends to lead to extensive separation into two phases, one of which is easily susceptible to corrosion or tarnishing, it may be desirable to compromise in workshop practice and use intermediate rates of cooling. This may be achieved, for example, by quenching in boiling water or by cooling the alloy on an iron plate, which is thick enough to take up its heat at an "intermediate" rate.

In the recovery of gold from wastes, it must be borne in mind that since nickel is ferromagnetic, filings of nickel white golds are removed along with iron filings when wastes containing them are exposed to the action of a magnet.

4.3 Noble metal white golds[1]

Au-Pt alloys have been reviewed by Darling [63] and by Hume-Rothery,[64] and the mechanical properties, constitution and fabrication of alloys in the Au-Pt system have been examined by Nowack.[65] Other alloy systems such as the Au-Pt-Rh system[66] have also been studied.

Palladium is, however, a more effective whitener of gold than is platinum, and according to one report[47] the 10% Pd alloy corresponds in colour with the 25% Pt alloy. Moreover, palladium is normally cheaper than platinum and has only about half the latter's specific gravity. The noble metal white golds which have been developed for use in jewellery are therefore almost exclusively Au-Pd alloys to which varying amounts of other metals such as silver, nickel, platinum and zinc have been added. White golds used in dentistry tend to be more complex (see Chapter IV).

Although palladium white golds are more expensive than nickel white golds, and have appreciably higher melting points than the latter, their malleability, ductility and corrosion resistance are much higher and they do not oxidise with change of colour on annealing or soldering. They can also be cast.

The compositions of some palladium white golds are listed in Table XXII.

Table XXII. Compositions of some Palladium White Gold Alloys

Au	Ag	Pd	Other metals	Reference
83.3	—	16.7	—	4
81.0	4.0	15.0	—	4
75.0	5.0	20.0	—	4
43.0	41.5	15.5	—	4
45.0	45.0	10.0	—	67–69
57.0	—	35.0	7.5 (Ni)	70
*60.0	—	30.0	10.0 (Pt)	71
74.0	—	21.5	4.5 (Pt)	72
†70–75	—	5–25	15–40 (Pt)	73
‡65–70	7–12	10–12	4.0 max. (Pt)	74
			6–10 (Cu)	
			1–2 (Zn)	
‡60–65	10–15	6 10	4–8 (Pt)	74
			9–12 (Cu)	
			1–2 (Zn)	
§75.0	9.9	5.4	5.1 (Cu)	
			3.5 (Zn)	
			1.1 (Ni)	
58.0	—	34.5	6.5 (Fe)	75
			1.0 (Al)	

*To produce an alloy having increased tensile strength and hardness 0.1—2.0% of Ru, Ir, Os or Rh may be added either alone or in combination. The addition of Ir is preferred.
†Can be hardened by a heat treatment.
‡Dental white gold casting alloys.
§A proprietary product.

4.4 White golds of other types

Alloys of gold with cobalt and iron have been studied[76] and compared with those with nickel. Young [7] has reported that cobalt whitens gold to a lesser extent than nickel, Perrault[78] has claimed the use in jewellery where hardness and ease of polishing are important of a 75Au-4Co-10Ag-11Cu alloy, and Sistaire and Chamer[28] have given details of a fine grained white gold alloy for jewellery containing 39-80Au/0-45Cu/0-15Zn/2.5Co/8-20Ni plus 0.005–0.05% Re as a grain refiner.

More recently[79] a number of metals have been compared in regard to their effectiveness as bleaching agents for gold. This has confirmed the outstanding activity in this respect of nickel, with other metals such as cobalt, silver, palladium, indium, tin, manganese and zinc having in comparison only subsidiary activity. Nevertheless it was found that there was advantage to be obtained from using cobalt together with nickel because of the more malleable nature of the resulting alloys. These observations are being followed up.

Of the other "subsidiary" bleaching agents, indium appears to have been least used, though a range of indium containing gold alloys, some of which are in the white category, have been described.[41] Manganese has apparently enjoyed some application.[31]

5. Violet Gold

Blue to violet coloured alloys are formed by gold and aluminium but are too brittle for use as such. The alloy corresponding to the intermetallic compound composition $AuAl_2$ is the most intensely violet in colour and a procedure has been patented[80] for the application to metallic substrates of violet coatings of this material. Powdered Au-Al alloy of the required composition is described as being entrained, either alone or mixed with gold or aluminium powder, in a gas jet directed at the substrate surface, and heated either by combustion or by an electric arc. The molten particles of alloy adhere to the surface to give highly decorative carat gold coatings.

Certain Au-Fe alloys, e.g. the 75Au-24Fe alloy, assume a blue colour, but have not found application. In the case of these alloys, the colour is apparently an interference colour resulting from the formation of a thin film of base metal oxide on the surface of the alloy.

No information has been released concerning a blue alloy, described as a blue carat gold, which has been developed by one manufacturer and which is used in jewellery and for watch faces.

The method by which purple films were produced on the Tutankhamun gold is unknown, but the presence of iron and of arsenic in such gold may not be without significance.[81]

6. Special Aspects of the Use of Carat Golds

6.1 Hardness and wearing properties: Age hardening of gold alloys

Introduction

Although as described above conventional yellow and red carat gold alloys based on the Au-Ag-Cu system are susceptible to age hardening by both precipitation and ordering mechanisms, the possible extent of such hardening is limited and may not always be realised in practice. The wear resistance of these carat golds, particularly those of higher caratages, therefore, normally leaves something to be desired. Their relative softness also results in most yellow and red golds having the disadvantage that they are not susceptible to machining.

It is therefore perhaps surprising that the potential of precipitation age hardening, ordering and grain refinement for producing strong hard wearing gold alloys, which has been widely exploited in the development of dental gold alloys and spinneret alloys, has not been developed in jewellery manufacture to the same extent. This is particularly so, since age hardening is such a feature not only of the Au-Ag-Cu system, which is the basis of most yellow and red carat golds, but also of the Au-Cu-Ni system which is the basis of nickel white gold alloys. Moreover, age hardening was first exploited in the case of a 64Au-12.5Pt-7.0Cu-16.5Ag alloy which was marketed as a dental alloy as early as 1906 by the S.S. White Dental Manufacturing Company.[82] This alloy could be softened by quenching from a red heat and hardened by air cooling from such a temperature. Age hardening processes have been the subject of a number of reviews.[83-90] In what follows extensive use has been made of the last of these reviews.

Hardening or strengthening of a metal or alloy results from any limitation of its deformation under stress. In the individual crystal, plastic deformation is a result of the propagation of dislocations in the crystal along slip planes which are defined by the structure of the crystal lattice. It is initiated under stress at positions at which imperfections or disclocations are present and is blocked and terminated by any irregularities in the crystal structure which are of such a nature as to prevent any further propagation of the dislocation. These phenomena are compounded in polycrystalline metal structures.

Propagation of dislocations, for example, is hindered in alloy solid solutions by the presence of atoms of the solute metal in the crystal lattice of the solvent metal. The greater the difference between the sizes of the atoms of the solute and solvent metal, the greater is this "solid solution hardening" effect. Alloy solid solutions therefore tend to be harder than their solvent metals. Propagation of dislocations is likewise increasingly blocked during cold working of metals by the irregularities in structure to which such working gives rise. Work hardening therefore results. It is blocked also at grain boundaries, and this may contribute towards making fine grained metals often harder and stronger than coarse grained ones. In metal composites in which inert materials such as oxides or carbides have been dispersed in a metal or alloy on the other hand, the propagation of dislocations is blocked by the particles of the dispersed phase, and so-called dispersion hardening results from this and from restraints which the disperse phase may exercise upon the deformation of the metal matrix.

The precipitation by means of ageing treatment of new phases in an alloy can be regarded as being in effect a special case of dispersion hardening. Where these processes involve dimensional changes and the creation of internal strains, however, hardening may result in part from such strains as well as from "dispersed phase" effects.

Hardening by ordering

The classical example of ordering, namely that which occurs in Au-Cu solid solutions (see Figure 6) has been extensively investigated.[14-27,82,90-94] At higher temperatures Au-Cu solid solutions are face centred cubic in structure with a random arrangement of gold and copper atoms. Below a certain temperature, however, the gold and copper atoms assume definite or ordered lattice positions,[22] which are particularly evident at compositions corresponding to AuCu and $AuCu_3$. In the first alloy in which ordering was recognised, namely that corresponding in composition to $AuCu_3$, the effect is that the gold atoms go to the corners of the cubes in the crystal lattice and the copper atoms to the centres of the cube faces. This involves little distortion of the parent lattice, and does not affect the hardness of the alloy significantly. In the alloy corresponding in composition to AuCu, however, the planes in the face centred cubic lattice arrange themselves into layers of all-gold and all-copper atoms. This causes a small change in the lattice dimensions, severe internal strains, and a consequent increase in hardness to a maximum of about 180 BHN at about 75% full ordering. Ordering in Au-Cu alloys, and the internal strains to which it gives rise, can be restricted by addition of small amounts of metals such as iron, cobalt and nickel.[27]

Ordering reactions such as these occur also in the copper rich solid phases of the Au-Ag-Cu system[17,23-27] and therefore contribute to age hardening in a wide range of carat gold alloys. Ordering is also responsible in part at least for the age hardening of the many dental alloys including that of the S.S. White Dental Manufacturing Company referred to above. It is also used (see later) to harden copper containing gold alloys for use as wipers in sliding electrical contacts, and has also to be taken into account as an influence in Au-Pd thermo-couple alloys and in Au-Ni electrical contact alloys.

Hardening by precipitation

Precipitation results normally from solubility changes with temperature, and its exploitation in age hardening of alloys results from the slowness with which diffusion processes occur in the solid state, and therefore the slowness with which equilibrium conditions in this state are normally attained. This implies that if an alloy, whose equilibrium state at ordinary temperatures is double phase, is heated at a temperature at which it is single phase and then quenched, the resulting alloy is obtained in a metastable single phase state. Heating at a temperature below which its equilibrium state is single phase, however, will result in its separation into two phases and an age hardening effect. This phase separation is promoted by lattice vacancies which may be present in the quenched alloy, because of their effects on diffusion rates. It may also in some cases be promoted by small additions of certain elements.[85]

The detailed processes may be complex and in any one system depend usually upon the ageing temperature, the composition of the alloy and the solution treatment and quenching procedures. Two basic mechanisms appear to be involved.

The first involves nucleation and growth of the precipitating phase. This occurs when the second derivative of the Gibbs free energy with respect to concentration or composition is positive, i.e. when

$$\left(\frac{\partial^2 G}{\partial c^2}\right)_{T,P} \geqslant 0.$$

It has been observed in many systems but occurs particularly where the separating phase differs markedly in composition from the solvent phase from which it crystallises out. It can occur for example in the separation of Au_4Ti from Au-Ti alloys.[95] It could be expected to occur also under certain conditions in some dental gold alloys which contain both iron and platinum, and in which age hardening is accompanied by separation of $FePt_3$.[96,97] Where precipitation of this type occurs preferentially at grain boundaries, i.e. by discontinuous precipitation, it can cause brittleness and in some instances increased susceptibility to corrosion at such positions, and steps may be necessary to prevent it either by changing the ageing conditions or by adding other elements.[85]

The second mechanism, namely, spinodal decomposition, occurs when the second derivative of the Gibbs free energy with respect to concentration is negative. It is of interest because it has been studied particularly in the Au-Pt (see Figure 30) and Au-Ni (see Figure 11) systems. It involves the breakdown of the metastable solid solution into a "modulated" structure consisting of alternating layers of the separating phases of continuously varying composition throughout the alloy. Because of the similarity between the solid phase relationships in these systems and those in the Au-Ag-Cu system, spinodal decomposition could be expected to occur under certain ageing conditions during the age hardening of many carat gold alloys. In each of these systems the hardening effects of the primary phase separation may be reinforced as a result of secondary changes with temperature which occur in the equilibrium compositions of the segregating phases during the final cooling of the alloy from the ageing temperature. They are also influenced by whether the separating phases are coherent or incoherent with the matrix lattice.

Precipitation in Au-Pt alloys

Spinodal decomposition was discovered only in 1957 when Tiedema *et al.*,[98] using electron diffraction, found periodic concentration perturbations in a decomposing Au-Pt solid solution. This was followed by other work which included thermodynamic calculations,[99] and X-ray studies to determine the nature of the precipitate[100] and the rate of precipitation near the spinodal curve,[101,102] i.e. under conditions where the value of the second derivative of the Gibbs free energy with respect to concentration is close to zero. Both coherent and incoherent precipitation were found to occur in this system,[103] and the solution treatment temperature was found to be important at short ageing times, since it affected the concentration of lattice vacancies.[104] The deformation and fracture of Au-Pt polycrystals hardened by spinodal decomposition were also studied.[105]

Precipitation in Au-Ni alloys

Parallel work which has been carried out on the Au-Ni system[20,82,106,107] illustrates the complexity of precipitation processes which have been found to vary greatly in alloys of different compositions and under different conditions. Thus discontinuous precipitation was found in a 50Au-50Ni alloy.[108] In a 70Au-30Ni alloy, however, a metastable precipitate was formed below 225°C and quenched in lattice vacancies were seen as controlling its rate of formation.[107] Such quenched in vacancies can influence the rate of spinodal decomposition.[109] At low temperatures, viz. below 165°C, ageing was found to lead to a modulated structure similar to that found in Au-Pt, but ageing at temperatures over 300°C led to a Widmanstatten structure.[110] Other investigators[111] have observed a modulated structure upon ageing below 225°C in gold alloys containing 20 – 80% Ni, whilst in a 64Au-36Ni alloy precipitation was found[112] to occur discontinuously with the formation of a two-phase lamellar structure above 225°C and homogeneously below this temperature. Discontinuous precipitation has also been observed with this system at temperatures above 400°C.[111] This is in contrast to observations on the Au-Fe system in which discontinuous precipitation occurred at lower ageing temperatures.[113–115]

Discontinuous precipitation has been exploited by Johnson and his colleagues[116–120] to develop a fibrous microstructure and so give high strength Au-Ni alloys, and other workers[121–124] have reported upon the effect of vacancies in the pre-precipation processes upon the mechanism of spinodal decomposition. Finally a distinct crystallographic orientation between the gold rich and nickel rich phases resulting from discontinuous precipitation has been observed.[125]

Precipitation in other gold alloy systems

These studies of the binary Au-Pt and Au-Ni systems have greatly increased understanding of the precipitation process and have demonstrated in particular the extent to which the control of solution treatment and ageing conditions can contribute to the avoidance of undesirable microstructures during age hardening, such as those resulting from discontinuous precipitation. They indicate that the potentialities and the techniques which are available for the development of age hardenable gold alloys suitable for use in jewellery manufacture may not so far have been fully exploited. Thus most of the early work[18–21,90,126] which has been done on the age hardening of gold alloys was carried out prior to the more recent work described above.

A variety of elements other than nickel and platinum have been described as giving rise to alloys with gold which are susceptible to age hardening. They include aluminium,[127] titanium,[95] chromium, iron, cobalt, germanium, indium and tin, and although each of these modifies the colour of gold when it is alloyed with it, and although some have been reported to give rise to brittle products with poor working properties, there are indications that some have the potentiality to produce useful alloys.

In regard to gold alloys containing chromium, recent developments which have been described include that of an Au-Cr-Co alloy[128] which was effective in sliding electrical contacts. In this, 0.05% Zr was incorporated in order to control discontinuous precipitation, and the D.P.H. (Diamond Pyramid Hardness) could be increased from 83 to 250 during ageing. They also include a number of patents[129–132] on age hardening gold alloys which contain besides chromium, such other elements as cobalt, platinum, iron and copper as co-hardeners. The claims describe the compositions and ageing properties of 46 alloys with chromium contents ranging from 1–12%, and aged hardnesses up to 450 D.P.H. Those of higher chromium content probably derive some of their age hardening properties from the fact that chromium has limited solubility (ca. 8%) in gold at ageing temperatures.

Iron has also limited solubility in gold at ageing temperatures and Nowack reported many years ago[19] on the age hardening properties of gold alloys containing 15–20% Fe. More recently these properties have been reported upon by Frebel and Predel[113–115] and others[133] and Japanese investigators have claimed the development of age hardening gold alloys containing up to 30% Fe, as well as such other elements as cobalt, chromium, nickel and platinum.[130–132,134,135] Aged hardnesses of up to 460 D.P.H. were obtained and in general hardness increased with iron content. Their corrosion resistance was excellent. Age hardenable Au-Pd-Fe alloys for use in electrical resistance wires are described elsewhere.

Cobalt is another metal which exhibits a great change in solubility in gold with temperature, its solubility falling from 8.4% at 995°C to about 0.1% at 400°C. Only recently, however, has its precipitation from gold alloys been reported upon to any extent.[136–142] Moreover there are Japanese patents claiming its use in amounts of 1–12% in multicomponent age hardening gold alloys containing chromium, iron and platinum as co-precipitating elements.[129,131,132]

The use of tin and indium in small concentrations in certain age hardening dental alloys[94] is probably effective because of the formation of In-Pt and Sn-Pt intermetallic compounds which precipitate out during ageing. Other precipitation processes are involved in other developments.[143,144]

There is clearly scope in terms of our new understanding of age hardening processes, for further studies not only of known age hardening gold alloys, but also aimed at the development of new alloys of this type. Particular interest attaches to the possible identification of a precipitating species which could be used for the age hardening of golds of caratages greater than 18. The use of $CuAl_2$ in the hardening of sterling silver[82] exemplifies this type of development. A patent claim by Sekine *et al.*[145] is of particular interest in his connection, since it describes 18 carat alloys which contain copper and aluminium together with a range of other alloying metals which are susceptible to age hardening, and which can be cast and worked.

To be at all widely applicable in jewellery fabrication, it will be appreciated that the necessary heat treatments required for age hardening should be simple.

6.2 Tarnish resistance, corrosion and stress corrosion cracking of gold alloys

Superficial corrosion

A useful and frequently quoted generalisation which stems from the extensive studies of Tammann[146] is that the resistance of gold to tarnish and corrosion at ordinary temperatures is not greatly reduced by the addition to it of silver and base metals, so long as the gold content of the resulting alloy is not below 50 atomic per cent. This corresponds to about 15.6 carat for Au-Ag alloys and to about 18 carat for Au-Cu alloys. It must be remembered, however, that this is a generalisation, and that a necessary condition is that the alloy is a homogeneous solid solution and does not undergo separation into two phases, one of which contains a lesser proportion of gold.

Below about 50 atomic per cent gold, the exposure of the base metal atoms in homogeneous Au-Ag-Cu solid solutions is such that corrosion and tarnish resistance tend to decrease significantly, and for Au-Cu and Au-Ag solid solutions, and for Au-Ag-Cu alloys containing certain ratios of Ag:Cu, Tammann has recorded[146] the limiting gold contents below which they become susceptible

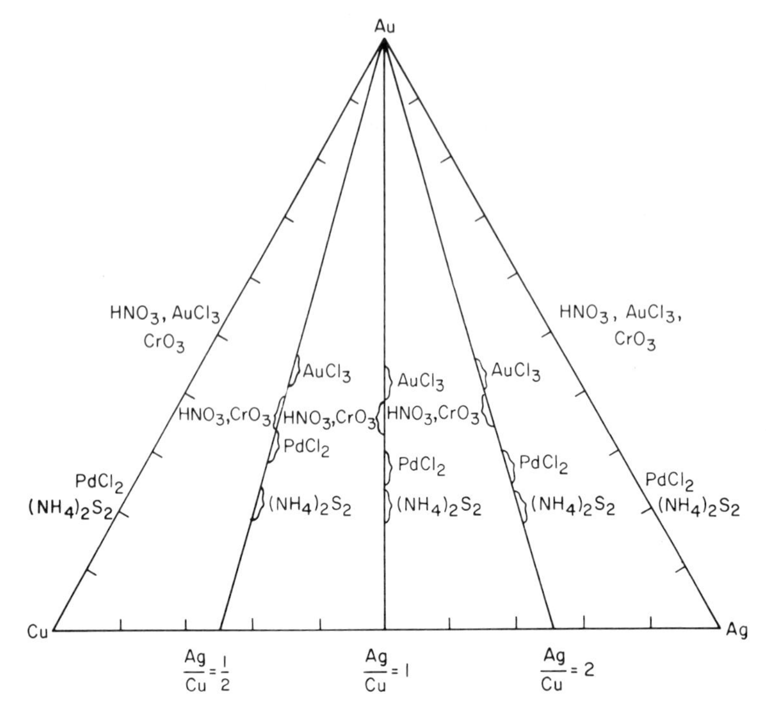

FIG. 12. Limiting Au contents of Au-Cu, Au-Ag and some Au-Ag-Cu alloys above which they are not attacked by various reagents. (Compositions in atomic per cent.) (After Tammann.[146])

to superficial attack by a variety of reagents. In general these limiting gold contents lie in the range 20–50 atomic per cent.

Tammann's findings are summarised in Figure 12. The fact that ternary Au-Ag-Cu alloys containing less than about 25% Au are attacked by ammonium sulphide solutions accords with the known susceptibility to tarnish of such low caratage alloys.

Tammann's observations on Au-Ag-Cu alloys are of limited significance, however, when considering the resistance to tarnish and superficial corrosion of low caratage gold jewellery alloys, since other metals and particularly zinc and nickel may be present in them in addition to silver and copper. Apart from actual alloy composition, an important factor is the physical state of the alloy, and in particular whether its equilibrium state at ambient temperatures is single phase or double phase. Alloys whose equilibrium state is single phase are generally more resistant to tarnish and corrosion, and their resistance is not affected significantly by the nature of the heat treatments to which the alloys are exposed. From this point of view zinc additions are important, since if zinc is added in high enough concentrates the equilibrium state of the alloy tends to be single phase. If, however, the alloy composition is such that its equilibrium state at ambient temperatures is double phase, then its condition will be affected by the nature of the heat treatment to which it has been exposed, and its tarnish and corrosion resistance affected accordingly. Samples which have been quenched in water for example may differ in their tarnish resistance from samples which have been air cooled. The overall picture in regard to the tarnish resistance of low carat gold jewellery alloys is therefore complex and it is not possible to generalise and relate such resistance directly to their gold content.

Stress corrosion cracking

Superficial corrosion or tarnishing is not, however, the only form of environmental attack to which gold alloys of caratages of 14 or less may be susceptible, since at positions where the alloys are highly stressed, localised corrosion especially at grain boundary sites may lead to what is known as stress corrosion or season cracking. Since much jewellery may have high local stress concentrations as a result of working or of other factors, this effect is of importance. Sufficient stress to induce corrosion cracking may be provided by even such an operation as burnishing.[3] Stress corrosion cracking can develop very rapidly (see Table XXIII), and in environments which cause no superficial attack on the alloy, and which might therefore be regarded as harmless.

Thus it may be induced not only by exposure to acids during pickling but also as a result of contact with reagents such as ink, traces of hydrochloric acid in the atmosphere, perspiration, etc. It has frequently been initiated at points of stress created in annealed low carat alloys by subsequent stamping. Articles such as fountain pen nibs, rings, chains, etc. are well known examples. Unless the stresses in such articles are fully relieved by annealing, they remain as focal points for attack by any reagent capable of inducing corrosion cracking. A case has been quoted[30] in which the collapse of hard rolled low caratage foil, stored as received from the supplier in its original packaging, was found to be a result of attack by fumes of hydrochloric acid from an adjacent workshop. In the same environment annealed foil remained stable. Great care should therefore be taken to relieve stress in low caratage gold alloys wherever feasible.

The processes involved in stress corrosion cracking have been the subject of much debate, and will be discussed here only with reference to gold alloys. It is usually characterised by attack along grain boundaries (intergranular corrosion) under conditions of tensile stress, which progressively opens up the grain boundary positions to attack, and leads to the formation of cracks and ultimate failure. In some cases, however, transgranular cracking may occur. In such cases attack tends to be focused at imperfections in the crystals. Thus Bakish and Robertson[147,148] found that deformation of $Au\text{-}Cu_3$ by cold rolling created a

Table XXIII. Dependence of the Sensitivity to Stress Corrosion of Homogeneous Au-Ag and Au-Cu Alloys on their Gold content when Exposed to 2% Aqueous $FeCl_3$ Solution*

Gold wt. %	Gold at. %	Stress in kg/mm²	Time to fracture of the specimen in minutes
(a) *Au-Ag alloys*			
0	0	15	>10,000
1.8	1.0	12	1,500–3,000
5.0	2.8	10	ca. 1,500
10.0	5.75	11	ca. 300
15.0	8.8	10	ca. 60
33.3	21.5	10	ca. 1
40.0	27.0	10	ca. 1
50.0	35.4	10	ca. 60
54.9	40.0	10	> 3,000
60	45.2	10	>10,000
(b) *Au-Cu alloys*			
33.3	14	18	ca. 1
45.1	21	18	ca. 1
50.0	24.3	15	ca. 1
55.0	28.3	18	ca. 1
67.4	40.0	22	>10,000

*After L. Graf.[152]

multiplicity of sites on the alloy surface at which attack by corrosion cracking reagents such as aqueous solutions of $FeCl_3$ occurred. Slip bands blocked by grain boundaries and coherent twin boundaries became sites of attack at small average strains, whilst with increasing strains the sites of reaction were observed to spread into the grains until they were covered with such sites. These authors related their observations to the known[148] decrease in susceptibility to stress corrosion cracking with increasing cold working, by suggesting that under such circumstances the intensity of reaction at grain boundaries is reduced because of the very large number of competing sites. In tune with this they found[149] that when plastically deformed Au-Cu alloys were annealed, stress corrosion attack reverted to grain boundary sites before microscopically visible re-crystallisation was observed. The two effects occurred close together both in respect of time and temperatures, however, and localised changes in slip bands preceding crystallisation were seen as a possible reason for such differences in the coincidence of the two effects as were observed.

Bakish and Robertson complemented the above studies of polycrystalline Au-Cu alloys, by detailed studies of $AuCu_3$ single crystals.[150] and by observation of potential differences between grain and grain boundary sites.[151] They concluded that in the systems which they studied stress corrosion cracking was caused by irreversible reactions initiated at unstable structural sites. Whereas in pure metal, chemical reaction at such sites ceased with the elimination of unstable grain edges, in an alloy the preferential oxidation of one component of the alloy at an unstable or active site was seen as resulting in the establishment of a galvanic cell, in which a residue of the more noble metal acted as cathode and in which equilibrium could not be attained. Reaction therefore proceeded along the active site until penetration was complete. Since the operation of the above process was dependent upon the alloy composition and upon the relative oxidation potentials of the alloy components in the corroding environment, the great variations in susceptibility to stress corrosion cracking of different alloys in different media were explained.

Most of the available information pertaining more specifically to the occurrence of stress corrosion cracking in Au-Ag, Au-Cu, Au-Cu-Ag and Au-Cu-Ni alloys has emerged, however, from the work of Graf and his co-workers.[152-159]

Au-Ag alloys are fully solid solution in type, even at low temperatures, and the same applies to Au-Cu alloys which have been heat treated at 700–800°C and quenched in order to prevent phase transformations at lower temperatures. Table XXIII shows the manner in which the sensitivity of such alloys to stress corrosion cracking was found by Graf[152] to vary with their gold content. In the Au-Ag alloys the sensitivity to stress corrosion, which is nil with pure silver, increases progressively with gold content to a maximum and then falls again to zero at a gold content of about 45 atomic per cent. The Au-Cu alloys, studied over a narrower range of composition, show a similar trend. Clearly at a certain level of gold content the alloy is protected from stress corrosion, just as it is from superficial corrosion (Tammann effect).[146] This is illustrated in Figure 13 for a variety of reagents which do not attack gold, and in Figure 14 for two reagents which do attack gold.

Au-Ag (and Au-Cu) alloys do not exhibit stress corrosion in all environments, however, and Table XXIV indicates why this may be so.

It will be noted that only where the reagent is such that a current actually flows between electrodes of Au and Ag placed in it, does corrosion cracking occur. Graf concluded therefore that only those reagents capable of promoting dissolution of silver in an Au-Ag cell are capable of producing stress corrosion cracking in Au-Ag alloys. In other words only those reagents produce stress corrosion cracking, which can by dissolution of the less noble metal in the alloy, create cathodic residues of gold.

In the case of exposure of Au-Ag and Au-Cu alloys to reagents which dissolve gold, such as aqua regia and potassium cyanide solutions, the findings of Graf and Budke[157] are summarisd in Figure 14. These indicate that with aqua regia susceptibility to stress corrosion cracking reaches a maximum

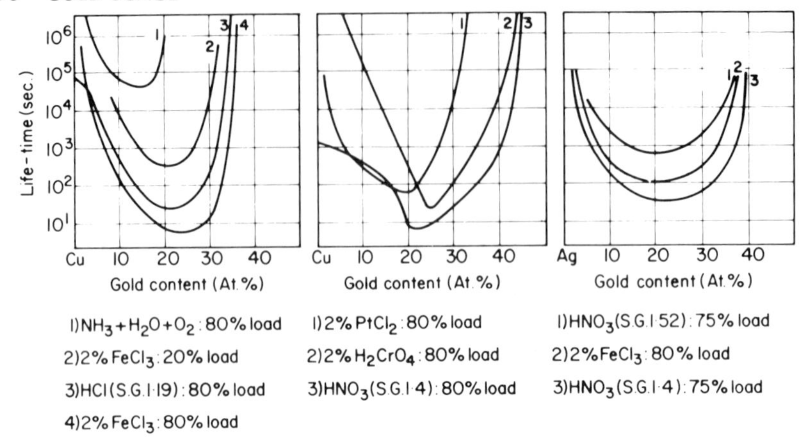

FIG. 13. Lifetimes of Au-Cu and Au-Ag alloys, exposed under stress to various reagents in which gold is not attacked, plotted against their gold contents. (After Graf.[152])

between 20 and 35 atomic % Au as it does with other reagents. Up to 40 atomic % Au, attack on these alloys clearly occurs both by superficial and stress corrosion mechanisms, whereas from 40–100% Au stress corrosion is not operative and failure is by superficial corrosion only. With potassium

Table XXIV. The Action of Different Reagents

Reagent	Concentration	Electrochemical action* Current in mA	Time to fracture of the specimen in minutes**
HNO_3	conc.	0.8–1.0	1
$HCl + HNO_3$	3:1 (conc.)	2.5	1
$K_2Cr_2O_7 + H_2SO_4$	conc.	2–3	60
$KMnO_4 + H_2SO_4$	—	2–3	1
CrO_3	2% aq.	0.3	13
HCl	conc.	0.2	40
H_2SO_4	conc.	0	>15,000
$FeCl_3$	2% aq.	0.8	3
$CuCl_2$	2% aq.	0.5	5
$ZnCl_2$	2% aq.	0	>15,000
$CrCl_3$	2% aq.	0	>15,000
$NaCl$	5% aq.	0	>15,000
$CuSO_4$	5% aq.	0	>15,000
$Fe_2(SO_4)_3$	2% aq.	0.1	160

(After Graf[152])

* In a gold–silver cell.
** For a 33.3 Au-66.7 Ag alloy under a tensile load of 10 kg/mm^2.

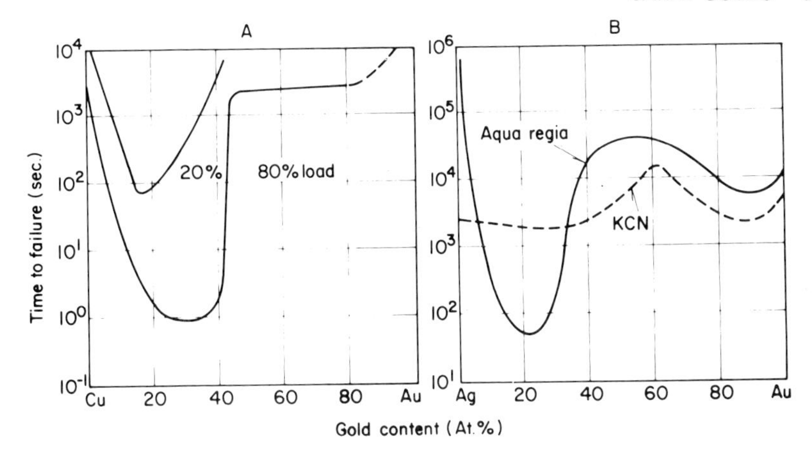

FIG. 14. Lifetimes of Au-Cu and Au-Ag alloys, exposed under stress to reagents which attack gold, plotted against their gold contents.

A. Au-Cu exposed to aqua regia under 80% and 20% loading.

B. Au-Ag exposed to aqua regia and to 10% aq. KCN under 90% loading.

(After Graf.[152])

cyanide solutions, however, Au-Ag alloys exhibit no susceptibility to stress corrosion cracking. This was attributed by Graf and Budke[157] to the fact that aerated cyanide solutions dissolve gold to form the very stable $Au(CN)_2^-$ complex, so that any gold cathodic areas created by initial attack on active sites tend to be dissolved irreversibly. With aqua regia as the corroding agent, however, it was concluded that gold dissolved at the corroding surface is reprecipitated on the walls of cracks or crevices by the action of exposed solute metal, so that the cathodic areas necessary for stress corrosion cracking are created.

Graf[159] have also established that in alloy solid solutions the reactivity of the grain boundaries and of disturbed areas on the grain surfaces is enhanced, and increases with the concentration of the solute metal up to a maximum of 50%. In accord with this they have observed that Au-Cu solid solutions which are susceptible to corrosion cracking under stress, also exhibit intercrystalline corrosion even in the absence of stress. Perhaps of greatest significance, however, has been their demonstration that this enhanced reactivity in the grain boundary areas of homogeneous solid solutions is further increased whilst they are undergoing flow or deformation. In the light of this the special role of tensile stresses in stress corrosion cracking of homogeneous alloys is easily understood. Such stresses produce highly activated flowing areas at the base of notches, so that intercrystalline or transcrystalline cracks develop rapidly. In polycrystalline material the cracks are predominantly intergranular, whilst with single crystals the cracks are transcrystalline.

On the basis of the work of his school on the stress corrosion of homogeneous alloys containing noble metal components, Graf[159] has attributed such corrosion to:

(a) Enhanced reactivity of such alloys at grain boundary sites, and increase in such reactivity with concentration of the solute metal up to 50 atomic %.

(b) Increase in reactivity of such alloys whilst they are undergoing deformation or flow.

(c) Exposure of the alloys under tensile stress to electrochemical attack, an essential feature of such attack being the formation on the walls of notched sites of cathodic areas, towards which the alloy becomes anodically polarised.

Whilst the above generalisations may be accepted as applying to homogeneous gold alloys, the position in regard to the susceptibility to stress corrosion cracking of heterogeneous gold alloys which contain more than one type of solid solution is more complex. This has been exemplified by Graf in a discussion of Au-Ag-Cu and Au-Cu-Ni alloys.

Aged Au-Ag-Cu alloys over wide ranges of composition contain two solid solution phases, which at low caratages and approximately equal percentages of silver and copper are respectively silver rich Au-Ag and copper rich Au-Cu.[15] Both would be expected to be susceptible to stress corrosion in 2% $FeCl_3$ solution. Nevertheless Graf has stated that alloys containing these phases are not susceptible to such corrosion, even though homogeneous Au-Ag-Cu alloys may be attacked. Apparently the less noble phase protects the more noble one, so that an incipient crack is blocked off as soon as it encounters the more noble phase.

In the early production of 14 carat fontain pen nibs, for example, Loebich[30] has stated that when ternary Au-Ag-Cu alloys were used, it was found desirable to age the fabricated nibs. In the aged condition they did not undergo stress corrosion cracking in use, whereas if heated to the point where they became homogeneous, cracking by the action of the ink became likely. When certain 14 carat quaternary Au-Ag-Cu-Zn alloys are used, however, such ageing is apparently unnecessary.

Graf has also referred (private communication) to heat treatment at 450°C in a salt bath, followed by air cooling, as a means of reducing the susceptibility of 8 carat gold alloys to stress corrosion cracking. The susceptibility of such alloys to stress corrosion cracking is, however, considerably influenced by their zinc contents. At zinc contents of less than 15%, they tend to be duplex in character and resistant to stress corrosion if aged. With zinc contents greater than 15%, however, they become single phase and sensitive to stress corrosion cracking even after ageing.

In the case of Au-Cu-Ni alloys, the two solid solution phases which separate from the homogeneous alloy on ageing are a nickel rich phase containing a little gold and copper, and a copper rich phase containing high gold but little nickel.[25] Graf points out that the latter phase might be expected to be susceptible to stress corrosion because the basis metal (copper) is less noble than the other main component (gold). On the other hand, the former phase should show no tendency to stress corrosion because nickel and copper (as a result of passivation of the nickel) have about the same electrochemical potentials, and this expectation is confirmed in practice. Thus aged Au-Cu-Ni alloys contain one phase which should be susceptible, and one phase which should not be susceptible to stress corrosion. The susceptible phase is, however, more noble than the non-susceptible one, so that its corrosion is inhibited and in practice the non-susceptible nickel rich phase is strongly attacked and leached out.

These two examples illustrate the complex interplay of factors which determine the behaviour of ternary gold alloys under stress in corroding environments. Since many low caratage gold alloys are quaternary or even more complex in character, any attempts to improve their resistance to stress corrosion cracking must inevitably be largely based upon empirical or intuitive trials.

That there may be factors additional to those discussed above which are operative in particular cases is illustrated by an observation by Chaston.[160] Chaston, in pursuing the idea that selective chemical attack at grain boundaries might be localised by the formation of a protective film of corrosion product over the faces of the alloy crystals, carried out experiments with a 9 carat Au-Ag-Cu-Zn alloy which had been cold rolled to about 80% reduction in thickness. In acid $FeCl_3$solutions, this material cracked in less than five minutes. If, however, the test sample was protected with a wax coating on one side and exposed to the action of a rotating brush on the other side, no cracking occurred after thirty minutes, though considerable general corrosion occurred.

From a practical point of view it must be emphasised that susceptibility to stress corrosion cracking is an inherent property of an alloy in respect of the environment to which it is exposed. It is not normally to be expected in Au-Ag-Cu alloys of caratages of 14 or higher, though some 14 carat alloys may be susceptible to it. It must always be borne in mind, however, especially in the processing, storage and use of gold alloys of caratages less than 14. These should not be exposed in pickling baths or in atmospheres containing hydrochloric acid fumes whilst in an unannealed or stress condition. The possibility that internal stresses created by cold working may, at higher levels, reduce the susceptibility of a gold alloy to stress corrosion has been referred to above.

No record of systematic attempts to develop low carat gold alloys resistant to stress corrosion cracking have been noted.

Closely related to stress corrosion cracking in aqueous environments is the cracking of low carat golds and of clad or plated golds which occurs as a result of their exposure under stress to mercury.

Except that cracking results from preferential diffusion of the liquid metal at activated grain boundary sites, rather than from electrochemical attack, stress corrosion creacking by mercury is very similar to stress corrosion cracking by conventional aqueous corroding agents.[156] Thus it does not occur with pure metals, and with homogeneous Au-Cu alloys the susceptibility to corrosion cracking by mercury increased with copper content up to a copper content of 50%, and then gradually decreases to zero again as the copper content is increased to 100%.

Low melting solders in the liquid state can induce stress corrosion cracking in the same way as mercury.

6.3 Staining of the skin and clothes by gold jewellery

Staining of the skin or of clothes by gold jewellery is an effect which is well known as a source of occasional embarrassment in the jewellery trade. It can be caused either by abrasion of the jewellery alloys or as a result of the formation upon them of surface films as a result of tarnishing or corrosion.[160,161]

Abrasion may occur either as a result of friction between parts of a piece of jewellery, or as a result of the presence on the skin of hard materials such as may be present in cosmetics, or in the stiffening or filling materials in textile fabrics. The result is the formation of very fine metallic particles which adhere to the skin or the clothing and stain it black. Such blackening may be avoided either by preventing the abrasion or, where this is feasible, hardening the surface of the jewellery by hard gold plating.

As indicated above, Au-Ag-Cu alloys are neither tarnished nor corroded at caratages which are higher than between 15.6 and 18 depending on their composition. Provided jewellery alloys are homogeneous, and provided that minor metal additions are not present in them which decrease their resistance, they will therefore not be susceptible to tarnishing or corrosion at these higher caratage levels. Jewellery alloys of lower caratages may, however, depending on their composition and their homogeneity, be susceptible. In such cases staining of the skin or of clothes may arise in two ways. Thus it may occur as the result of tarnishing of the alloy whilst the jewellery is in storage. The tarnish films are usually so thin as not to be readily noticed, and consist of sulphides and oxides formed by reaction of Ag, Cu or other base metals in the alloy, with oxygen, moisture and sulphurous impurities in the air. These dark coloured tarnish products can be one source of dark stains on skin and fabrics when the jewellery is worn. Another source of such stains, however, can be products formed by corrosion of the jewellery alloy by perspiration, which contains such substances as sodium chloride, amino-compounds and fatty acids. Such corrosion may even be accelerated by electrolytic effects where different metals and alloys are present in a piece of jewellery. Traces of perspiration which remain on low caratage jewellery when it is stored must also have an effect on the subsequent development of tarnish films upon it.

It has been suggested that variations in the quantity and composition of the sweat gland secretions by different individuals and even by the one individual may be a factor where staining arises from corrosion of gold jewellery alloys by sweat. Whilst no evidence to support these suggestions has been recorded, it will be apparent that high sweat secretion on occasions when low carat gold jewellery is worn could well increase the risk of staining as a result of corrosion.

The above comments relate to yellow and red carat gold alloys only. Staining effects are apparently not observed with the harder and more corrosion and tarnish resistant palladium white golds. As has already been described, rhodium-plated low-carat nickel white golds may give rise to problems during use as a result of abrasion or local corrosion.

6.4 Pickling and "colouring" or surface enrichment of carat golds

In jewellery manufacture the normal pickling of carat golds to remove surface films of oxide formed during heat treatments is carried out[162] with preparations based on either sodium hydrogen sulphate ("safety pickle") or sulphuric acid (5–10%), or in the case of white 18 carat golds on which nickel sub-oxides are apparently formed under mild oxidising conditions, by preparations containing sulphuric acid (10%) plus an oxidising agent such as potassium dichromate.[163] Detailed procedures are described in a number of publications.[164–170]

More active reagents can be used, however, to produce special effects on surfaces cleaned by these methods. In the "colouring" of carat golds for example, the alloy is exposed to media which — in effect — preferentially dissolve base metal components from the alloy. In this way the surface layers of the alloy are enriched in gold and on polishing a product of more gold-like appearance is obtained. This "colouring" process has a long history and was apparently practiced in Mycenae, Troy and Japan and by the Incas in South America, who used for the purpose the juices of a sort of wood sorrel containing oxalic acid. In more modern times hydrochloric acid, sulphuric acid (1:1) and aqua regia reagents have been used. Using the latter, there is always some gold dissolved in addition to base metals, but this is largely redeposited on the alloy surface by electrochemical replacement reactions.[164,171] Different compositions of solution may be used, however, for treating silver and gold-rich alloys and special procedures followed if more than superficial surface enrichment is desired.[172] The most marked effects are obtained with copper-rich alloys of caratage 14 or less.

6.5 Electropolishing and electroetching

Electropolishing is a selective electrolytic dissolution of the metal which leads to smooth and brilliant surfaces being obtained. In the jewellery industry this technique is often used to obtain a final polish of massive gold pieces, and is particularly useful for polishing recessed portions of objects not accessible to polishing by mechanical means. Even for soldered gold alloy objects this method is most useful, since a simultaneous polish of the alloy and soldered areas is possible. The technique is not recommended for thin gold electroplate, which is attacked too severely.

Electroetching of gold is often used to obtain a dull matt finish on gold and gold alloys for decorative purposes; patterns can be etched on the surfaces of gold objects by using photoresists to screen parts of the metal from electrolytic attack. This gold or gold alloy foil can even be pierced in this way, thus often avoiding the need for manual methods.

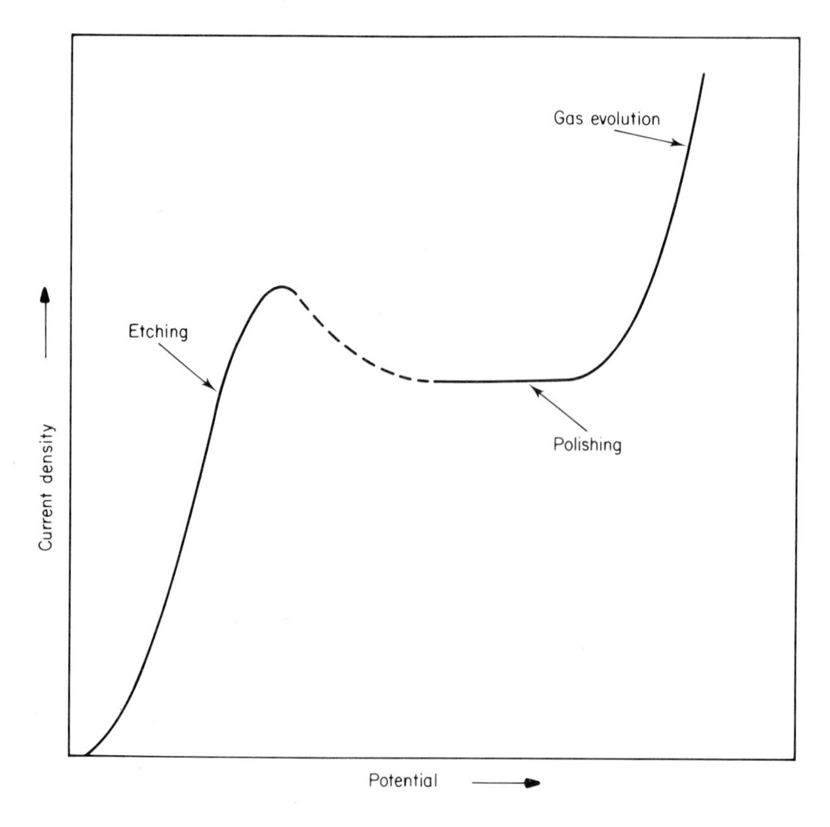

FIG. 15. Typical anodic current-potential curve.

Both of these techniques rely upon the anodic dissolution of the metal,[173–175] and can be conveniently described in terms of the typical anodic current-potential curves (Figure 15) in which the current initially increases with increasing potential until a peak or plateau is reached; in this electrochemical domain the surface of the anode usually becomes etched. Upon further increase of the potential, passivation of the surface sets in, and the current values become unstable and decrease. This is followed by a stable plateau at these lower current density levels over which polishing of the anode surface occurs; optimum polishing usually occurs at the lowest ratio of current:potential. When the potential is increased further, the current increases once more, and dissolution of the metal is accompanied by the evolution of gas; pitting of the surface of the anode occurs, which can be accompanied by polishing action at very rapid rates of gas evolution.

It has been found that the quality of the polished finish depends upon both the temperature of the solution and the concentration of the reagents, and that the height of the polishing plateau in the current-potential curve is affected by the viscosity and extent of agitation of the solution. This clearly demonstrates the diffusion controlled nature of the polishing action. The original condition of the surface is however also a controlling factor in the quality of the final polish.

The exact chemical reactions and mechanisms of reactions which occur on the surface of the anode are not clear. The functions of an ideal polishing process must include[176] both "smoothing" by elimination of the large scale irregularites ($>1\ \mu m$) and "brightening" or "optical smoothing" by removal of the smaller irregularities (down to about $0.01\ \mu m$).

Electropolishing can be distinguished in that it combines both smoothing and brightening, whereas smoothing of the surface is also observed in electroetching.

Current theories ascribe[174] the smoothing effect to the formation of a viscous layer of reaction products around the anode which is relatively thick compared to the surface irregularities, and has an approximately plane interface between the layer and the bulk of the solution. Smoothing during electropolishing can be accounted for qualitatively by the differences in concentration gradient in the viscous layer over the peaks and valleys of the rough surface of the anode; preferential dissolution of the peaks results in a smoothing of the surface.

Brightening, however, does not always occur when a viscous layer is present, and there is now considerable evidence suggesting that the formation of some type of film on the surface of the anode occurs during the polishing process.[174] This film is probably only a few atomic layers thick, and its nature is not certain; it is probably not identical to bulk oxides or insoluble metal salts. Thus it is likely that the film undergoes a continuous process involving dissolution by the electrolyte and renewal by metal from the anode. Such a film would dissolve readily if exposed directly to the electrolyte, and it appears that a further function of the viscous layer is to assist in the maintenance of the film by restricting chemical attack.

The first electropolishing effects on gold were reported at the start of this century,[173] but the first industrial applications are reported[176] to have occurred during the 1940's. Historically the laboratory methods were developed first and are fairly well documented. While no doubt the industrial processes are based upon similar principles much of this information remains unpublished.

Slightly different conditions appear to be necessary for polishing pure gold and gold alloys.[177,178] Recently an electropolishing bath containing $K_4Fe(CN)_6$, alkali tartrates, $Na_2B_4O_7$ and KCN was patented [179] for gold which operates at temperatures $\geq$70°C and current densities of $\leq$120 A/cm^2. The behaviour of cyanide baths, adjusted to different concentrations and pH's with dilute HCl, in the polishing of a 14 carat alloy has been reported upon,[180] and a bath for electropolishing Au-Cu alloys described.[181]

When electopolishing gold in these cyanide media, stainless steel, carbon or platinum have been used as cathode materials;[178] cathodes should have an area at least ten times greater than that of the anode surface to be polished.[176] The anodes should be agitated rapidly in the electrolyte.[174,176,178] These solutions are toxic, and should be used under a ventilated hood, since cyanogen compounds are evolved.[174,176] The life of such a solution is of the order of 10 to 15 minutes,[178] and consequently has to be replenished frequently with fresh cyanide; as a result the performance of these solutions is often erratic.[174]

Gold and gold alloys of a fineness of at least 8 carat can also be electropolished in acidic solutions containing thiourea.[182] Improved electropolishing of gold alloys was reported in acidic solutions of thiourea containing metallic ions common with the alloying additions in the gold alloy.[183]

It has also been reported[184] that gold can be electropolished at 80°C in ferric chloride, which melts in its own water of crystallisation.

A solution consisting of 100 g KCN, 30 g KCl, 50 g $K_4[Fe(CN)_6]$ and 1 000 m*l* H_2O has been recommended[178] for deep etching at room temperature of a variety of gold alloys; a stainless steel cathode is used, and a current density low enough to prevent gas evolution at the anode is suggested.

Other electrolytes suitable for electroetching of gold have been reviewed.[176] They include:–

(i) a solution of Au(III)-chloride;
(ii) warm 20 to 25% KCN solutions, which produce fine-grained matt surfaces;
(iii) solutions containing Fe(III)-chloride and concentrated HCl, using a current density of 0.25 A/cm^2; and
(iv) a solution prepared by mixing equal parts of 10% solutions of KCN and NH_4-persulphate, which has apparently been used in industrial production.

Table XXV. Cyanide Solutions for Electropolishing Gold and Gold Alloys

Solution No.	Composition of electrolyte	Current density (A/dm^2)	Voltage drop across cell (V)	Polishing time (minutes)	Temperature (°C)	Reference
1	67.5 g/l KCN 15 g/l $K_4[Fe(CN)_6]$ 15 g/l Na-K-tartrate 18.50 g/l H_3PO_4 2.5 ml/l NH_3(aq) 1 000 ml water	100	4.5–5.0	1–4	55–62	178, 181
2	90 g/l KCN 22.5 g/l $K_4[Fe(CN)_6]$ 18.75 Na-K-tartrate 15 g/l H_3PO_4 3.75 g/l Cu-cyanide 15 ml/l NH_3(aq)	150	12	1–2	88	173
3	67.5 g/l KCN 15 g/l Na-K-tartrate 18.5 ml H_3PO_4(d=1.69) 2.5 ml NH_3(liq) (d=0.9) 1 000 ml water	100–150	5–10	1–4	55–62	173
4	200 g KCN 20 g KF 20 g K_2CO_3 800 ml glycerine 120 ml H_2O	93	12	1–2	93–99	178
5	150 g KCN 120 g KF 400 ml triethanolamine 100 ml H_2O	78	13	1–2	88	178
6	40 g/l KCN 10 g/l $K_4[Fe(CN)_6]$ 20 g/l Na_2CO_3 5 g/l NaOH	—	15–20	—	—	176
7	20 ml dilute HCl 80 ml 10% KCN(aq)	90	1.3–1.7	—	—	180

Table XXVI. Thiourea Solutions for Electropolishing Gold and Gold Alloys

Solution No.	Composition of electrolyte quantity per litre	Current density (A/dm^2)	Cell voltage (V)	Polishing time (minutes)	Temperature (°C)	Type of alloy (%)
1	25 g thiourea 3 ml H_2SO_4 10 g tartaric acid	1.5–3.5	2.0–4.5	2–5	20–45	Au 58.5; Ag 10; Cu 27.9; Zn 3.6
2	25 g thiourea 100 ml H_2SO_4	1.3–2.3	—	—	25	— do —
3	50 g thiourea 5 ml H_2SO_4	3.5–6.5	3–7	< 2	50–60	— do —
	— do —	2.8–9.2	3–7	—	47	Au 83.3; Ag 10; Cu 6.7
	— do —	2.8–9.2	3–7	—	47	Au 75; Ag 10; Cu 15
	— do —	2.8–9.2	3–7	—	47	Au 75; Ag 4; Cu 4; Zn 4; Cd 13
4	25 g thiourea 10 ml H_2SO_4	1.3–7.3	1.5–5.2	—	60	Au 33.3; Ag 12; Cu 37.8; Zn 16.4; Ni 0.5
5	50 g thiourea 25 g Cu SO_4 Acidic medium	6	3	5–10	—	Au 83.3; Ag 10; Cu 6.7
6	75 g thiourea 10 g $AgNO_3$ Acidic medium	5	3	10	—	Au-Ag alloys
7	50 g thiourea 15 g $NiSO_4$ Acidic medium	6	—	5–10	60	White 8 to 18 carat Au-Ni alloys with minor Ag-Cu content

Low current densities are recommended for most of these electrolytes.[176]

The compositions of various cyanide-containing solutions, as well as the experimental conditions under which it has been claimed they can be used for electropolishing gold are summarised in Table XXV. Similar data for patented thiourea solutions are listed in Table XXVI. These solutions are not generally effective for the polishing of nickel or palladium white golds.

6.6 Chemical polishing and chemical etching

Chemical etching of gold is easily achieved via solutions containing a complexing ligand for gold and a suitable oxidant.[176,178]

Aqua regia (3 to 4 parts concentrated HCl to 1 part concentrated HNO_3) rapidly etches gold and most gold alloys, but is not really suitable for gold alloys with a major proportion of silver. An alkaline solution of cyanide containing an oxidising agent such as H_2O_2 or dissolved O_2 is recommended for metallographic and decorative etching, provided that the resist used tolerates alkaline solutions. A very useful solution for metallography[185] is an aqueous mixture of equal volumes of 10% NH_4 persulphate and 10% KCN; the individual solutions are stable, but the mixture has a life of a few minutes only. This solutions should be used under a fume hood.[178] Metallographic etching of gold alloys is also possible by means of a 50% solution of I_2 in aqueous KI;[185] the film which is formed on the surface can be removed with a KCN solution.

A procedure for polishing or brightening alloys by chemical means with KCN and H_2O_2 has been patented.[186] A concentrated (20%) solution of KCN is used, to which an equal volume of 50% H_2O_2 is added. This technique is highly hazardous and should be carried out in an efficient fume chamber. It may be applied in order to restore the colour of white gold alloys which have developed a yellow shade.

6.7 Investment casting of gold alloys

The melting and casting of gold alloys into ingots is described in some detail in most texts and handbooks on the working of gold. In contrast, investment casting of gold alloys, which has expanded in recent years to the point where it is probably the most important single production process in jewellery manufacture, is the subject of very few technical publications. Of these, most are concerned with the casting of dental gold alloys,[187] which is better documented than that of gold jewellery alloys. Those that do relate to jewellery alloys tend to be directed more to the individual craftsman than the jewellery manufactuer.

In the adaption of the process for jewellery fabrication considerable effort has been devoted to the development of high quality rubber, wax and investment materials, and a wide range of equipment for casting has become available in recent years. The number of parameters which can affect the quality of final castings is so great, however, that optimum efficiency in the investment casting of gold alloys is apparently not always achieved, even with sophisticated equipment, particularly with low caratage alloys. Whilst the nature of the alloys used is a factor in this situation, it is only one of many factors, and cannot be considered in isolation.

Features of the casting of gold jewellery are the smallness of the amount per pour even when "large" casting equipment is used, the small and intricate nature of many castings and their high surface to volume ratios. The fluid pressure distribution and opposing surface tension effects play a significant role in determining the speed with which moulds may be fed, and many of the defects encountered in casting are associated primarily with incorrect feeding during solidification. If patterns are not correctly sprued, temperatures are not correct or the alloys cast unsuitable, good results cannot be obtained even with the best equipment. Consistency in control of the various casting parameters is of vital importance.

Investment materials usually consist of silica (quartz or cristabolite) bonded with hydrated calcium sulphate, which becomes dehydrated during firing. Removal of such investment from castings has been found[188] to be best effected, after preliminary high pressure water spray or ultrasonic[189] treatment, by a boiling solution of ammonium sulphate (150 g/litre). With certain alloys (see below) to which small amounts of Al have been added, such removal is facilitated.

Apart from other factors, the presence of grain refining agents in the casting alloys contributes greatly to improvement of the surface finish and the strength of castings — an effect which has been exploited particularly in the making of dental gold alloy castings, in which mechanical strength is of particular importance. They do not appear to enjoy, as yet, widespread use in gold jewellery casting alloys, though Handy and Harman[190] have claimed that Au-Ag-Cu and Au-Ag-Cu-Zn alloys containing small additions of nickel or palladium ($<0.1\%$) and of ruthenium ($<0.05\%$) are particularly useful for casting of jewellery.

During casting shrinkage occurs in three ways; by thermal contraction of the liquid metal between the temperature to which it is heated and its liquidus temperature, by contraction during passage from the liquid to the solid state, and by thermal contraction as it cools from the solidus point to room temperature. Contraction in the liquid state is not of great importance since as the liquid metal contracts in the mould, more molten metal can flow into the mould to compensate for the contraction. Contraction during solidification and cooling is more important, however, and for a range of dental alloys cast in different dimensions and shapes has been reported to be $1.25 \pm 0.1\%$[191] though higher figures than this (1.42–1.67) have been reported.[192] These figures are less than the percentage linear thermal contraction of gold alloys and gold from their solidus points, as determined experimentally, which may be in the range of 1.5–2.0%. This anomaly can be explained[187] if it is assumed firstly that solidification occurs first at the walls of the mould, and secondly that the first layer of metal to solidify adheres to the mould until it gains sufficient strength as it cools to pull away from the mould, and to contract independently. Castings with a high ratio of surface to volume, in which the onset of shrinkage could be expected to be delayed significantly could therefore be expected to show diminished casting shrinkages.

In practice the shrinkages which occur in the preparation of the wax pattern and in the casting alloy must be compensated for where precision casting such as is necessary for dental restorations is called for. This may be done normally by taking advantage of one, or a combination of more than one, of the following effects:

(a) Thermal expansion of the wax pattern;
(b) Setting expansion of the investment; and
(c) Thermal expansion of the investment.

Thermal expansion of the wax pattern may be exploited by maintaining the temperature of the water-investment mixture at a suitable temperature (say 40–42°C) after the pattern has been invested. Distortion of the wax investment under these conditions can cause difficulties, however, so that most precision casting exploits one or other of the latter two effects.

In (c), reliance for compensation is upon the expansion of the mould during heating to the temperature (say 700°C) at which casting is carried out. In (b) reliance is mainly upon the expansion of the mould which occurs when it is heated to say 427–482°C, during which the expansion is determined in large measure by the changes caused by the dehydration of the investment which takes place.

The chief among a variety of causes for rejection of gold alloy castings is porosity, which can arise either in the course of cooling and solidification or as a result of gas inclusions. Voids at the ingate area may be a result of the supply of molten metal being cut off before the mould is completely filled. Under such circumstances voids tend to form in the zone of dendrite formation, the gas inclusions developing around the dendrites in positions which are usually filled as solidification proceeds, if liquid metal supply is adequate. Voids or gas inclusions may also result from the trapping or entrainment of air in the mould during pouring and solidification. In such cases a common cause is a too turbulent flow of the molten metal when poured.

As distinct from the above, microporosity may develop if the molten alloy is too close to its liquidus temperature when poured, and solidification therefore so rapid that shrinkage develops throughout the whole casting. It is characterised by the presence of small irregular voids distributed throughout the casting.[193]

A further type of defect arising during cooling and solidification is sub-surface porosity, which occurs near the surface of castings, and which is thought to arise when solidification of the central part of the casting is delayed, as may be the case if the molten metal is fed to the mould too rapidly and at too high a temperature. The initially formed solidified "skin" to the casting is locked to the walls of the mould and the central core of the casting later pulls away from this skin as it solidifies, creating sub-surface voids. Surface defects may arise if gas or water vapour is generated by the mould in contact with the molten metal. In such cases the fault may be sought in the procedures followed in preparation of the mould.

Finally pinhole porosity may result from the presence of hydrogen in the melt. Any action which could bring moisture (the commonest source of hydrogen) into contact with the melt should therefore be avoided. Where easily reduced oxides such as CuO are present in the melt, hydrogen may be converted either wholly or in part into steam. This reaction is avoided where a deoxidiser such as silicon is present in the casting alloy, since this converts the oxygen into a form (SiO_2) which does not react with hydrogen. It does not, however, eliminate the hydrogen.

Remelting of gold casting alloys can also give rise to casting difficulties, if it is done often enough to bring about changes in the alloy composition. One cause of trouble in such cases is the loss of zinc; because of its volatility zinc containing casting alloys cannot be used in vacuum casting systems. Another cause may be a build-up of oxides in the casting alloy, especially where this has been melted without adequate protection from oxidation. With the higher melting nickel white golds *(quo vide)* difficulties may also arise from hot shortness caused by sulphur or even silicon generated during the melting operation, particularly under reducing conditions, as a result of the formation of low melting eutectics.

6.8 Continuous casting and hot working

Continuous casting

The need to increase productivity in the large scale fabrication of gold alloys makes the development of procedures which are potentially less labour intensive than traditional ones a matter of considerable interest.

Although no account of the continuous casting of gold alloys has been noted, it is apparently practised in both North America and Europe.[194]

The continuous casting of fine silver alloys on a large scale, however, has been described[195] and in the absence of published information on the similar casting of gold alloys, seems worth mentioning. It does away with casting in individual moulds, and is claimed to yield better quality castings, and to improve the economy of the production of various silver and silver alloy intermediates, through better yields, and the processing of larger batches.

The metals (see Figure 16) are melted in induction furnaces (1) and fed from a casting ladle (2) to a holding furnace (3), whose lower end has a flanged-on, water cooled graphite mould (4) through which the metal is extruded. Before casting begins, this mould is closed by a rod, which is withdrawn by the rollers of an extractor machine (5). The melt solidifies on passing through the mould and can be taken off continuously. By the use of different moulds, castings of different cross sections can be produced as strips, bolts, rods or piping. These castings are cut into desired lengths by a heavy-duty circular saw (6) and passed on for further processing, such as extrusion pressing, wire rolling and drawing.

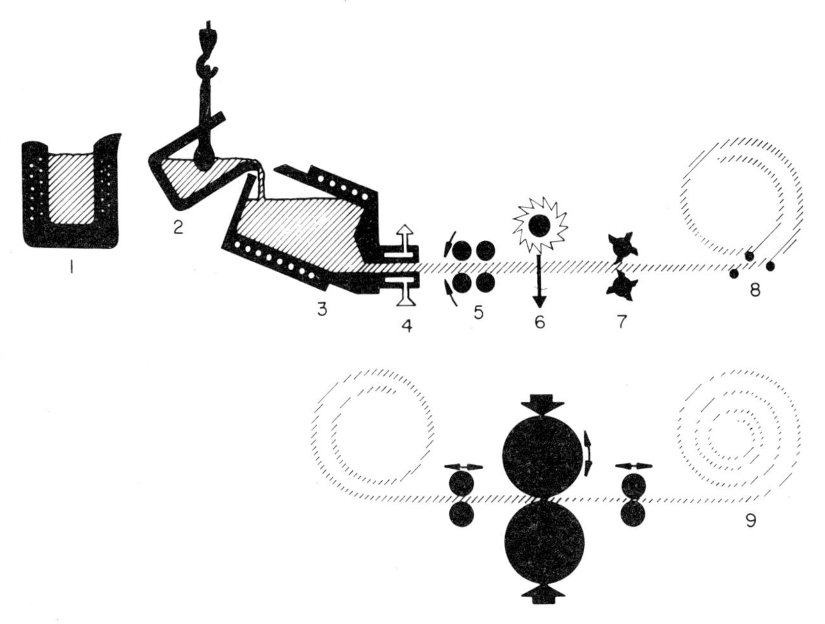

FIG. 16: Schematic drawing of the continuous casting and rolling plant for silver and silver alloys of Degussa's Technical Metal Products Division.

Hot working

Because of the hardening which occurs when metals are cold worked, there is a limit to the extent of deformation to which gold alloys may be subjected before they must be annealed in order to soften them. In hot working, however, metals are deformed at temperatures at which no work hardening occurs. Re-softening by recrystallisation and recovery processes takes place continuously during working. Moreover, since at hot working temperatures metals are softer, larger deformations are possible and power costs and tool wear are reduced.

Despite these potential advantages of hot working, the facts that wrought gold alloys can usually be cold worked and that they are usually processed in small amounts have reduced interest in hot working on the part of the processors of gold alloys. The result is that very little information has been published concerning the hot working properties of gold alloys, and most gold alloys in use have been developed primarily for fabrication by cold working. In fact only one report has been noted on studies of the hot working of gold alloys, and this relates to the behaviour during hot working of Au-Cu-Zn alloys with shape memory.[196]

Nevertheless some hot working processes might possibly be applied with advantage to gold alloys. In particular, precision hot stamping is an excellent way of producing small components in large numbers, and at least one company hot stamps carat gold alloys to produce watchcases.[194] Hot forging of gold alloy knife blades has also been used as a means of getting increased toughness and wear resistance.[7] Hot rolling is also more efficient than cold rolling, requiring less time for a given reduction and no intermediate annealing, and hot extrusion is an ideal way of breaking down the structure of a cast ingot and forming it into stock for subsequent sizing by rolling and drawing. By extruding to near the finished size, much of the annealing needed at intermediate stages in cold rolling or drawing can be avoided.

Considerations such as these indicate that systematic studies of the hot working properties of gold alloys might well be rewarding.[197]

6.9 Contamination of carat golds

Certain contaminants of gold and its alloys can prevent their coherence, so that they show brittleness and hot shortness. The most important of these is lead,[198] which may find its way into the alloys from soft solder residues, or from the lead underlays of anvils. It forms — inter alia — the brittle compound Au_2Pb, which deposits in the grain boundaries and which melts as low as 418°C. In its presence even in small concentrations (0.005–0.06%) wires break up and sheets crack during cold working, and in hot working there is little coherence of the metal and hot shortness is evident. As little as 0.2% Pb in coinage alloys is sufficient to cause cold shortness.[199] Other observations in this area have been summarised by Grover and Grimwade.[200]

Analogous effects are produced by phosphorus and sulphur in gold alloys.[30,201] In these instances the embrittling agents are the phosphides and sulphides of such metals as copper, silver or nickel. Sources of phosphorus contamination are few, but sulphur may be derived from a variety of sources such as sulphur containing gas used for heating, crucibles, or even plaster of Paris.

The most "normal" source of contamination, however, is oxygen which can give rise to the formation of oxides both within and on the surface of the lower caratage alloys during melting and annealing unless appropriate precautions are taken.[201] Because of these effects, melting and annealing operations should not be prolonged without protective measures. In the case of nickel bearing alloys, such measures may involve coating with fluxing agents since the oxide films which are formed on them are difficult to remove by normal pickling methods.

Silicon can be troublesome, since it may be generated at high temperatures under reducing conditions if a melt is in contact with silica. It is especially serious in Au-Pt and Au-Pd alloys and in nickel white golds in which it forms brittle compounds.

Not all minor components of gold alloys, however, have deleterious effects. Thus aluminium has been incorporated[202,203] in small amounts in casting gold alloys, on wich it is claimed to form an oxide layer at high temperatures, so that castings are cleaner and brighter and more easily stripped. Moreover, as has already been mentioned, certain elements such as rhenium, iridium, ruthenium and rhodium have highly beneficial effects on certain alloys because of their grain refining action. Other elements (iron, cobalt, nickel) can be used to control Au-Cu ordering reactions in gold alloys, whilst as is described later, still other elements such as yttrium and the rare earth metals can be used to modify the recrystallisation temperature of gold and therefore to control its creep behaviour when exposed to heat under stress.

6.10 Colour standardisation and matching

Colour control and matching is an important feature not only of the fabrication of carat gold jewellery, but also of the production of decorative gold coatings.

Although no international standards have been agreed upon, standards which have been set up in Switzerland,[204] France[205] and in particular Germany [206] are widely used. The DIN standard colours, which enjoy code designations, are based on those of a series of reference alloys prepared by melting. They are applied both to wrought gold and to electroplated gold alloys, though in the latter case, as a result of structural differences, coatings of the same composition as a standard alloy may not match it in colour. Standard colours used in watchcase plating, together with the compositions of the reference gold alloys as illustrated by Massin and Reid[207] are given in Table XXVII.

Comparisons of colour with the standards are usually made subjectively using diffuse reflected light — the procedure involving placing a tissue paper over the alloy surface in order to avoid confusing specular reflections of other objects and colours in the vicinity.

Objective methods[208–210] for describing colours of metals in terms of their optical constants as determined by one or other form of reflectance spectroscopy are difficult, and the equipment used for more physical studies of the optical constants of gold, and its alloys with such metals as Ag, Cu and Zn[211–215] has been unsuitable for routine industrial use. Gardam, however, has

Table XXVII. Standard Colours in Watchcase Plating

Colour coding	Colour description	Carat	Corresponding reference gold alloys (Composition, parts per thousand)				
			Gold	Silver	Copper	Nickel	Zinc
1N-14*	Pale yellow	14	585	265	150	—	—
2N-18*	Pale yellow	18	750	160	90	—	—
3N*	Yellow	18	750	125	125	—	—
4N*	Rose (pink)	18	750	90	160	—	—
5N*	Red	18	750	45	205	—	—
ON**	Yellow-green	14	585	340	75	—	—
8N**	White	14	590	—	220	120	70

* Standards common to Germany, France and Switzerland.
** Standards common to Germany and France.

described[208] the use of a simplified commercial spectrophotometer for the determination of spectral reflectivity curves such as those in Figure 17.

Although such curves completely define the efficiency of light reflection for metals, Gardam has pointed out that they are less convenient for comparison and discussion than "chromacity coefficients", which can be calculated[208] from the spectral reflectivity curve, and plotted on a colour triangle. Certainly any system which may be adopted for defining objectively the colours of metals must take account of their overall reflectivities or luminances. This is well illustrated in the case of nominally "white" metals such as sterling silver, rhodium plate, chromium plate and tin-nickel plate. The "whiteness" of these metals is a result of their approximately horizontal (see Figure 17) spectral reflectivity curves. Their actual colours as recorded by the eye, however, are determined as much by the average heights of these curves (i.e. the overall spectral reflectivities to which they correspond) as by their departures from the horizontal. With decreasing overall reflectivity, the metals appear increasingly grey, and with greater reflectivity in certain parts of the spectrum overtones of colour appear in this greyness — e.g. faint pink overtones in the grey colour of tin-nickel plate (see Figure 17).

The transformation of instrumental readings into a form which corresponds to visual perception requires definition and agreement upon the colour detecting characteristics of the average eye. In promoting such definition and agreement, through the adoption of colour mix data representative of the normal observer, and of reference stimuli in terms of which the amounts of the three (red, green and blue) stimuli required to match colours over the whole spectrum can be expressed, the C.I.E. (Committée Internationale de l'Eclairage)[216] has achieved considerable success. The time may therefore be near when it will be possible to describe the colours of carat golds in terms of internationally accepted colour specifications derived by accepted procedures from spectral reflectivity curves.

In the assessment of metal colours, whether subjectively or objectively, the important effects of surface structure, surface finish and tarnish films must be emphasised. Thus in respect of finish, the multiple reflections which occur in roughened gold surfaces cause such surfaces to assume a reddish hue, because of the higher reflectivity of gold surfaces for light of longer wavelengths.[217] In respect of structure, it is of interest that deformation of certain gold alloys during working produces colour changes.[218–221] Thus Au-Ag alloys containing 58.5–75% Au tend to be greenish white in the annealed condition, but yellow to yellow-green after cold working. Similar development of yellow colour is apparently also observed in cold working of certain Au-Ag-Cu alloys. Thus a 52Au-22Ag-26Cu alloy is reddish white in the soft condition but yellowish white when hardened, a 55Au-28Ag-17Cu alloy is white when soft but yellow-green when work hardened, and a 58Au-32Ag-10Cu alloy is whitish-green when soft and yellowish green when hard. Effects of cold working on colour are apparently not observed, however, with all Au-Ag-Cu alloys.[222]

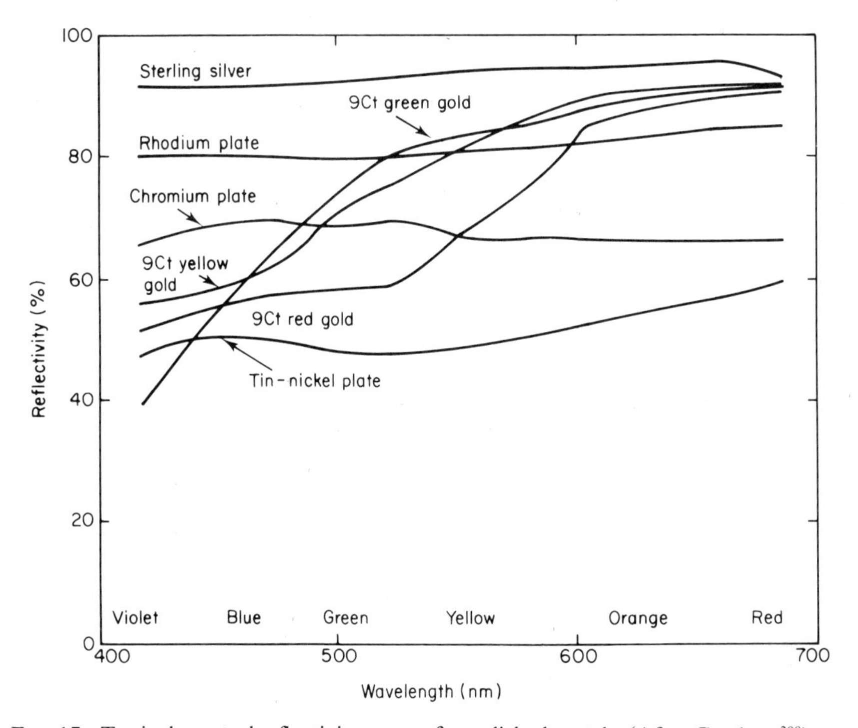

FIG. 17. Typical spectral reflectivity curves for polished metals. (After Gardam.[208])

In relation to the above, it may be noted that electron microprobe analyses which have been made[223] across castings of certain low caratage Au-Ag-Cu-Zn alloys have indicated that the dendritic centres in such castings are enriched in gold and that the interdendritic areas are eutectic in character with both copper-zinc and silver-zinc rich regions. The problems involved in relating the colours and compositions of such heterogeneous alloys will be apparent.

These problems are almost certainly increased in the case of polished surfaces. Thus microhardness determination established that the gold-rich dendritic centres were significantly softer (Hv98) than the interdendritic (Hv147) and eutectic structures (Hv180). The possibility of gold enrichment of the surfaces of alloys of this type occurring during polishing as a result of a smearing of the gold-rich phase over the surface must therefore be taken into account. The fact that electron probe microanalyses of the polished surfaces of one wrought and two cast low carat alloys all revealed an increase in the gold content of the surface which extended to a depth of about 1 μm indicates that this possibility cannot be ignored. The enrichment was greater in the case of the polished wrought specimen. Effects such as these would be expected to be enhanced in the case of speciments from the surfaces of which base elements or phases have been leached out in pickling or colouring treatments.

The factors affecting the colour of gold alloys have been reviewed by Rivlin and Brook,[223] and by Saeger and Rodies.[223 a]

7. Carat Gold Solders

7.1 Caratage, colour and other requirements

In each application in this field, and apart from the normal requirements in respect of melting range (or working temperature) and other soldering or brazing properties, the colour and caratage of the solder or braze must be appropriately matched with the colours and caratages of the parts being joined. Additionally the behaviour of the solder or braze during subsequent polishing and wear must be such that no discontinuities in the appearance of the finished article become apparent during these processes. The hardness, texture, and resistance to tarnish of the solder in situ must therefore not differ too greatly from the corresponding properties of the joined parts. These are signifcant and important considerations.

7.2 Some observations on current practice

In practice many makers of jewellery purchase their soldering and brazing alloys and the fluxes to be used with them, in the form in which they require them, as proprietary products from one or other of the major suppliers of precious metals. Apart from specifying the caratage, colour, melting range, flow characteristics, and the recommended flux, no other information is normally available from most suppliers concerning their proprietary products. In each caratage supplied, and for each colour where feasible, each supplier endeavours, however, to make available a number of alloys with melting ranges covering a range of temperatures, which can be used in appropriate succession for first, second, third, etc. brazes in the fabrication of any one article. In some cases a recommended working temperature within the melting range is also specified.

Standard specifications for carat gold solders have not so far been introduced.

Where the parts being joined are not of compositions which are too close to those with the minimum melting point for the alloy systems involved, the individual jeweller can choose, however, to make up his own solder by altering the composition of a portion of the alloy being joined so as to reduce its melting range appropriately. Where the alloys are Au-Ag-Cu alloys this can be achieved in each caratage by increasing the ratio of Cu:Ag in the alloy. The effect of this on the colour of the alloy must be taken into account however. Tables have been presented (see Table XVIII) which reflect the melting behaviour of the different caratage Au-Ag-Cu alloys with variations in their Cu:Ag ratios.

A recent and important development is that of brazing alloy pastes, which are available in a range of carat gold alloys, and with various flux contents. These pastes flow freely, do not settle out on standing and can be fed from automatic dispensers in automated brazing machines. Such systems apparently have advantages over systems based on the use of brazing alloy wires. The alloys are incorporated in the pastes in finely divided form.

7.3 Compositions and melting ranges of some typical carat gold solders

The compositions and melting ranges of some typical carat gold solders, assembled from various sources, are listed in Table XXIX. From these a number of points emerge.

Typical working temperatures for commercially available "yellow" gold solders as taken from one supplier's catalogue, for example, are as follows:

9 carat	793°C	14 carat	780°C
	760°C		720°C
	715°C		670°C
	615°C		
18 carat	820°C		
	750°C		
	700°C		

Table XXVIII. Showing the Effect of Varying Silver Contents on the Melting Points of Au-Ag-Cu Carat Gold Alloys

Composition per cent			Ratio		Melting point (°C)	Flow point (°C)
Gold	Silver	Copper	Silver	Copper		
22 carat gold						
91.6	6.3	2.1	3	1	1024	1035
91.6	4.2	4.2	1	1	971	1003
91.6	2.1	6.3	1	3	954	979
18 carat gold						
75.0	21.4	3.6	6	1	976	1005
75.0	17.0	8.0	2	1	934	968
75.0	12.5	12.5	1	1	882	905
75.0	8.0	17.0	1	2	882	893
75.0	3.6	21.4	1	6	881	902
15 carat gold						
62.5	32.0	5.5	6	1	966	1004
62.5	25.0	12.5	2	1	890	925
62.5	18.75	18.75	1	1	860	877
62.5	12.5	25.0	1	2	875	895
62.5	5.5	32.0	1	6	910	925
12 carat gold						
50.0	43.0	7.0	6	1	900	939
50.0	33.3	16.7	2	1	809	838
50.0	25.0	25.0	1	1	807	827
50.0	16.7	33.3	1	2	851	865
50.0	7.0	43.0	1	6	881	905
9 carat gold						
37.5	53.5	9.0	6	1	905	948
37.5	41.5	21.0	2	1	800	820
37.5	31.25	31.25	1	1	825	863
37.5	21.0	41.5	1	2	875	900
37.5	9.0	53.5	1	6	915	950

It will be noted that no solders are supplied with working temperatures below 600°C. Solders of this type which are used in the electronics industry are too brittle to use in jewellery. Red and white carat gold solders are usually available with more limited ranges of working temperatures.

It is perhaps significant that a number of the alloys contain cadmium, which is not desirable because of the toxicity of oxide fume to which it gives rise. As a result there is a tendency, particularly in the U.S.A., to use zinc and tin in preference to zinc and cadmium in many gold solders, and there is a possibility that incorporation of cadmium in carat gold solders may be forbidden in many countries in the future.

Table XXIX. Compositions of some Typical Carat Gold Solders together with their Solidus and Liquidus Temperatures

Au	Ag	Cu	Zn	Cd	Sn	Ni	Solidus tempera-ture (°C)	Liquidus tempera-ture (°C)
*80.0	—	—	8.0	—	—	12.0	782	871
75.0	12.0	8.0	—	5.0	—	—	826	887
75.0	9.0	6.0	—	10.0	—	—	776	843
75.0	9.0	6.0	10.0	—	—	—	730	783
75.0	2.8	11.2	9.0	2.0	—	—	747	788
75.0	—	15.0	1.8	8.2	—	—	793	822
*66.6	10.0	6.4	12.0	—	—	5.0	718	810
66.6	15.0	15.0	3.4	—	—	—	—	—
58.5	25.0	12.5	—	4.0	—	—	788	840
58.5	10.3	24.2	—	7.0	—	—	792	831
58.5	8.8	22.7	—	10.0	—	—	751	780
58.5	11.8	25.7	4.0	—	—	—	816	854
58.5	25.7	11.8	4.0	—	—	—	786	818
58.5	24.2	10.3	7.0	—	—	—	765	808
58.5	4.9	25.6	2.0	9.0	—	—	738	760
58.5	8.0	22.0	2.1	9.4	—	—	744	776
*58.3	15.0	5.7	15.0	—	—	6.0	—	—
*50.0	25.0	10.0	9.0	—	—	6.0	—	—
**41.7	24.0	16.3	9.0	9.0	—	—	—	—
41.7	35.0	21.9	1.4	—	—	—	—	—
33.3	30.0	16.7	—	20.0	—	—	635	709
33.3	30.0	16.7	20.0	—	—	—	695	704
33.3	40.5	17.0	6.6	2.6	—	—	722	749
33.3	1.8	49.4	2.3	10.2	3.0	—	689	776
33.3	31.0	28.0	7.7	—	—	—	737	808
*33.3	42.0	10.0	9.7	—	—	5.0	738	807
25.0	35.0	20.0	10.0	10.0	—	—	—	—
*25.0	58.0	—	17.0	—	—	—	—	—

* A white gold.
** A yellow gold recommended for soldering green golds.

7.4 Metallurgical aspects

Perhaps the most striking feature of the alloys whose compositions are presented in Table XXIX is the fact that they can almost all be regarded as modified Au-Ag-Cu carat gold alloys. Modification is effected in each caratage range firstly by variation of the Ag:Cu ratio. Apart from its effects on the melting range, such modification also affects the colour of the alloy. It is therefore almost always accompanied by incorporation of one or more of the metals zinc, cadmium, nickel and tin. This is done in such a way as to achieve both the caratage, colour and melting range which is desired.

The rationale underlying the use of zinc in carat gold solders in the amounts indicated in Table XXIX will therefore be apparent. Not only can it be used to assist in the adjustment of the melting range of a solder by modification of the Ag:Cu ratio, but it can also be used to compensate for changes in the colour of the solder caused by such modification.

The same applies in general terms to the use of cadmium or of zinc plus cadmium and to that of zinc plus tin. In these complex systems, however, the phase relationships are not well defined and any attempts to understand the formulation and properties of the alloys in terms of phase relationships are unrewarding.

Nevertheless it is a matter of experience that higher Ag:Cu ratios in the solder tend to promote greater fluidity. Solders with lower Ag:Cu ratios on the other hand tend to be less fluid in the molten state. Alloys of the former type are more suited to joining because of their better penetration, whilst alloys of the latter type are more suited to joins where a build up of an area is called for.

As will be noted from Figure 9 also, high Ag:Cu ratios give rise to paler shades of alloys. White gold solders of a "non-palladium containing type", which are discussed further below, are formulated by the addition of nickel and zinc to these high AG:Cu ratio alloys.

In a similar manner, red gold solders are based upon alloys of high Cu-Ag ratio. For soldering of green golds, certain gold solders in the yellow range may be recommended.

Grain refinement, which is as necessary in these carat gold solders as in the carat gold themselves, does not appear to present problems.

White carat gold solders could possibly also be looked upon in formulating white carat gold solders of this type, as based upon the Au-Cu-Ni system, with the necessary reductions in melting range being achieved, and the white colour retained by reducing the nickel content to about 5% and incorporating in its place appropriate percentages of such metals as silver, copper and zinc. In practice, however, and as will be seen from Table XXIX the amounts of silver which are incorporated by at least one supplier in his products are such that they can better be regarded as alloys based on the Au-Ag-Cu system, which have been modified by addition of nickel and zinc.

Whether any of the white gold solders which are supplied commercially are based on modification of palladium white gold compositions is not clear. White gold solders of the type mentioned by Gee[7] (75Au-15Ag-10Pt) would need to offer particular advantages to compete in price today with white gold solders such as those listed in Table XXX.

Alloys based more directly on the Au-Ni-Cu system have been described by Johnson, Matthey & Company[224] who have claimed alloys containing 30–80% Au, 1.0–67.5% Cu and 2–12% Ni, which may be modified by inclusion of 0.5–7.0% Cr and a trace to 0.5% B, as suitable for jewellery fabrication. These alloys were developed primarily, however, for special applications in the aircraft industry.

Table XXX. The Compositions of some Nickel Containing White Gold Solders*

Au	Ni	Cu	Zn	Ag	Melting range or working temperature (°C)
80.0	12.0	—	8.0	—	782–871
66.6	5.0	6.4	12.0	10.0	718–810
58.3	6.0	5.7	15.0	15.0	704–736
50.0	6.0	10.0	9.0	25.0	724–782
41.7	5.0	8.3	15.0	30.0	702–732
33.3	5.0	10.0	9.7	42.0	721–788

* The compositions listed are those given by one supplier for products in his catalogue.

The incorporation of indium and gallium in carat gold solders has been reported upon,[225–227] and low melting gallium bearing gold alloys have been described[228] as useful in both dental and jewellery work, which contain 1–10% Ga, 0–25% Cu, 0–33% Ag, 0–10% Zn and the rest Au.

REFERENCES

1. Forbes, J. S., Hallmarking gold and silver, *Chem. Br.* **7** (1971), 98–102.
2. van Heerden, J., "Hallmarking and South Africa". Paper presented to an International Jewellery Symposium, Mbabane, Swaziland, September, 1974 (cf. *CHEMSA* **2** (1976), 80–82).
3. Wise, E. M. (Ed.), "Gold: Recovery Properties and Applications". Van Nostrand Company Inc., Princeton, N.J., 1964.
4. Sterner-Rainer, L., "Die Edelmetall-Legierungen in Industrie und Gewerbe". Verlag Wilhelm Diebener G.m.b.H., Leipzig, 1930.
5. Smith, E. A., "Working in Precious Metals". N.A.G. Press, London, 1933.
6. Anon., "Kleine Edelmetallkunde: Kursus für Gold- und Silberschmiede". *Gold, Silber, Uhren, Schmuck.* (This course appears in a series of articles extending over a number of years.)
7. Gee, G., "Gold Alloys". London, 1929.
8. "Der Junggoldschmied: Ein Lehr- und Werkstattbuch für den Nachwuchs des Goldschmiedehandwerks". Rühle-Diebener-Verlag-KG, Stuttgart, 1956.
9. "Diebeners Handbuch des Goldschmieds", Vols. I–IV. Rühle-Diebener-Verlag-KG, Stuttgart, 1963–1969.

10. Brepohl, E., "Theorie und Praxis des Goldschmieds", Third Edition. VEB Fachbuchverlag, Leipzig, 1973.
11. Raynor, G. V., The alloying behaviour of gold, *Gold Bulletin*, **9** (1976), 12–19 and 50–54.
12. "Gmelin's Handbuch der Anorganischen Chemie", 8th Edition, System No. 62, "Gold", 1954, pp. 954–956.
13. Hansen, M., "Constitution of Binary Alloys", Second Edition. McGraw Hill Book Company, Inc., New York, 1958.
14. McMullin, J. G. and Norton, J. T., On the structure of gold–silver–copper alloys, *Trans. Am. Inst. Min. Metall. Eng.* **185** (1949), 46–48.
15. Sterner-Rainer, L., Die Verbindung Au–Cu in Goldlegierungen, *Z. Metallk.* **17** (1925), 162–165.
16. Masing, G. and Kloiber, K., Ausscheidungsvorgänge im System Kupfer-Silber-Gold, *Z. Metallk.* **32** (1940), 125–132.
17. Hultgren, R. and Tarnapol, L., Effect of silver on the gold–copper superlattice, Au–Cu, *Trans. Am. Inst. Min. Metall. Eng.* **133** (1939), 228–237.
18. Spanner, J. and Leuser, J., Der Einfluss von Zusatzmetallen, besonders von Silber auf die Umwandlungshärtung der Gold-Kupfer-Legierungen, *Metallwirtsch.* **14** (1935), 319–322.
19. Nowack, L., Vergütbare Edelmetall-Legierungen, *Z. Metallk.* **22** (1930), 94–103.
20. Wise, E. M., High strength gold alloys for jewellery and age-hardening phenomena in gold alloys, *Trans. Am. Inst. Min. Metall. Eng.* **83** (1929), 384–404 (cf. U.S. Patent 1,577,995 (1926)).
21. Carter, F. E., "Gold, Silver, Copper Alloys", *Trans. Am. Inst. Min. Metall. Eng.*, Technical Publication No. 86. Vol. 78, 1928, pp. 786–803.
22. Chaston, J. C., Heat treatment of gold alloys: the mechanism of disorder-order hardening, *Gold Bulletin*, **4** (1971) 70–71.
23. Shashkov, O. D., Syutkina, V. I. and Rudenko, V. K., Effect of atomic ordering on the decomposition process in a gold–copper–silver alloy, *Fiz. Metal. Metalloved.* **37** (4) (1974), 782–789 (*C.A.* **81** (1974), 157100c).
24. Raub, E., Der Einfluss des Silbers auf die Umwandlungen des Systems Gold-Kupfer, *Z. Metallk.* **40** (1949), 47–54.
25. Raub, E. and Engel, A., Das Driestoffsystem Gold–Kupfer–Nickel. II. Die Umwandlungen der Gold–Kupfer–Legierungen im Dreistoffsystem Gold–Kupfer–Nickel, *Metallforschung,* **2** (1947), 147–158.
26. Raub, E. and Walter, P., Die AuCu-Umwandlung in ternären Legierungen, *Z. Metallk.* **41** (1950), 240 243.
27. Johnson, Matthey & Company Ltd., "An Improved Gold Alloy", U.K. Patent 681,484 (1951).
28. Sistare, G. H. and Chamer, E. S. (Handy and Harman), "Fine Grained White Gold Alloy". U.S. Patent 3,512,961 (1970).
29. Wahlbeck, H. G. E., Alloying material for jewellery gold. Swedish Patent 347,020 (1972) (*C.A.* **78** (1973), 46943y).
29a. Leuser, J. and Wagner, E., Degussa aus Forschung und Production, *Degussa Festschrift,* (1953), 47–58.
30. Loebich, O., Metallkundliche Probleme bei der gewerblichen Goldverarbeitung, *Z. Metallk.* **44** (1953), 288–292.
31. Anon., Kleine edelmetallkunde: Kursus für Gold- und Silberschmiede, *Gold, Silber, Uhren, Schmuck,* **0** (1976), January, 76.
32. Andryushchenko, I. A. *et al.,* "Gold alloy", U.S.S.R. Patent 448,239 (1974) (*C.A.* **82** (1975), 159594m).

33. Andryushchenko, I. A. *et al.,* "Gold alloy", U.S.S.R. Patent 448,238 (1974) *(C.A.* **82** (1975), 159593k).
34. Andryushchenko, I. A. *et al.,* "Gold alloy", U.S.S.R. Patent 453,443 (1974) *(C.A.* **82** (1975), 174409y).
35. Leuser, J. and Wagner, E., Uber den Einfluss verschiedenartiger Wärmbehandlung auf die Eigenschaften von Schmuckgold, *Z. Metallk.* **44** (1953), 282–286.
36. Barshtein, N. P. *et al.*, "Gold Alloy", U.S.S.R. Patent 455,158 (1974) (*C.A.* **82** (1975), 174411t).
37. Belousova, T. P. *et al.*, "Gold Based Alloy", U.S.S.R. Patent 377,380 (1973) (*C.A.* **79** (1973), 69630d).
38. Andryushchenko, I. A. *et al.,* "Gold alloy", U.S.S.R. Patent 462,874 (1975) *(C.A.* **83** (1975), 102209f).
39. Anon, Kleine edelmetallkunde: Ein Kursus für Geschäft und Werkstatt, *Gold, Silber, Uhren, Schmuck,* (1975), November, p. 68 (cf. earlier issues also).
40. Jackson, R. S., "The Development of an Improved 9-carat Yellow Gold Casting Alloy". Report from the University of Aston, Birmingham, to the Chamber of Mines of South Africa, 1976.
41. Atkinson, B., "The Bleaching of Gold to Produce White Gold". Project Report No. 6a/2 to the Worshipful Company of Goldsmiths, London, 1972.
42. Starchenko, I. P. and Lifshits, V. A., Effect of alloying elements on the colour of gold alloys, *C.A.* **83** (1975), 31923w.
43. Belais, D., U.S. Patent 1,330,231 (1930).
44. Belais, D. and Bondy, A. R., "White Gold". U.S. Patent 1,391,449 (1921).
45. Koch, T. U.S. Patent 1,340,451 (1920).
46. Belais, D., "White Gold". U.S. Patent 1,584,352 (1921).
47. Anon., Kursus für Gold- und Silberschmiede: 2.1.2.2.1 Weissgolde, *Gold, Silber, Uhren, Schmuck,* 1963 (11), 45; 1976 (1), 76; (2) 75; (3) 138.
48. Jarrett, T. C., "Effect of Composition on Colour and Melting Point of 10 ct., 12 ct. and 14 ct. Gold Alloys", Am. Inst. Min. Metall. Eng., Technical Publication No. 109, 1939.
49. Wise, E. M., "High Strength Gold Alloys for Jewellery". Am. Inst. Min. Metall. Eng., Technical Publication No. 147, 1928 (cf. U.S. Patent 1,577,995 (1926)).
50. Belousova, T. P. *et al.*, Ductility of Z IMNTs 750 white gold, *C.A.* **81** (1974), 15821q.
51. Raub, E. and Engel, A., Das Dreistoffsystem Gold–Nickel–Kupfer, I, *Metallforschung,* **2** (1947), 11–16 (cf. Ref. 12, pp. 931–932).
52. Stagno, E. and Pinasco, M. R., Structure métallographique de l'or blanc 750% à l'état brut de fusion et après traitements thermiques, *Mém. Sci. Rev. Métall.* **73** (1976), 35–55.
53. Stagno, E. and Ienco, M. G., Structure d'un certain nombre d'alliages binaires et ternaires riches en or à l'état brut de fusion et après traitements thermiques, *Mém. Sci. Rev. Métall.* **73** (1976), 117–140.
54. Carter, F. E., U.S. Patent 1,355,811 (1920).
55. Liebkecht, O., "Gold Alloy". U.S. Patent 1,523,026 (1925).
56. Gardner, L. A. (American Telephone & Telegraph), "Alloy for Electrical Contacts". U.S. Patent 1,565,358 (1924).
57. Piccard, J., U.S. Patent 1,926,313 (1933).
58. Brenner, B. (Sigmund Cohn Corp.), "Novel Gold Alloys and Potentiometer Wires Produced from them". U.S. Patent 2,840,468 (1958).
59. Williams, R. V., "Spinnerette". U.S. Patent 1,647,822 (1925).

60. Donnelly, R. G., Gilliland, R. G. and Slaughter, G. M., "Brazing Alloys". U.S. Patent 3,079,251 (1963).
61. Gamer, N. T., "High Temperature Brazing Alloys". U.S. Patent 3,577,233 (1971).
62. Comptoir Lyon-Alemand Louyot et Cie., White gold. French Patent 2,102,510 (1972) *(C.A.)* **77** (1972), 167816x),
63. Darling, A. S., Gold–platinum alloys: a critical review, *Platinum Met. Rev.* **6** (1962), 62–67 and 106–111.
64. Hume-Rothery, W., The platinum metals and their alloys: a review of their electronic structure and constitution, *Platinum Met. Rev.* **10** (1966), 94–100.
65. Nowack, L., Weissgolde, *Metallwirtschaft*, **7** (1928), 465–466.
66. Raub, E. and Falkenburg, G., Das system Gold–Platin–Rhodium und die binären Randsysteme, *Z. Metallk.* **55** (1964), 392–397.
67. Fahrenwald, F. A., U.S. Patent 1,296,938 (1919).
68. Fahrenwald, F. A., U.S. Patent 1,339,505 (1920).
69. Fahrenwald, F.A., U.S. Patent 1,415,233 (1922).
70. Richter, K. G. P., U.S. Patent 1,165,448 (1915).
71. Peschko, R. J., U.S. Patent 1,169,753 (1916).
72. Dufour, G. H. (Marshall Field & Company), "Noble Metal Alloy". U.S. Patent 1,282,055 (1918).
73. Powell, A. R. and Box, E. R. (Johnson, Matthey & Company, Ltd.), "Precious Metal Alloys". U.S. Patent 2,071,216 (1937).
74. Crowell, W. S., "Metal Handbook. Vol. I. Properties and Selection of Metals". American Society for Metals, 8th Edition, 1961.
75. Zwingmann, G. (Degussa). German Patent 1,236,207 (1967).
76. Raub, E. and Walter, P., Alloys of gold with cobalt and iron, *Z. Metallk.* **41** (1950), 234–238.
77. Young, R. S., "Cobalt". Am. Chem. Soc. Monograph No. 108, Rheinhold Publishing Corp., 1948.
78. Perrault, R., "Le Cobalt". Dunod, Paris, 1946.
79. O'Connor, G. P., "Development of White Gold Alloys: Optimisation of". BNF Metals Technology Centre, Confidential Report No. PR288/5 to the South African Chamber of Mines, October 1975.
80. Derouwaux, P. and Hofman, C., "Procédé pour déposer une couche d'un alliage d'or violet sur un objet". Swiss Patent 533,691 (1973).
81. Wood, R. H., The purple gold of Tut'ankhamun, *J. Egypt. Archeol.* **20** (1934), 62–65.
82. Vines, R. and Wise, E., "Age Hardening of Metals". American Society Metals, Cleveland, 1940.
83. Merica, P. D., The age hardening of metals, *Trans. Am. Inst. Min. Metall. Eng.* **99** (1932), 13–54.
84. Hardy, H. K. and Heal, T. J., Report on precipitation, *Prog. Met. Phys.* **5** (1954), 143–278.
85. Kelly, A. and Nicholson, R., Precipitation hardening, *Prog. Mater. Sci.* Pergamon Press, New York, **10** (1963), No. 3.
86. Martin, J., "Precipitation Hardening". Pergamon Press, New York, 1968.
87. Fine, M., "Strength of Metals". Edited by D. Pecker. Reinhold Publish. Corp., 1964, pp. 161 ff.
88. Kelly, A., "Electron Microscopy and the Strength of Crystals". Edited by G. Thomas. Interscience, 1963, pp. 988 ff.

89. Brown, L. and Hamm, K., "Strengthening Mechanisms in Crystals". Edited by A. Kelly and R. Nicholson. Applied Science Publishers, 1971.
90. Fehrer, F. C., "Age Hardening of Gold Alloys". Report to the Chamber of Mines of South Africa, Johannesburg, 1976.
91. Wise, E. M., Crowell, W. S. and Eash, J. T., The role of platinum metals in dental alloys. I and II. *Trans. Am. Inst. Min. Metall. Eng.* **99** (1932), 363–412.
92. Dowson, A., Precipitation hardening of metals, *Br. Dent. J.* **90** (1951), 205–211.
93. Leinfelder, K. F., O'Brien, W. J. and Taylor, D. F., Hardening of dental gold–copper alloys, *J. Dent. Res.* **51** (1972), 900–905.
94. Kurnakow, N. *et al.*, The transformations in alloys of gold with copper, *J. Inst, Met.* **15** (1916), 305–332.
95. Graham, M., "Precipitation Strengthening of an Au+4 Atomic % Ti Alloy. Ph.D. Thesis, Northwestern University, Evanston, Ill., 1975.
96. Leinfelder, K. F. *et al.,* Hardening of high fusing gold alloys, *J. Dent. Res.* **45** (1966), 392–396.
97. Smith, D. L. *et al.*, Iron platinum hardening in casting golds for use with platinum, *J. Dent. Res.* **49** (1970), 283–288.
98. Tiedema, T. J., Bouman, J. and Burgers, W. G., Precipitation in gold platinum alloys, *Acta. Metall.* **5** (1957), 310–321.
99. van der Toorn, L. and Tiedema, T., Precipitation in gold–platinum alloys. I. Thermodynamics, *Acta. Metall.* **8** (1960), 711–714.
100. Carpenter, R. W., Growth of a modulated structure in gold–platinum alloys, *Acta. Metall.* **15** (1967), 1567–1572.
101. van der Toorn, L., Precipitation in gold–platinum alloys. II. Influence of the spinodal curve on the rate and mechanism of the precipitation process, *Acta. Metall.* **8** (1960), 715–723.
102. Carpenter, R., "Strengthening Effects of Spinodal Decomposition in Au-Pt Alloys and the Kinetics of the Reaction. Ph.D. Thesis, Univ. Berkeley, Calif., 1967.
103. Weise, J. and Volkman, G., Coherent–incoherent precipitation in the gold–platinum system, *Z. Metallk.* **59** (1968), 904–909.
104. Kralik, G., Decomposition kinetics of gold platinum alloys by means of electrical resistivity, *Z. Metallk.* **61** (1970), 751–756.
105. Carpenter, R. W., Deformation and fracture of gold–platinum polycrstals strengthened by spinodal decomposition, *Acta. Metall.* **15** (1967), 1297–1308.
106. Borelius, G., Spinodal hypotheses and precipitation in gold–platinum and gold–nickel alloys, *Phys. Scr.* **4** (1971), 127–131.
107. Sivertsen, J. and Wert, C., Aging in gold–nickel alloys, *Acta Metall.* **7** (1959), 275–282.
108. Gerlach, W., Heterogeneous precipitation in the system gold–nickel, *Z. Metallk.* **40** (1949), 281–289 (*C.A.* **44** (1950), 101b).
109. Gibala, R., Precipitation in gold/nickel alloys, *Trans. Am. Inst. Min. Met. Eng.* **230** (1964), 255–256 (*C.A.* **60** (1964), 10317a).
110. Fukano, Y., Precipitation phenomena in thin films of supersaturated gold–nickel alloy, *J. Phys. Soc. Japan*, **16** (1961), 1195–1204 (*C.A.* **55** (1961), 20871).
111. Fisher, R. and Embury, J., "Precipitation in Gold–Nickel Alloys. Proc. Europ. Reg. Conf. Electron Microscopy, 3rd, Prague, 1964, pp. 149–150.
112. Sivertsen, J. and Sundahl, R., X-ray study of aged Au-Ni alloys, *Acta. Metall.* **9** (1961), 162–163.
113. Frebel, M. and Predel, B., Two variants of discontinuous precipitation in gold–iron solid solutions, *Z. Metallk.* **64** (1973), 913–920.

114. Frebel, M. and Predel, B., Kinetics of discontinuous lamellar precipitation of gold–iron solid solutions, *Arch. Eisenhuettenw.* **45** (1974), 483–490.
115. Frebel, M. and Predel, B., Volume diffusion controlled discontinuous precipitation in gold–iron mixed crystals, *Mater., Sci. Eng.* **15** (1974), 221–230 (*C.A.* **83** (1975), 14440u).
116. Johnson, A. *et al.*, Age hardening in a gold–nickel alloy, *Nature*, **201** (1964), 1020–1021.
117. Ordonez, J., Johnson, A. A. and Mukherjee, K., Preparation of fibre reinforced nickel–gold alloys using a solid state reaction, *Nature*, **217** (1968), 442–444.
118. Fesolowich, A. *et al.*, Fibre composite alloys: preparation by controlled dissociation of metallic solid solutions, *Science,* **167** (1970), 1374–1376.
119. Johnson, A., Hughes, E. J. and Barton, P. W., Precipitation hardening in gold–nickel alloys, *Int. Dent. J.* **18** (1968), 655–667.
120. Johnson, A. A. and Masson, D. B., Preparation and properties of fiber reinforced alloys, *J. Sci. Ind. Res.* **32** (1973), 657–663.
121. Cohen, J. B. *et al.*, Evidence for vacancies associated with pre-precipitates, *Acta. Metall.* **14** (1966), 545–550.
122. Cohen, J. B., Kimball, O., Meshii, M. and Rundman, M., Evidence for vacancies associated with pre-precipitates. Reply, *Scr. Metall.* **2** (1968). 83–86.
123. Kimball, O., "Investigation of the Pre-precipitation Process in the Au-Ni System". Ph.D. Thesis, Northwestern University, Evanston, Ill., 1967.
124. Cohen, J. B. and Kimball, O. F., Preprecipitation in the gold–nickel system, *Trans. Metall. Soc. Am. Inst. Min. Metall. Eng.* **245** (1969), 661–669.
125. de Keijzer, A. and de Groot, C., Orientation relations during precipitation in a gold–nickel alloy, *Proc. Kon. Ned. Akad. Wetensch., Ser. B.* **73** (1970), 46–63 (cf. also 139–146 and 209–227).
126. Wise, E. M. and Eash, J. T., The role of platinum metals in dental alloys. III. The influence of platinum and palladium and heat treatment upon the microstructure and constitution of basic alloys, *Trans. Am. Inst. Min. Metall. Eng.* **104** (1933), 277–307.
127. Sekine, E. *et al.*, "High Strength Gold Alloy Containing Cu and Al Together With Optional Other Constituents to Enhance Properties". Jap. Patent 49–48, **813** (1974) (*C.A.* **83** (1975), 14732r).
128. Brandes, E. A., A gold–chromium–cobalt alloy for sliding contacts, *Gold Bulletin,* **8** (1975), 73–79.
129. Harigatani, H., Kawanishi, I. and Asahina, M., "Gold Alloy". Jap. Patent 48–38, 534 (1973) (*C.A.*, **81** (1974), 5630n).
130. Hariya, H., Kawanishi, K. and Asahina, M., "Gold Alloy". Jap. Patent, 48–24, 931 (1973) (*C.A.* **80** (1974), 18514g).
131. Hariya, H., Kawanishi, K. and Asahina, M., "Gold Alloy". Jap. Patent 48/39, 131 (1973) (*C.A.* **80** (1974), 136657d).
132. Harigaya, H., Kasai, K. and Asahina, M., "Gold Alloy". Jap. Patent, 49–23, 971 (1974))*C.A.* **82** (1975), 128417b).
133. Hornbogen, E., Lead-induced discontinuous precipitation in Fe-Au, *J. Mater. Sci.* **9** (1974), 518–521.
134. Kasai, K., "Age Hardenable Gold Alloy". Jap. Patent, 49–123, 927 (1974) (*C.A.* **82** (1975), 102104x).
135. Kasai, K., "Gold Alloys". Jap. Patent, 50–25,425 (1975) (*C.A.* **83** (1975), 102200w).

136. Clipstone, C. J. and Gaunt, P., Precipitation in gold–cobalt alloys, *Philos. Mag.* **20** (1969), 173–179.
137. Healy, T. and Leak, C., Proc. Int. Conf. Strength Metals and Alloys, Vol. 9, 1967, pp. 521 ff.
138. Gaunt, P., A magnetic study of precipitation in a gold–cobalt alloy, *Philos. Mag.* **5** (1960), 1127–1145.
139. Campbell, R. and Muldawer, L., Dislocation decoration by precipitation in gold–cobalt alloy, *Philos. Mag.* **6** (1961), 531–534.
140. Gaunt, P. and Silcox, J., An electron microscope examination of small cobalt particles precipitated in a gold matrix, *Philos. Mag.* **6** (1961), 1343–1345.
141. Ardell, A. and Hovan, M., Precipitation hardening of a Cu_3Au-Co alloy, *Mat. Sci. Eng.* **9** (1972), 163–174.
142. Gettleman, L. *et al.*, Studies on a new dental casting alloy, *J. Dent. Res.* **46** (1967), 595–601 (cf. *J. biomed. Mater. Res.* **6** (1972), 25–32).
143. Harigaya, H., Kasai, K. and Asahina, M., "Gold Alloy". Jap. Patent, 49–23,735 (1974) (*C.A.* **82** (1975), 63180x).
144. Azuma, S., Anzai, K. and Sato, A., "Gold Alloys Age-hardenable Rapidly at Low Temperatures". Jap. Patent 45–08,551 (1970) (*C.A.* **73** (1970), 38065v).
145. Sekine, E. *et al* "High Strength gold Alloy Containing Copper and Aluminium with Optional other Constituents to Enhance Properties". Jap. Patent 49–48813 (1974) (*C.A.* **83** (1975), 14732r).
146. Tammann, G., Die chemischen und galvanischen Eigenschaften von Mischkristallreihen und ihre atomverteilung, *Z. Anorg. Allg. Chem.* **107** (1919), 1–240. (cf. *Z. Metallk,* **44** (1953), 298–301).
147. Robertson, W. D. and Bakish, R., "Structural Factors Associated with Stress Corrosion Cracking of Homogeneous Alloys", *in* "Stress Corrosion Cracking and Embrittlement". Ed. by W. D. Robertson, John Wiley & Sons Ltd., New York, 1956, pp. 32–47.
148. Bakish, R. and Robertson, W. D., Structure dependent chemical activity of polycrystalline Cu_3Au-experiments relation the mechanism of stress corrosion cracking of homogeneous solid solutions, *Trans. Am. Inst. Min. Metall. Eng. (J. Metals),* **206** (1957), 1277–1282.
149. Wilson, T. C., Edmunds, G., Anderson, E. A. and Pearce, W. M., "Symposium on Stress Corrosion Cracking of Metals", *ASTM–AIME,* 1944, p. 173.
150. Bakish, R. and Robertson, W. D., Structure-dependent chemical reaction and nucleation of fracture in Cu_3Au single crystals, *Acta. Metall.* **4** (1956), 342–351.
151. Bakish, R. and Robertson, W. D., Galvanic potentials of grains and grain boundaries in copper alloys, *Trans. Electrochem. Soc.* **103** (1956), 319–325.
152. Graf, L., Stress corrosion cracking in homogeneous alloys, *in* "Stress Corrosion Cracking and Embrittlement". Ed. by W. D. Robertson, John Wiley & Sons Ltd., New York, 1956, pp. 48–60.
153. Graf, L., Die Ursache der Spannungskorrosionempfindlichkeit homogener Legierungen, *Z. Metallk.* **38** (1947), 193–207.
154. Graf, L., Die Spannungskorrosion bei homogenen Legierungen, ihre Ursachen und ihr Mechanismus, *Werkst. Korros.* **8** (1957), 329–344.
155. Graf, L., Spannungskorrosion heterogener Legierungen, *Z. Metallk.* **38** (1947), 207–212.
156. Graf, L. and Klatte, H., Zum Problem der Spannungskorrosion homogener Mischkristalle. IV. Das Festigkeitsverhalten von Kupfer–Gold und

Kupfer–Zink–Mischkristallen unter Einwirkung von Quecksilber und Chemischen Agentien, *Z. Metallk.* **46** (1955), 673–680.

157. Graf, L. and Budke, J., Zum Problem der Spannungskorrosion homogener Mischkristalle. III. Abhängigkeit der Spannungskorrosions-Empfindlichkeit von Kupfer–Gold- und Silber–Gold–Mischkristallen von Goldgehalt und Zusammenhang mit dem "Mischkristall-Effekt", *Z. Metallk.* **46** (1955), 378–385.
158. Graf, L., The causes and the mechanism of stress corrosion cracking of homogeneous non-supersaturated alloys as derived from experimental work with alloys containing noble metal components, *in* "The Theory of Stress Corrosion Cracking in Alloys—the Proceedings of a Research Evaluation Conference". Edited by J. C. Scully, NATO Scientific Affairs Division, Brussels, 1971, pp. 399–417.
159. Graf, L. *et al.,* Was spricht gegen das Ausfreissen von Deckschichten und gegen eine Versprödung als Ursachen der Spannungsrisskorrosion?, *Z. Metallk.* **66** (1975), 749–954.

159a. Chaston, J. C., Stress corrosion, *Nature,* **161** (1948), 891–892.

160. Gardam, D. E., "Why Jewellery Sometimes Blackens the Skin or the Clothing". Special Report No. 4 to the Worshipful Company of Goldsmiths, London, 1969.
161. See reference (6), Sections 2.12.1.2 and 2.1.2.3.7, and specially 1976 (4), p. 91.
162. Sell, G., "Pickling Agents for Precious Metals: a Comparison of the Efficiencies of Various Pickles in Removing Oxide from Silver and Gold Alloys". Worshipful Company of Goldsmiths, Technical Advisory Committee Project Report No. 29/2, London, September, 1973.
163. Sell, G., "Pickling Agents for Precious Metals: a Comparison of the Efficiencies of Various Pickles for 18 Carat White Gold". Worshipful Company of Goldsmiths, Technical Advisory Committee Project Report No. 29/3, London, September, 1973.
164. Krause, H., "Metallfärbung", 3 Auflage, München, 1951, p. 30, 2 Auflage, Berlin, 1937, pp. 118–119.
165. Machu, W., "Nichtmetallische anorganische Uberzüge", Wien, 1952, p. 190.
166. Buss, G., "Mitt. Forschungsinst. Probieramts Edelmetalle Schwäbisch-Gmünd", Vol. 11, 1937–1938, pp. 93–94.
167. Rechenberg, *Metallwaren-Ind. Galvanotech,* **30** (1932), 57–58.
168. Anon., "Diebeners Handbuch des Goldschmieds", Leipzig, 1929, pp. 484–485.
169. Oldham, P. A., *Metal Ind. New York,* **37** (1939), 72–75.
170. Buchner, G., "Die Metallfärbung", 4 Auflage, Berlin, 1935, pp. 118–119.
171. Busser, A., *Metallwaren-Ind. Galvanotech,* **33** (1935), 162–163.
172. Anon., See reference (6), Section 2.1.4.1 and 1976 (8), p. 84.
173. Jacquet, P. A., Electrolytic chemical polishing, *Metall. Rev.* **1** (1956), 157–211.
174. Tegart, W. J. McG., "The Electrolytic and Chemical Polishing of Metals in Research and Industry". Pergamon Press, 1959.
175. Fedot'ev, N. P. and Grilikhes, S. Ya., "Electropolishing, Anodizing and Electrolytic Pickling of Metals", Moscow, 1957. Translated by A. Behr, 1959. Published by Robert Draper Ltd.
176. Gmelin's Handbuch der Anorganischen Chemie, 8th Edition, System No. 62, *Gold* (1954), 402–403.
177. Kushner, J. B., Modern gold plating, Part 27, *Prod. Finish.* (Cincinnati), **7** (1942), 42–44.
178. Wise, E. M., "Gold: Recovery, Properties and Applications". D. van Nostrand Co. Inc., 1964, pp. 359–361.

179. Mulnet, G., Bath for electrolytic polishing, *Fr. Demande,* 2,222,451 (1974).
180. Gleekman, L. W., Evans, G. E. and Grove Jr., C. S., An improved cell for electrolytic polishing, *Met. Prog.* **61** (6) (1952), 92–93.
181. Bakish, R. and Robertson, W. D., Metallographic techniques for Cu-Au alloys, *J. Met.* **7** (1955), 424.
182. Reichert, M., "Electropolishing of Gold and Gold Alloys. U.S. Patent 2,712,524 (1955).
183. Fischer, J., "Electropolishing of Gold Alloys". U.S. Patent 2,712,525 (1955).
184. Milazzo, G., "Electrochemistry: Theoretical Principles and Practical Applications". Translated by P. J. Mill, Elsevier, 1963, p. 493.
185. Blazy, A. and Mohler, J. B., Metallography for the electroplater, *Met. Finish.* **45** (1947), 68–71.
186. Mulnet, G., "A Procedure for Chemical Polishing of Alloys of Gold and Silver Used in Jewellery". French Patent 2,137,296 (1972).
187. Phillips, R. W., "Skinner's Science of Dental Materials", 7th Edition. W. B. Saunders Company, Philadelphia, 1973.
188. Sell, G. C. E., "An Evaluation of Various Chemical Methods for Removal of Investment from Castings". Worshipful Company of Goldsmiths, Technical Advisory Committee Report No. 29/1, London, September, 1973.
189. Geckle, R. A., Ultrasonic cleaning for the manufacturing jeweller, *Am. Jewell. Manuf.* **21** (1976) (6), 10–12, 26, 28, 30, 32, 33.
190. Handy and Harman, "Fine Grained Gold Alloy". British Patent 1,197,778 (1970).
191. Souder, W., Fifteen years of dental research at the National Bureau of Standards, *J. Am. Dent. Assoc,* **21** (1966), 58–65.
192. Hollenback, G. M. and Skinner, E. W., Shrinkage during casting of gold and gold alloys, *J. Am. Dent. Assoc.* **33** (1946), 1391–1399.
193. Ryge, G., Kozak, S. F. and Fairhurst, C. W., Porosities in dental gold castings, *J. Am. Dent. Assoc.* **54** (1957), 746–754.
194. Gainsbury, P., private communication.
195. Degussa, Frankfurt. Press release, 1976.
196. Brook, G. B. and Iles, R. F., Gold–copper–zinc alloys with shape memory, *Gold Bulletin,* **8** (1975), 16–21.
197. BNF Metals Technology Centre, "An Assessment of the Hot Working of Gold Alloys, and Proposals for Research". Report to the Chamber of Mines of South Africa, August 1976.
198. Nowack, L., Über den Einfluss geringer Bleizusätze auf Gold, *Z. Anorg. Allg. Chem.* **154** (1926), 395–398.
199. Rose, T. K., 33rd Annual Report of the Royal Mint, 1902, p. 73.
200. Grover, R. K. and Grimwade, M. F., "The Effect of Lead on the Workability of Commercial Carat Gold Alloys". Report to the Chamber of Mines of South Africa, 1975.
201. Richards, E. Th., Deutsch. *Goldschm.-Ztg.* **39** (1941), 34 ff (from ref. 30).
202. Jordan, D. R., "Gold Alloy for Jewellery". U.S. Patent 3,810,755 (1974) (*C.A.* **81** (1974), 53577x).
203. Duerrwaechter, E. and Denzer, D., "Centrifugal Casting of Metallic Jewellery". German Offen. 2,024,071 (1971) (*C.A.* **76** (1972), 49121y).
204. Standard specification NIHS-03-SO (July 1961), Switzerland.
205. Standard specification CETEHOR-07-70 (March 1966), France.
206. Standard specifications DIN 8-322 (1966) and DIN 8-238, Germany.

207. Massin, M. and Reid, F., "Watch Industry Applications" *in* "Gold Plating Technology". Ed. by Reid, F H. and Goldie, W., Electrochemical Publications, Ayr, Scotland, 1974, p. 587.
208. Gardam, G. E., The colours of some metals and alloys, *Trans. Inst. Met. Finish,* **44** (1966), 186–188.
209. Gardam, G. E., Study of the dyeing of anodized aluminium using colour measurements, *Trans. Inst. Met. Finish,* **41** (1964), 190–199.
210. Christie, J. S., "Instruments for the Measurement of Metallic Appearance of Metal Surfaces", Am. Soc. Test. Mater. Spec. Tech. Publ. No. 478, 1970, pp. 59–78.
211. Köster, W. and Stahl, R., Effects of alloying deformation, recrystallisation and short and long range order on the optical constants of noble metals and their alloys, *Z. Metallk.* **58** (1967), 768–777.
212. Pepperhof, W., "Temperaturstrahlung". Darmstadt, 1956.
213. Stahl, R., Spranger, H. J. and Aubauer, H. P., Optical properties of the ordered phase $AuCu_3$, *Z. Metallk.* **60** (1969), 933–938.
214. Rivory, J., Optical properties of ordered and disordered Au-Cu alloys, *J. Phys. (Paris),* **35** (1974), May, C4–51 to C4–56.
215. Saeger, K. E., Einige Bemerkungen zur Farbe der Edelmetalle und ihrer Legierungen, *in* "DODUCO, 1922 to 1972". Ed. by Dr. E. Durrwachter, DODUCO KG, 7530 Pforzheim, West Germany.
216. C.I.E. (Committée Internationale de l'Eclairage), "International Lighting Vocabulary", 3rd Edition, C.I.E., Paris, 1970.
217. Roberts, E. I., private communication to the Chamber of Mines of S.A.
218. Tammann, G., Chemische und galvanische Eigenschaffen von Mischkristallreihen usw., *Z. Anorg. Chem.* **107** (1919), 1–239.
219. Tammann, G. and Wilson, C., Die Anderung des galvanischen Potentials der Metalle usw., *Z. Anorg. Chem,* **173** (1928), 156–163.
220. Boas, W., "Dislocations and Mechanical Properties of Crystals". Ed. by Fisher *et al,* Wiley, New York, 1957, pp. 406–407.
221. Yates, E. L., The change in colour of a silver–gold alloy on plastic deformation, *Aust. J. Phys.* **16** (1963), 40–46.
222. Anon., Kleine Edelmetallkunde, *Gold, Silber, Uhren, Schmuck* (1975) (12), 47.
223. Rivlin, V. G. and Brook, G. B., "Factors Affecting the Colour of Gold Alloys". Report by the Fulmer Research Institute to the Chamber of Mines of South Africa, August 1976.
223a. Saeger, K. E. and Rodies, J., The colour of gold and its alloys: The mechanism of variation in optical properties, *Gold Bulletin* **10** (1977), 10–14.
224. Sloboda, M. H. and Boughton, J. D. (Johnson, Matthey & Company, Ltd.), "Brazing Alloys". British Patent 1,280,460 (1972). cf. U.S. Patent 3,658,997 (1972).
225. Steltman, L., "Fusible Noble Metal Alloys". South Africa Patent 69 06,938 (1970) (*C.A.* **75** (1971), 24583s).
226. Andryushchenko, I. A. *et al.*, "Brazing Alloy for Gold Articles". U.S.S.R. Patent 451,509 (1975) (*C.A.* **83** (1975), 32206b).
227. Andryushchenko, I. A. *et al.,* "Solders for Gold Articles". U.S.S.R. Patent 451,510 (1974) (*C.A.* **83** (1975), 14846f).
228. Hatswell, J. S. and Sloboda, H. (Johnson, Matthey & Co. Ltd.), "Gold Dental Alloys". U.S. Patent 3,892,564 (1975).

4
Dental Golds

1. Introduction

Gold is used in dentistry as the pure metal, in the form of foil, powder or the thin flake-like crystals known as mat gold. It is also employed in wrought alloy wires and plates, in dental gold alloy solders, and in two types of casting alloys. Alloys of the first type are used as such in restorations, and are based upon the Au-Ag-Cu system, whilst alloys of the second type are used in enamel veneered form, and are based upon the Au-Pd-Pt system.

These applications have been described by Wise[1] in 1964, and in the biennial editions of the American Dental Association's "Guide to dental materials and devices".[2] They are also discussed in texts on dentistry by Greener, Harcourt and Lautenschlager[3] and by Phillips,[4] and in a technical handbook produced by Degussa.[5]

2. Gold Foil, Mat Gold and Gold Powder

Gold foil for restorative purposes is normally about 60μm thick and is crumpled into small pellets before use. It may also be used in corrugated or rope like forms. Mat gold, on the other hand, is a tenacious structure of fine loosely packed dendritic gold crystals prepared by electro-deposition. It is compressed and sintered together to form thin cylinders or strips, sections of which are cut as needed. It is commonly wrapped in gold foil before application, or used in the form of mat foil, in which a layer of mat gold is sandwiched between layers of gold foil. Both gold foil[6] and mat golds[7-9] have long histories of use in dentistry.

Gold powder[10,11] on the other hand is a more recent introduction into dentistry. It is prepared (see later) either by atomisation of molten gold, or by chemical precipitation. The average particle size is about 15 μm with the largest particles less than 75 μm in diameter. It is difficult to apply directly and is usually used wrapped in gold foil.

All three forms of gold are commonly applied in "cohesive" form, i.e. in a state in which they readily undergo cold bonding. They are therefore made from gold of high purity, and are gently heated — usually in an alcohol flame — just before use, in order to remove surface contamination and to ensure effective cold welding when they are impacted in the tooth cavity. A temperature of at least 315°C is necessary for this purpose. Heating must not be unduly prolonged, however, if excessive grain growth in gold foil or excessive sintering of mat gold or gold powder are to be avoided.

These "cohesive" golds all give fillings which — from a clinical point of view — do not differ very greatly in strength, density or surface hardness, irrespective of whether they are compacted by manual or pneumatic malleting or by other means, though individual findings have varied.[12–17] The hardness (Knoop) of the gold is increased to as high as 75, and the tensile strength to as high as 45,000 psi (309 MN/m^2) during placement, provided its surface has been adequately decontaminated by the initial heat treatment. Nevertheless, pure gold fillings are best suited for areas not subject to high stress. In such situations the stronger cast cold alloys give better service. The specific gravity of the pure gold fillings (14.3 – 15.9) is significantly less than that of pure gold (19.3 at 20°C), and indicates that compaction in the tooth cavity is never complete.

There is apparently no recrystallisation of compacted and work hardened gold fillings of this type at oral temperatures.

Gold-platinum foil (platinum centred gold foil) yields stronger fillings[2] but is more difficult to manipulate. It is made[1] either by sandwiching platinum between two sheets of gold, and then rolling and beating, or by depositing gold on both sides of a platinum sheet and then rolling and beating. It may contain up to 40% Pt. An analogous cohesive product has been made using for the centre sheet a gold alloy containing 0.1 – 0.5% Ca.

The cohesive or cold bonding properties of dental golds can be destroyed by treatment with such gases as ammonia, and so-called "non-cohesive" gold foil is made in this way for use in the lining of prepared cavities, before completing their filling with cohesive gold.

The incorporation of small amounts of other metals, and in particular calcium, in powdered gold has been found[14] to be advantageous. It does not affect the welding of the powder in the cavity adversely, and with electromalleting significantly increased hardnesses are achieved. The compositions of the alloy powders studied were not revealed, and the precise mechanisms responsible for the increased hardnesses which were observed are not clear. The powders were sintered under controlled conditions before use.

3. Casting Alloys based upon the Au-Ag-Cu System

3.1 Specifications

Most alloys of this type in use today conform to American Dental Association

Specification No. 5 for dental casting gold alloy[2] which has been accepted as Specification No. 7 of the Fédération Dentaire Internationale and is also the basis of the International Standards Organisation (I.S.O.) recommendation R150 for such alloys.

The specification is based principally on the classic work of Coleman[18] and of Taylor, Paffenbarger and Sweeney[19] on the physical properties of inlay casting golds and on that of Brumfield[20] on the testing of such materials. Four types of alloy are specified, for which the composition and physical property limits are set out in Table XXXI.

The limits of composition ensure adequate resistance to corrosion in the mouth environment, whilst the specifications in respect of physical properties are designed to facilitate choice of an alloy of appropriate strength and abrasion resistance for each casting. Colour is left for negotiation between purchaser and supplier. It will be clear that a range of alloy compositions conforming to the specification for each type is possible, and alloys ranging in colour from yellow through to white are marketed.

3.2 Metallurgical aspects

The compositions of the dental gold alloys which are supplied commercially as conforming to the American Dental Association specification referred to above are reflected in analyses[21,22] carried out on 127 samples. The results are summarised in Table XXXI. From these it will be seen that the alloys in use are all essentially Au-Ag-Cu alloys, which are modified by incorporation of varying quantities of platinum, palladium and zinc.

The effects of replacing gold in Au-Ag-Cu alloys by palladium or platinum have been studied by Wise, Crowell and Eash[23] and by Wise and Eash.[24] They found that replacement with palladium was effective in bleaching the colour of the Au-Ag-Cu alloys, the introduction of over 15 atomic per cent palladium resulting in alloys of increasing whiteness. Platinum was a less effective bleaching agent. Both palladium and platinum, and particularly platinum, caused a considerable increase in the hardness of the alloy solid solutions. Moreover the strengths of these solid solutions could be further increased by age hardening by substantially the same ratios as could be achieved by similar treatment of the Au-Ag-Cu alloys. Age hardening was found to occur over a wide range of conditions, the hardening effects in each case being accompanied by considerable increases in the melting points of the alloys. The introduction of both platinum and palladium also reduced the rate of grain growth in the alloys, platinum being the more effective additive in this respect. Platinum additions also resulted in the development of a duplex fibrous structure in the resulting alloys when they were wrought.

In the alloys as used, special grain refining agents (see below) are usually

Table XXXI. Composition and Physical Properties of Dental Gold Casting Alloys as laid down in American Dental Association Specification No. 5

Type	Gold and metals of the platinum group	Vickers number			Tensile strength	Elongation 5 cm gauge length		Fusion temperature
		Quenched		Hardened	Hardened	Quenched	Hardened	
	Min %	Min	Max	Min	Min (kg/cm²)	Min (%)	Min (%)	Min (°C)
I	83	50	90	—	—	18	—	930
II	78	90	120	—	—	12	—	900
III	78	120	150	—	—	12	—	900
IV	75	150	—	220	6.350	10	2	870

Table XXXII. Range of Percentage Composition of Dental Casting Gold Alloy

Type of alloy	Component					
	Gold	Silver	Copper	Palladium	Platinum	Zinc
I	80.2–95.8	2.4–12.0	1.6– 6.2	0.0– 3.6	0.0–1.0	0.0–1.2
II	73.0–83.0	6.9–14.5	5.8–10.5	0.0– 5.6	0.0–4.2	0.0–1.4
III	71.0–79.8	5.2–13.4	7.1–12.6	0.0– 6.5	0.0–7.5	0.0–2.0
IV	62.4–71.9	8.0–17.4	8.6–15.4	0.0–10.1	0.2–8.2	0.0–2.7

incorporated. The lower hardness types (Types I and II of Table XXXII), containing the higher percentages of gold and the lower percentages of platinum group metals respond little to heat treatments, whereas the harder alloys (Types III and IV) are softened by holding at 700°C briefly and then quenching. Subsequent hardening, with loss of ductility, can be achieved usually by cooling uniformly from 450° to 250°C over thirty minutes, though appropriate softening and hardening procedures are usually recommended by suppliers for each of their alloys.

It is a matter of interest that age hardening was first observed in 1905 at the S.S. White Dental Manufacturing Company in an Au-Ag-Cu-Pt alloy which was later placed on the market. The mechanism of hardening of typical present day casting alloys[25-29] has been reported as involving not only Au-Cu ordering, but also precipitation, as in the case with the carat gold alloys discussed in Chapter III. Barton, Eick and Dickson[30] have shown that the relationship between the Brinell and Vickers hardness numbers for dental casting alloys is a linear one, and that addition of 19 to the Brinell numbers gives a good approximation to the Vickers number.

In view of its beneficial effects on the properties of castings, grain refinement may be promoted in dental alloys by incorporation in them of small quantities of certain metals. Those which are most effective for gold alloys are of relatively high melting point, and show eutectic or peritectic points close to the edge of the phase diagram for the binary system with gold.[31] They are metals with high cohesive strength and low solubility in the liquid phase, which have the property of lowering the surface free energy of the liquid metal, and thereby stabilising embryo crystal nuclei. Metals in this category are iridium and ruthenium (see Figures 18 and 19), and rhenium[32] and rhodium.[33]

Hypoeutectic Au-Co alloys have been reported[34] as possessing properties which would make them suitable for dental casting alloys.

The corrosion of various dental gold alloys in artificial saliva has been studied by electrochemical methods and the influence of various alloying elements discussed.[35]

4. Casting Alloys for Coating with Enamel Veneers

From about the middle 1950s there has been, for aesthetic reasons, considerable clinical application[35] of dental gold alloys reaction bonded to dental porcelain coatings. These must meet certain requirements. Thus they must bond firmly on firing with the porcelain, which must not crack nor be discoloured in the process. Discoloration of the porcelain on firing tends to exclude the use of copper or silver except in small amounts and alloys of the Au-Ag-Cu system can therefore not be used. In practice the alloys found most

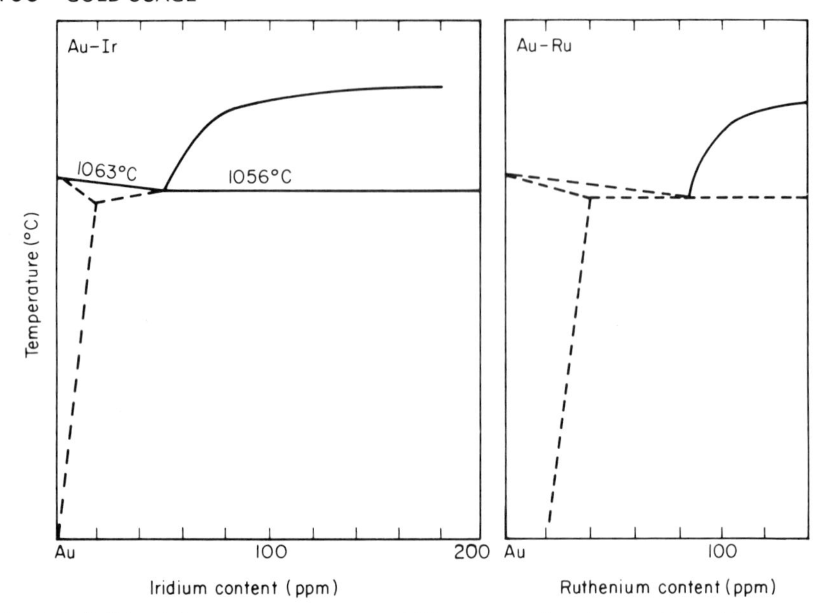

FIG. 18. Solid solution ranges for ppm additions for grain refining. (After Nielsen and Tuccillo.[31])

suitable are essentially Au-Pd-Pt alloys to which one or more of the metals iron, indium and tin are added in amounts of up to about 1%, together with grain refining agents of the types referred to above. The iron, indium and tin (see below) function as age-hardening agents. Together with the small quantities (< 1%) of copper and silver which are usually added, they also promote reaction bonding of the metal to the porcelain.[37–38] Additions of silver,[39] aluminium[40] silicon,[40] zinc[45] or titanium[41] may also be made to promote, *inter alia*, reaction bonding.

As Table XXXIII indicates, the alloys as used in practice may contain a large number of components. How essential this multiplicity of components may be is difficult to assess, but the properties of Au-Pd-Sn alloys containing 100 ppm of iridium indicate that platinum-free alloy with satisfactory properties could probably be developed.[42]

Alloys such as those listed in Table XXXIII, and other types,[45] can be satisfactorily matched in thermal expansion coefficient with dental porcelains. Moreover they undergo age hardening to the point where such structures as bridges do not flex sufficiently under stress to crack off the porcelain veneer. In practice such age hardening is effected if possible during the porcelain firing cycle.[46]

The mechanisms by which these alloys age harden have been studied,[47–49] and have been shown to be associated with the presence of platinum and of the

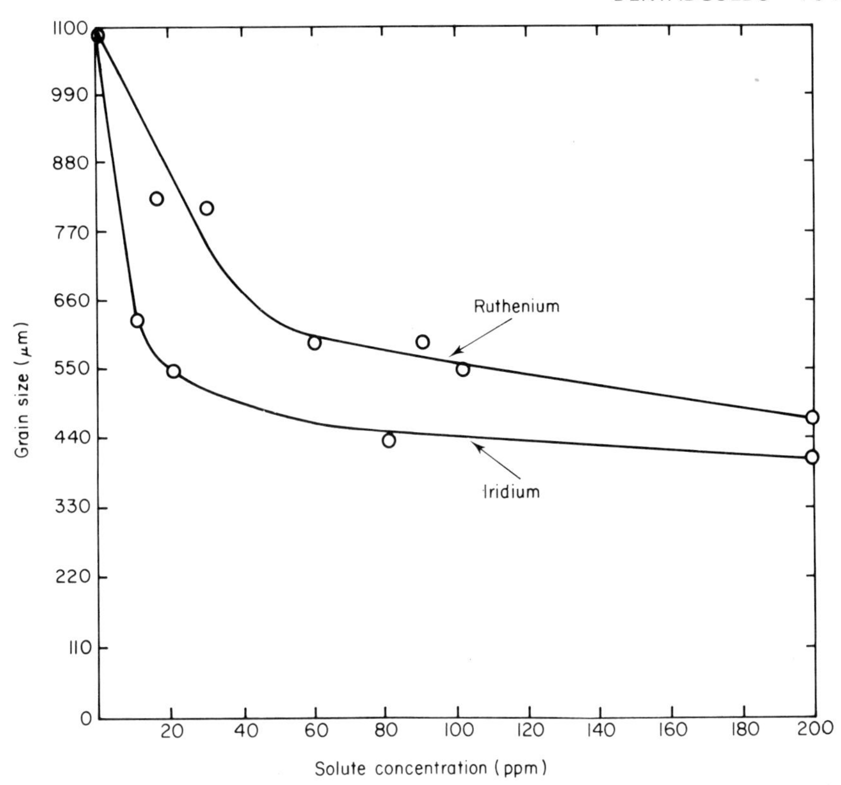

FIG. 19. Grain size versus solute concentration in gold. (After Nielsen and Tuccillo.[31])

small percentages of iron, indium and tin in the alloys. Thus age hardening effects are not observed in the Au-Fe-Pd, Au-Pd-Sn and Au-In-Pd systems, but are a feature of the Au-Fe-Pt, Au-Pt-Sn and Au-In-Pt systems. In the case of the Au-Fe-Pt system, age hardening involves the formation of $FePt_3$ at casting temperatures and its subsequent precipitation. Analogous intermetallic compound formation is almost certainly involved in the Au-In-Pt and Au-Pt-Sn systems.

Where iron is present in alloys of this type it has been found to "fire refine" out of the alloys on melting.[50] Losses in the content of iron, and of certain other minor base metal components, may become significant if the alloys are subjected to repeated melting, particularly if the gas-oxygen torch method is used, and temperatures rise far above the liquidus point.

The other important and interesting aspect of alloys suitable for bonding to dental porcelains is that of the mechanisms by which bonding occurs between the alloy and the ceramic. In the course of a recent review of the development of these alloys, Sperner[51] has discussed these mechanisms in some detail.

Table XXXIII. Composition (Weight %) of some Gold Casting Alloys reported as suitable for Bonding to Ceramic Porcelains

	A[32]	B[32]	C[39]	D[41]	E[43]	F[44]
Au	84.0	77.5	72.0	86.0	67.7	78–85
Pd	4.9	8.9	10.0	2.1	16.7	8–13
Pt	7.8	9.8	13.0	6.9	11.6	4–8*
Ag	1.2	0.9	3.4	0.5	2.0	0.0–2.0
Cu	0.3	0.3	—	0.9	0.1	—
Fe	—	0.5	—	1.06	—	0.7–1.0*
In	1.0	1.5	—	—	—	0.0–1.0
Sn	0.5	0.5	—	—	0.6	0.9–1.4
Ir	0.1	0.1	—	0.1	—	—
Re	0.2	—	0.5	—	—	small amount
Rh	—	—	1.0	—	—	—
Si	—	—	0.1	—	—	—
Ti	—	—	—	2.44	—	—
Ru	—	—	—	—	1.3	—

*Fe:Pt ratio kept at 0.4–0.6

Although mechanical interlocking between the surfaces of ceramics and alloys may contribute to bonding between such materials,[52-58] it is now generally accepted that wetting of the alloy surface by the ceramic is an essential preliminary to bonding.[59-61] Once contact between the two materials has been established, diffusion across the interface results in chemical or reaction bonding between them. To the extent that chemical interaction across the interface may promote wetting these two processes cannot be regarded as independent of one another. The importance of oxide layers on the surface of the metal phase was early recognised.[62-66]

It was King, Tripp and Duckworth,[66] however, who first concluded from their studies of the enamelling of steel that during firing there was a continuous exchange of an oxidation-reduction type between metal ions in the ceramic and metal atoms in the alloy.

Understanding of the mechanisms of bonding of gold alloys to dental porcelains developed against the background of these advances in knowledge achieved with other alloy-ceramic systems. The significance of mechanical bonding was studied,[67-68] as well as that of oxide formation. The difficulties caused by discoloration of the ceramic as a result of reaction with base metal oxides from the alloy phase attracted considerable attention,[69] as did also the influence of various factors on the ease of wetting of the alloy by the ceramic and on the strength of bonding.[70-71] In respect of the latter studies, the conclusion was reached that in the wetting by the ceramic of pure metals such as gold, platinum and palladium, as well as certain of their alloys, Van der Waal's forces played a major role. In the case of gold alloys containing tin, however, it was concluded that exchange of metal atoms of the alloy and metal ions of the ceramic across the interface along the lines proposed by King *et al.*66 was operative during bonding.

Experimental evidence for the occurrence of such interchange has recently been provided by Sperner.[51] Using microprobe analysis, Sperner was able to demonstrate not only the migration of easily oxidisable metals such as indium and tin from dental alloys into the ceramic phase during firing, but also the movement of silicon, aluminium and oxygen across the interface into the alloy phase. He drew the conclusion that the model for the adhesion of enamel to steel as proposed by King, Tripp and Duckworth[66] can be applied also to explain the bonding of dental porcelain to gold casting alloys.

The work of a number of other investigators[73–83] who have studied different aspects of the bonding of dental gold casting alloys to dental porcelains should be read against the background of these observations. In particular, Nally *et al.*,[78–79] Lautenschlager *et al.*[82] and Radnoth[83] had already observed a concentration at the alloy-porcelain interface of the more easily oxidisable components of dental casting alloys.

In practice various factors affect the strength of bonding which is achieved through this mechanism.[84,85] Evolution of gases from the alloy surface during firing is one such factor which has been studied.[84] It can be eliminated by avoiding contamination of the alloy by handling it with the fingers, and by heating it to about 1000°C prior to enamelling. By far the most important factor determining the strength and stability of bonding, however, is the extent to which the contractions of the porcelain and the alloy which occur on cooling after firing are matched. If the contractions are significantly different, this can lead either to spalling or cracking of the porcelain veneer. The problems of producing alloys and porcelains which are compatible with one another have been solved mainly by commercial enterprise.

Where very thin porcelain layers are applied to a casting alloy, and there is danger of the enamel appearing grey as a result of base metal oxidation during bonding, a layer of gold may be interposed between the casting alloy and the porcelain. Studies of the effects of this upon the strength of the bond which is established on firing have given varying results.[73,75,86,87] This is understandable in terms of the factors that are involved. Sperner's observations[51] indicate, however, that under appropriate conditions such a gold layer does not constitute a barrier to the diffusion of such metals as tin and indium from the dental alloy through to the ceramic phase.

A recent study[88] has explored the possibility of applying a dental gold alloy veneer to artificial teeth. Most artificial teeth are made of acrylate polymer, which can cause allergic changes in the mucous membranes of the mouth in some cases. The dental gold alloys were applied by sputtering and their adhesion to the acrylate and their hardness were measured. No indication was given as to whether it was considered that the results justified clinical trials of acrylate teeth gold coated in this manner.

5. Wrought Gold Alloys for Plates and Wires

The basic requirements in these are high strength and resistance to corrosion in the mouth, and both demands are met by a range of alloys which may contain gold, platinum, palladium, silver, copper and zinc. Most of those in use are susceptible to age hardening, and high values of hardness and strength can be obtained.

American Dental Association Specification No. 7[89] prescribes two grades of wires, depending upon their precious metal contents (see Table XXXIV). These largely determine their tarnish and corrosion resistance. As in the case of dental casting alloys, the specification of colour is left in the hands of the purchaser, and suppliers endeavour to supply alloys in each type with a range of colours. Type I alloys are, however, more commonly white or platinum coloured, whilst Type II alloys are more frequently offered in gold shades. Ranges of composition for these two types of alloys for plates and wires are given in Table XXXV. These omit fraction percentages of iridium, indium and rhodium which may be present.

Wrought structures, although stronger than cast structures of the same composition, are tending to be replaced by the latter, because of the development of precision casting techniques.

6. Dental Gold Solders

Like the carat gold solders used in fabrication of jewellery, dental gold solders are essentially Au-Ag-Cu alloys. The composition and properties of typical such solders as recorded by Wise[90] are listed in Table XXXVI. Small amounts of tin (2–3%) and zinc (2–4%) are usually added to reduce the fusion temperatures. Where white solders are required, copper may be replaced by nickel (cf. carat gold solders).

As in jewellery manufacture, solder compositions must be selected to meet actual requirements in each instance. Thus solders with higher silver content flow better than solders which are higher in copper content. They penetrate joint areas better, therefore, and are preferable for joining parts together. Solders higher in copper content on the other hand are particularly well suited, because of their reduced tendency to flow, for building up of an area.

The low carat solders, with their higher hardness and strength, are used

Table XXXIV. Composition and Physical Properties for Dental Wrought Gold Wire Alloy according to American Dental Association Specification No. 7

Type	Gold and metals of the platinum group	Yield strength oven-cooled	Tensile strength oven-cooled	Elongation 50.8mm. gauge length		Fusion temperature
	Min	Min	Min	Quenched Min	Oven-cooled Min	Min
	(%)	(kg/cm^2)	(kg/cm^2)	(%)	(%)	(°C)
I*	75	8,800	9,500	15	4	950
II**	65	7,000	8,800	15	2	870

* **High precious metal**

**Low precious metal

Table XXXV. Range of Percentage Composition of Wrought Gold Wire Alloys

Type of alloy	Component						
	Gold	Silver	Copper	Palladium	Platinum	Zinc	Nickel
I	53.6–63.2	8.5–12.4	10.2–15.2	0.0– 8.2	6.8–17.6	0.0–0.6	0.0–1.9
II	60.0–67.1	8.4–21.4	10.2–19.6	0.0–10.3	0.0– 6.5	0.0–1.7	0.0–6.2

where joints are exposed to high stress. They are, however, more susceptible to tarnish and corrosion. Porosity has been noted as a factor affecting the strengths of joints made between dental casting alloys using hard gold solders.[91]

Table XXXVI. Composition and Properties of Precious Metal Solders

Class of solder	Composition			BHN (as cast)	Melting range (°C)
	Gold	Silver	Copper		
Low carat	45	30–35	15–20	140	816–691
General purpose	60	12–22	12–22	110	835–724
High carat	80	3–8	8–12	80	871–746

REFERENCES

1. Wise, E. M., Editor, "Gold: Recovery, Properties and Applications". D. van Nostrand Company Inc., New York, 1964.
2. American Dental Association, "Guide to Dental Materials and Devices". 7th Edition, 1974–1975.
3. Greener, E. H., Harcourt, J. K. and Lautenschlager, E. P., "Materials Science in Dentistry". Williams and Wilkins Co., Baltimore, 1972.
4. Phillips, R. E., "Skinner's Science of Dental Materials", 7th Edition. W. B. Saunders Company, Philadelphia, 1973.
5. Degussa, "Edelmetall-Taschenbuch". Frankfurt-am-Main, 1967, pp. 160–167.
6. Black, A. D., Dent. Rev. **29** (1915), 14–29.
7. Trueman, W. H., Dent. Cosmos, **10** (1868), 128 ff.
8. Koser, J. R. and Ingraham, R., Mat gold foil with a veneer cohesive gold foil surface for class V restoration, *J. Am. Dent. Assoc.* **52** (1956), 714–727.
9. Myers, L. E., Filling a class V cavity with a combination mat and cohesive gold foil, *J. Prosthet. Dent.* **7** (1957), 254–258.
10. Lund, M. R. and Baum, L., Powdered gold as a restorative material, *J. Prosthet. Dent.* **13** (1963), 1151–1159.
11. Baum, L., Gold foil (filling golds) in dental practice, *Dent. Clin. North Am.* (1965), Mar., 199–212.
12. Hollenback, G. M., Lyons, N. E. and Shell, J. S., A study of some of the physical properties of cohesive gold, *J. Calif. Dent. Assoc.* **42** (1966), 9–11.
13. Richter, W. A. and Cantwell, K. R., A study of cohesive gold, *J. Prosthet. Dent.* **15** (1965), 722–731.
14. Xhonga, F., Direct golds, *J. Am. Gold Foil Oper.* **13** (1970), 17–22 and **14** (1971), 5–15.
15. Mahan, J. and Charbeneau, G. T., A study of certain mechanical properties and the density of condensed specimens made from various forms of pure gold, *J. Am. Gold Foil Oper.* **8** (1965), 6–12.
16. Rule, R. W., A further report on physical properties and clinical values of

platinum-centred gold foil as compared to pure gold filling materials, *J. Am. Dent. Assoc.* **24** (1937), 583–595.
17. Hodson, J. T., Structure and properties of gold foil and mat gold, *J. Dent. Res.* **42** (1963), 575–582.
18. Coleman, R. L., Physical properties of dental materials (gold alloys and accessory materials), Research Paper No. 32, *J. Res. Nat. Bur. Stand.* **1** (1928), 867–938.
19. Taylor, N. O., Paffenbarger, G. C. and Sweeney, W. T., Inlay casting golds: Physical properties and specifications, *J. Am. Dent. Assoc.* **19** (1932), 36–53.
20. Brumfield, R. C., Tentative standard methods of testing of precious metal dental materials, *J. Am. Dent. Assoc.* **49** (1954), 17–30.
21. Eick, J. D., Caul, H. J., Smith, D. L. and Rasberry, S. D., Analysis of gold and platinum group alloys by X-ray emission and corrections for interelement effects, *Appl. Spectrosc.* **21** (1967), 324–328 (*C.A.* **67** (1967), 104852d).
22. Eick, J. D., Caul, H. J. and Dickson, G., Chemical composition of dental gold casting alloys and dental wrought gold wires, *Int. Assoc. Dent. Res. Abstracts*, **46** (1967), 21.
23. Wise, E. M., Crowell, W. S. and Eash, J. T., The role of platinum metals in dental alloys. I and II, *Trans. Am. Inst. Min. Metall. Eng.* **99** (1932), 363–412.
24. Wise, E. M. and Eash, J. T., The role of platinum metals in dental alloys. III. The influence of platinum and palladium and heat treatment upon the microstructure and constitution of basic alloys, *Trans. Am. Inst. Min. Metall. Eng.* **104** (1933), 277–307.
25. Leinfelder, K. F., O'Brien, W. J. and Taylor, D. F., Hardening of dental gold—copper alloys, *J. Dent. Res.* **51** (1972), 900–905.
26. Lüthy, H. and Tissot, P., A study of segregation in gold-based dental alloys by electron probe analysis, *Chimia,* **28** (1974), 391–393.
27. Lüthy, H. and Tissot, P., Thermal differential analysis of order-disorder transformations of dental alloys based on gold and copper, *Thermochim. Acta*, **10** (1974), 279–283 (*C.A.* **82** (1975), 64465f).
28. Kanzawa, Y., Yasuda, K. and Metahi, H., Structural changes caused by age-hardening in a dental gold alloy, *J. Less-Common Met.* **43** (1975), 121–128.
29. Bergman, M., Holmlund, L. and Ingri, N., Structure and properties of dental casting gold alloys. I. Determination of ordered structures in solid structures in solid solutions of gold, silver and copper by interpretation of variations in the unit cell length, *Acta. Chem. Scand.* **26** (1972), 2817–2831 (*C.A.* **78** (1973), 21319m).
30. Barton, J. A., Eick, J. D. and Dickson, G., Compression of Brinell and Vickers hardness tests on dental casting gold alloys, *J. Dent. Res.* **52** (1973), 163–169.
31. Nielsen, J. P. and Tuccillo, J. J., Grain size in cast gold alloys, *J. Dent. Res.* **45** (1966), 964–969.
32. Wagner, E., "Goldlegierung zum aufbrennen von Porzellan für zahnärtzliche Zwecke". German Patent 1,533,233 (1970) (cf. Wagner, E. and Ludwig, W. E., German Auslegeschrift 1,183,247 (1966).
33. Schmid, H., Grain size and hardening of gold–platinum alloys, *Metall.* **12** (1958), 612–619.
34. Gettleman, L., Harrison, J. D. and Brasunas, A. de S., Hypoeutectic gold–cobalt alloys for dental castings, *J. Biomed. Mater. Res.* **6** (1972), 25–32.
35. Brugirard, J. *et al.*, Study of the electrochemical behaviour of gold dental alloys, *J. Dent. Res.* **52** (1973), 828–836.
36. Ryge, G., Current American research on porcelain-fused-to-metal restorations, *Int. Dent. J.* **15** (1965), 385–392.

37. Vickery, R. C. and Badinelli, L. A., Nature of attachment forces in porcelain–gold systems, *J. Dent. Res.* **47** (1968), 683–689.
38. McLean, J. W. and Sced, I. R., Bonding of dental porcelain to metal. I. The gold alloy/porcelain bond, *Trans. Brit. Ceram. Soc.* **72** (1973), 229–233.
39. Yamaguchi, S., Tsuchiya, T. and Nishikawa, M., Study of a dental alloy based on gold–platinum alloy, *Werkst. Korros.* **26** (1975), 356.
40. Ingersoll, C. E. (Williams Gold Refining Co. Inc.), "Bright Cast Alloy and Composition". U.S. Patent 3,769,006 (1973).
41. Dr. Th. Wieland Scheideanstalt., *German Offen.* 2,302,837.
42. Leinfelder, K. F. and Servais, W. J., Platinum-free high fusing gold alloys for enamel veneering, *J. Dent. Res.* **49** (1970), 884.
43. Hirschhorn, L., "Porcelain Bonding Dental Gold Alloy". U.S. Patent 3,679,402 (1971).
44. Burnett, A. P., "Gold Base Alloys for Use in Dentistry and Industry". U.S. Patent 3,666,540 (1972) *(C.A.* **77** (1972), 65564w).
45. Nally, J. N., Chemico-physical analysis and mechanical tests of the ceramo-metallic complex, *Int. Dent. Assoc. J.* **18** (1968), 309–325.
46. Fairhurst, C. W. and Leinfelder, K. F., Heat treating porcelain-enamelled restorations, *J. Prosthet. Dent.* **16** (1966), 554–556.
47. Leinfelder, K. F., O'Brien, W. J., Ryge, G. and Fairhurst, C. W., Hardening of high-fusing gold alloys, *J. Dent. Res.* **45** (1966), 392–396.
48. Smith, D. L., Burnett, A. P., Brooks, M. S. and Anthony, D. H., Iron platinum hardening in casting golds for use with porcelain, *J. Dent. Res.* **49** (1970), 283–288.
49. Fuys, R. A., Fairhurst, C. W. and O'Brien, W. J., Precipitation hardening in Au-Pt alloys containing small quantities of Fe, *J. Biomed. Mater. Res.* **7** (5) (1973), 471–480.
50. Tuccillo, J. J., Lichtenberger, H. and Nielsen, J. P., Composition stability of gold base dental alloys for different melting techniques, *J. Dent. Res.* **53** (1974), 1127–1131.
51. Sperner, F., Edelmetallegierungen für die Dental-Keramik, *Z. Metallk.* **67** (1976), 289–295.
52. Dietzel, A., *Ceram. Abstr.* **13** (1934), 250.
53. Dietzel, A., *Ceram. Abstr.* **14** (1935), 107.
54. Cole, S. S., Jnr. and Sommer, G., Glass migration mechanism of ceramic-to-metal seal adherence, *J. Am. Ceram. Soc.* **44** (1961), 265–271.
55. Borom, M. P. and Pask, J. A., Role of "adherence oxides" in the development of chemical bonding at glass-metal interfaces, *J. Am. Ceram. Soc.* **49** (1966), 1–6.
56. Kautz, K., Effect of iron surface preparation upon enamel adherence, *J. Am. Ceram. Soc.* **20** (1937), 288–295.
57. Kautz, K., Reply to Staley's discussion of A critical analysis of some statements and experiments on the adherence of sheet steel ground coats, *J. Am. Ceram. Soc.* **21** (1938), 311–315.
58. Kautz, K., Oxide film between fired ground-coat enamels and iron, *J. Am. Ceram. Soc.* **22** (1939), 247–250.
59. Fulrath, R. M., Mitoff, S. P. and Pask, J. A., Fundamentals of glass-to-metal bonding. III. Temperature and pressure dependence of wettability of metals by glass, *J. Am. Ceram. Soc.* **40** (1957), 269–274.
60. Pask, J. A. and Fulrath, R. M., Fundamentals of glass-to-metal bonding. VIII. Nature of wetting and adherence, *J. Am. Ceram. Soc.* **45** (1962), 592–596.

61. Volpe, M. L., Fulrath, R. M. and Pask, J. A., Fundamentals of glass-to-metal bonding. IV. Wettability of gold and platinum by molten sodium disilicate, *J. Am. Ceram. Soc.* **42** (1959), 102–106.
62. Douglas, G. S. and Zander, J. M., X-ray diffraction study of the oxidation characteristics of nickel-pickled sheet iron as related to enamel adherence, *J. Am. Ceram. Soc.* **34** (1951), 52–59.
63. Healy, J. H. and Andrews, J. A., The cobalt-reduction theory for the adherence of sheet-iron ground coats, *J. Am. Ceram. Soc.* **34** (1951), 207–214.
64. Harrison, W. N., Richmond, J. C., Pitts, J. W. and Benner, S. G., A radioisotope study of cobalt in porcelain enamel, *J. Am. Ceram. Soc.* **35** (1952), 113–120.
65. van Houten, G. R., A survey of ceramic-to-metal bonding, *Ceram. Soc. Bull.* **36** (1959), 301–307.
66. King, B. W., Tripp, H. P. and Duckworth, W. H., Nature of adherence of porcelain enamels to metals, *J. Am. Ceram. Soc.* **42** (1959), 504–525.
67. Silver, M., Klein, G. and Howard, M., An evaluation and comparison of porcelains fused to cast metals, *J. Prosthet. Dent.* **10** (1960), 1055–1064.
68. Swartz, M. L. and Phillips, R. W., A study of adaptation of veneers to cast gold crowns, *J. Prosthet. Dent.* **7** (1957), 817–822.
69. Lyon, D. M., Cowger, G. T., Woycheshin, F. F. and Miller, Ch. B., Porcelain fused to gold—evaluation and esthetics, *J. Prosthet. Dent.* **10** (1960), 319–324.
70. O'Brien, W. and Ryge, G., Relation between molecular force calculations and observed strengths of enamel-metal interfaces, *J. Am. Ceram. Soc.* **47** (1964), 5–8.
71. Shell, J. and Nielsen, J. P., Study of the bond between gold alloys and porcelain, *J. Dent. Res.* **41** (1962), 1424–1437.
72. Ryge, G., Current American research in porcelain-fused-to-metal restorations, *Int. Dent. J.* **15** (1965), 385–392.
73. Leone, E. F. and Fairhurst, C. W., Bond strength and mechanical properties of dental porcelain enamels, *J. Prosthet. Dent.* **18** (1967), 155–167.
74. Anthony, D. H., Burnett, A. P., Smith, D. L. and Brooks, M. S., Shear test for measuring bonding in cast gold alloy-porcelain composites, *J. Dent. Res.* **49** (1970), 27–33.
75. Eichner, K., Radnoth, M. Sz. V., Riedel, H. and Vahl, J., Micromorphological investigation of the bond in different gold-ceramic systems, *Dtsch. zahnärtzl. Z.* **25** (1970), 274–280 (*Oral Res. Abstr.* **6** (1971), 679).
76. Wagner, E., *Zahnärtzl. Welt,* **66** (1965), 343.
77. Nally, J. N. and Berta, J. J., Recherches expérimentales sur les propriétés mécaniques des céramiques cuite sur alliages (1), *Schweiz. Maschr. Zahnheilk.* **75** (1965), 93–104 (Oral Res. Abstr., Vol. 1, 1966, 505a).
77a. Johnston, J. F., Porcelain veneers bonded to precious metal castings, *Can. Dent. Ass. J.* **26** (1960), 657–663.
78. Nally, J. N., Monnier, D. and Meyer, J. N., Distribution topographique de certains éléments de l'alliage et de la porcelaine au niveau de la liaison céramco-metallique, *Schweiz. Maschr. Zahnheilk.* **78** (1968), 868–878 (cf. *Int. Dent. J.* **18** (1968), 309–325).
79. Meyer, J. N. and Nally, J. N., 7th Annual Meeting of Int. Assoc. Dent. Res., Continental European Section, 19th–20th Sept., 1970, Erlangen (from reference 50).
80. Radnoth, M. Sz. V. and Lautenschlager, E. P., Metal surface changes during porcelain firing, *J. Dent. Res.* **48** (1969), 321–324.
81. Radnoth, M. Sz. V. and Lautenschlager, E. P., Morphology of the interspace

between fine metal alloys and fired ceramics in crowns, *Dtsch. zahnärtzl. Z.* **24** (1969), 1029–1036.
82. Lautenschlager, E. P., Greener, E. H. and Elkington, W. E., Microprobe analyses of gold porcelain bonding, *J. Dent. Res.* **48** (1969), 1206–1210.
83. Radnoth, M. Sz. V., Electron microscope investigation of oxide layer formation in metal-ceramic systems, *Dtsch. zahnärtzl. Z.* **25** (1970), 259–264 (*Oral Res. Abstr.* **6** (1971), 682).
84. Leinfelder, K. F., Fairhurst, C. W., O'Brien, W. J. and Ryge, G., Evolution of gases from high-fusing gold alloys, *J. Dent. Res.* **45** (1966), 1154–1158.
85. Knap, F. J. and Ryge, G., Study of bond strength of dental porcelain fused to metal, *J. Dent. Res.* **45** (1966), 1047–1051.
86. Freyberger, P., Fehlermöglichkeiten in der Metallkeramik, *Zahnärtzl. Welt*, **78** (1969), 1082 ff.
87. Lavine, M. H. and Custer, R., Variables affecting the strength of bond between porcelain and gold, *J. Dent. Res.* **45** (1966), 32–36.
88. Nenadović, T., Bibić, N., Kraljević, M. and Adamov, M., Mechanical properties of dental gold thin films, *Thin Solid Films*, **34** (1976), 211–214.
89. Reference 2, pp. 188–189.
90. Reference 1, p. 234.
91. Lautenschlager, E. P., Marker, B. C., Moore, B. K. and Wildes, R., Strength mechanisms of dental solder joints, *J. Dent. Res.* **53** (1974), 1361–1367.

5 Gold and Gold Alloys in Instrumental Applications

1. Introduction

The use of gold and of gold alloys in a wide variety of instruments is based for the greater part upon the extremely wide range of thermal, electrical and mechanical properties which gold and its alloys can be made to exhibit. Important among the electrical properties are electrical resistivity (or conductivity), contact resistance, temperature coefficient of resistance, and thermal emf against copper. Important among the mechanical properties are mechanical strength, malleability and resistance to wear, as well as low coefficient of friction. Moreover, its high thermal conductivity ensures rapid dissipation of heat when gold is used for contacts or in situations where constriction resistance arises. Resistance to tarnish and corrosion is also a notable feature of many gold alloys. It ensures constancy of surface properties, such as contact resistance, over extended periods.

In regard to resistivity Table XXXVII illustrates the range of resistivities which can be achieved, and how gold alloys can be used both as conductor and resistor materials.

From a market point of view use of gold in conductors and contacts is by far the most important. Quite apart from the extensive use of gold in the form of fine wires or surface coatings in electronic circuitry, gold and its alloys are used in many kinds of separable contacts for low and medium energy circuits, including relay make-break contacts, slip ring sliding contacts, wipers for potentiometer wires and connectors. Moreover, the nobility, softness, ductility and ease of bonding of pure gold makes it the preferred materials at sites where permanent joints have to be made by thermal or ultrasonic compression bonding in electronic devices. The major proportion of the industrial gold used in the U.S.A. in 1968 was reported to be for electrical contacts, with world usage for this purpose in 1973[1] possibly exceeding 100 tons.

From the scientific and technical point of view, however, the resistor and other applications of gold alloys, although they do not consume large amounts of gold, are of great importance. Thus the wide range in resistivities,

Table XXXVII. Electrical Resistivities of Gold and some Gold Alloys compared with those of some Pure Metals*

Composition	Resistivity in μ ohm cm at approximately 20°C
Pd (pure)	10.8
Pt (pure)	10.6
Cu (pure)	1.7
Ag (pure)	1.6
Au (pure)	2.1
10Au-90Ag	3.64
20Au-80Ag	5.22
90Au-10Ag	6.7
67Au-33Ag	10.8
50Au-28Ag-20Cu-2Zn	13.1
69Au-25Ag-6Pt	17.5
40Au-60Pd	25.0
10Au-30Ag-14Cu-35Pd-10Pt-1Zn	31.6
75Au-1Cu-18.5Ni-5.5Zn	35.0
59.4Au-1.0Cr-39.6Pd	35.1
97.5Au-2.5Cr	36.9
20Au-4Ag-26Pd-50Pt	41.3
20Au-20Ag-40Pd-20Pt	48.9
58.2Au-3.0Cr-38.8Pd	52.2
38Au-4Cr-58Pd	58.9
57Au-5Cr-38Pd	71.5
50Au-45Pd-Mo	100.0
40Au-50Pd-10Fe	158.0 **
49.5Au-40.5Pd-10Fe	183.0 **
70Au-20Pd-10V	>200.0

* Data from Wise,[2] and from Degussa and Doduco handbooks.
** Value dependent upon treatment of the alloy.

temperature coefficients of resistance, and thermal emfs of gold alloys have created applications for them in such instruments as standard resistances, potentiometer wires, high-pressure gauges, resistance thermometers and thermocouples. Moreover, even in those instances where similar electrical properties can be achieved with base metals or their alloys, gold and its alloys are often the preferred materials because of their tarnish resistance.

In the description of the uses of gold in instruments which follows, the emphasis will be primarily upon the properties of the gold alloys which recommend their use in each instance.

2. Gold Alloy Resistors for Standard Resistances and Potentiometer Wires

2.1 Gold-palladium alloys

A basic consideration in the development of gold alloys as resistors is information such as that in Figure 20 which is reproduced from an article by Sperner[3] in which the effects upon the resistivity of gold of additions of up to 1 atomic % of a number of other elements are plotted. From this it will be seen that the effects of palladium additions to gold are small compared with those of a number of other metals. Nevertheless many important resistance alloys are based upon the Au-Pd system; these alloys are employed not only in wire form, but also along with Au-Pt alloys in the form of surface coatings on insulating substrates to which they are applied by either thick or thin film techniques in the production of printed circuits.

The variations in specific resistance and temperature coefficient of resistance of Au-Pd alloys with composition are illustrated in Figure 21 taken from a review by Darling.[4]

The Au-Pd system is characterised by the formation of a continuous series of solid solutions in which it will be noted that resistivity reaches a maximum, and temperature coefficient of resistance a minimum at compositions corresponding to a high degree of alloying. Relationships such as these were the subject of early interpretations by Mathiessen and Vogt,[5] which have recently been examined by Whall *et al.*[6,7] Studies of the resistivity and thermoelectric power of Au-Pd alloys have also been reported by Rowland *et al.*[8] Not shown in Figure 21 are the thermal emfs of the alloys versus copper, the values of which are important if wires of the alloys are to be used in precise measurement of dc. Also not shown in Figure 21 is the fact that the electrical properties of Au-Pd alloys undergo slow changes with time, which have been associated with ordering reactions.[9] These can be accelerated, and subsequent changes in electrical properties suppressed by rapid (10^5 degrees C/sec) cooling which apparently quenches large vacancy concentrations into the alloys and allows ordering processes to proceed to equilibrium quite rapidly. In practice this suggests that Au-Pd resistances should be stabilised not by ageing, but rather by winding them from quenched wire.[10]

As mentioned above the addition of palladium to gold does not give rise to alloys of high resistivity. The alloys of this type which have proved most effective in practice are alloys of gold (cf. Figure 20) with iron, chromium, cobalt, molybdenum and vanadium. Iron, molybdenum and vanadium, however, can most appropriately be added not to gold itself, but to Au-Pd alloys.

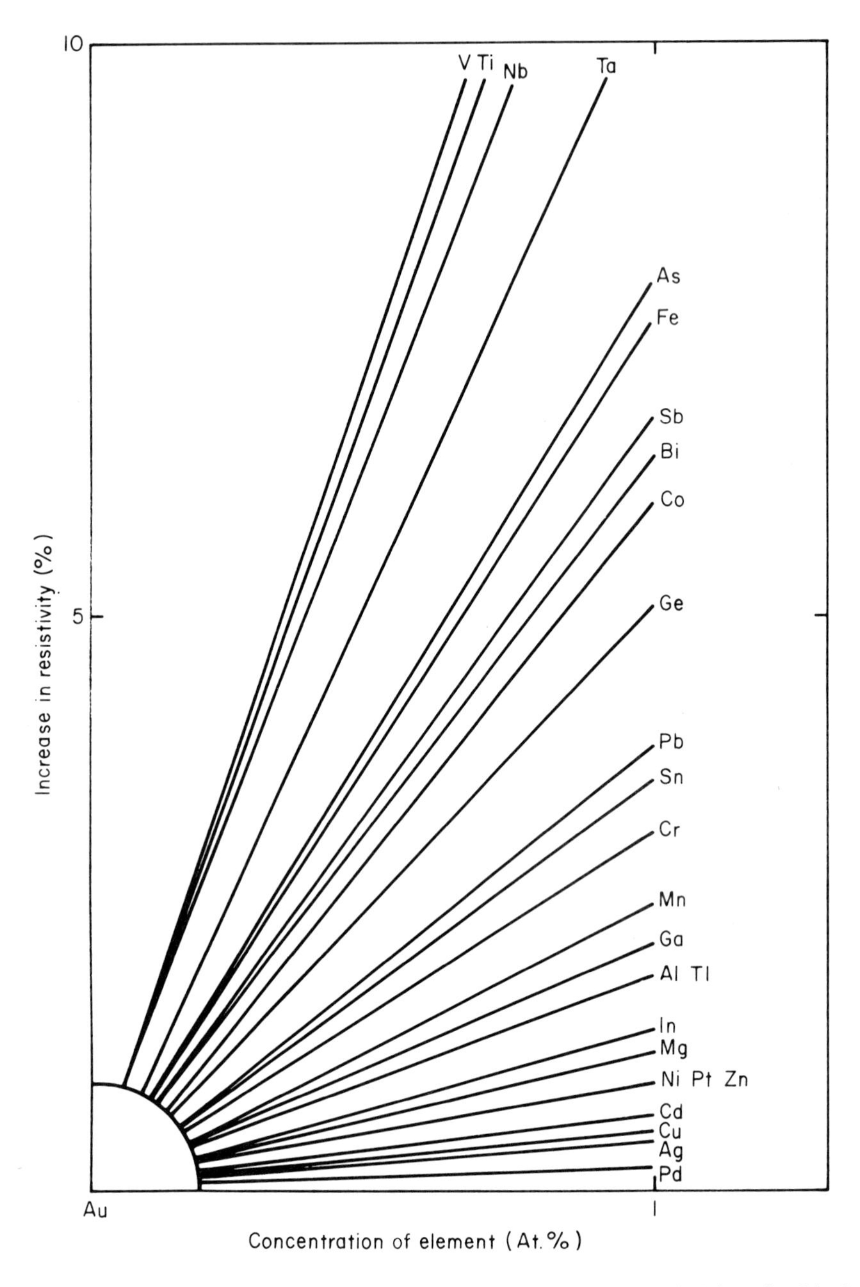

FIG. 20. A graphic representation of the effects on the electrical resistivity of gold of adding 1 atomic per cent of various elements. (After Sperner.[3])

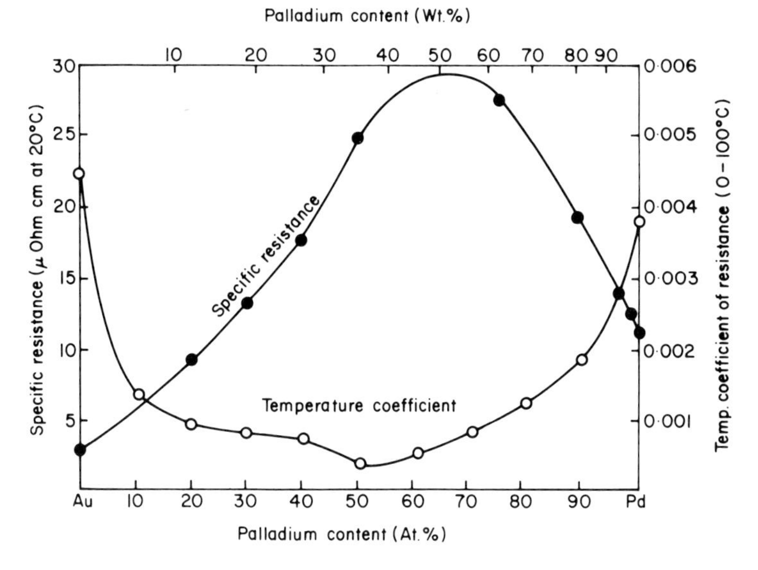

FIG. 21. The Au-Pd system is generally regarded as a classical example of complete mutual solubility. Although the electrical resistivity and temperature coeficient curves indicate deviations from this ideal state, no discontinuous variation of properties with composition occurs and many important resistance alloys are based on this system. (After Darling.[4])

2.2 Gold-iron and gold-palladium-iron alloys

Binary Au-Fe alloys containing up to 9% Fe, which have resistivities of 76-117 μ ohm cm, and which are easily worked[11] have not found favour because of their tendency to instability and the formation on them of surface oxide films. Better materials result from the addition of iron to Au-Pd alloys, the best being the 50Au-10Fe-40Pd ternary alloy, which is ductile and very corrosion resistant after solution treatment at 800°C and which has a specific resistance of 75 μ ohm cm. Subsequent ageing at 400-500°C and which has a specific resistance to about 180 μohm cm, with positive temperature coefficients as low as 10 ppm per °C being claimed. The increase in resistance with ageing is unexpected, since heat treatment orders the alloy;[12] this effect is usually associated with a decrease in resistivity. The response of the alloy to heat treatment is critically dependent on its composition and over ageing can cause rapid falls in resistivity. Cold working after ageing also reduces the resistivity because it destroys the ordered structure. Because of these effects, the applications of these alloys are limited to situations where high stability and constancy of resistance are not of primary importance.

2.3 Gold-chromium alloys

The work of Linde[13,14] in the 1930's indicated that these alloys could be considered as corrosion resistant alternatives to Manganin and Constantan, and their use as such was reported upon[15] by the U.S. Bureau of Standards in 1934. Experience of these[16] and in particular of their low temperature resistance characteristics, has led to their being used extensively. The alloy most used contains 2.0-2.1% Cr, and hard drawn wires of this alloy have a positive temperature coefficient and a specific resistance of 33 μohm cm. They are wound into coil form and then heat treated at low temperatures until the temperature coefficient has been reduced to acceptable values, which can be as low as 130 ppm over the range 20-100°C.[17] These alloys are not hard enough for use as slide wires, but serve well as stable fixed resistors. They find considerable application in the measurement of high pressures (see below).[18]

Au-Cr resistors can also be vacuum deposited as thin films on an insulating substrate.[19] Reports suggest that the stability of Au-Cr resistance wires is of a high order.[20,21]

2.4 Gold-cobalt alloys

An alloy containing 2.2% Co is readily drawn to wire and has a specific resistance of 30 μohm cm. Moreover, a temperature coefficient of almost zero can be obtained by low temperature ageing. The high thermal emf against copper of alloys of this type of about 47 micro-volts per °C,[22] however, makes them unsuitable for many uses.

2.5 Gold-palladium-molybdenum and gold-palladium-vanadium alloys

Until the 1960's, there were no gold-based alloys with high resistivity and low temperature coefficients of resistivity other than the Au-Fe-Pd alloys referred to above, in which resistivities of over 100 μohm cm could be developed. Moreover, as described above the Au-Fe-Pd alloys proved to have unsatisfactory features. Systematic studies have, however, established that both Au-Mo-Pd and Au-Pd-V alloys have the electrical properties, the stability, and the abrasion and tarnish resistance required for high resistance potentiometer wires.

The 50Au-5Mo-45Pd alloy is reported to be the most effective of this type.[23,4] It has a specific resistance of 100 μohm cm which is not affected appreciably by either heat treatment or working. Its temperature coefficient at 0.00012 per °C from 0-100°C is acceptably low for most purposes, and its thermal emf versus copper is −1.8 V per °C at room temperature. The fact that its resistance is not affected by deformation of the alloy during working is of particular interest.[24,25]

The physical properties of Au-V alloys were described by Köster and co-workers in 1961[26] and those of the Au-Pd-V alloys in 1962.[27] The latter were claimed[28] to be suitable for use in potentiometer wires, and in 1973 the 70Au-20Pd-10V alloy was reported[3] as being an alloy of this type chosen for commercial development. At that stage it was undergoing development trials in a number of applications. Its characteristics can be interpolated from Figure 22, from which it will be seen that it has a resistivity of more than 200 μohm cm, and that other electrical requirements are met.

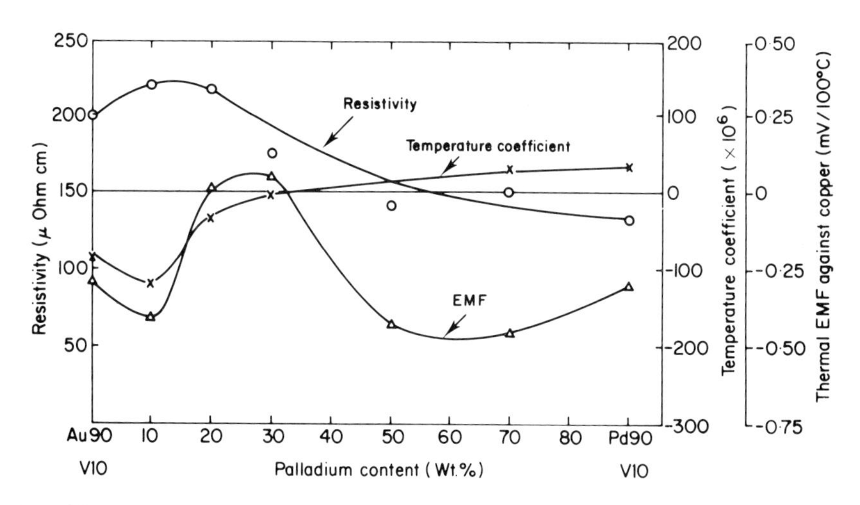

FIG. 22. The resistivity, temperature coefficient and thermal emf's of alloys in the ternary Au-Pd-V system with 10% V. At 20% Pd it will be seen that an alloy is obtained having an excellent combination of properties. (After Sperner.[3])

3. Gold Films in Resistance Thermometry

Gold films deposited on plastic foil by evaporation techniques have been found to be superior to wires when used in resistance thermometers for the measurement of small (0.01°C) temperature fluctuations in air streams.[29] They offer a large cooling area to the gas flow so that higher measuring currents can be applied. They also have low thermal inertia for which reason they have found application in aerodynamic studies where high sensitivity has been called for.[30]

4. Gold Alloys for Low Temperature Resistance Thermometry

4.1 Gold-manganese alloys

Platinum resistance thermometry presents problems at low temperatures because the resistance to be measured is so small. The presence of traces of certain other metals in such metals as copper, zinc and the noble metals, however, causes an anomaly — the Kondo effect — in their changes of resistance with temperatures. Typically in the presence of the impurities the electrical resistivity does not fall steadily with temperature to 0°K, but dips to a minimum at low temperatures and then rises again. The impurity metals which are active in causing this low temperature resistance minimum come typically from the first transition series, and include such metals as cobalt, iron, manganese, chromium and nickel. A very large research effort over the past 10 to 15 years has been aimed at achieving an understanding of the Kondo effect which was first observed in a nominally pure gold some forty years ago.

A new type of cryogenic resistance thermometer developed by Degussa[31] exploits the anomalous effect in the resistivity of a 99.6Au-0.4Mn alloy, which has proved to be stable at low temperatures and which can be drawn into fine wire. The effect of the manganese addition to this alloy is that the resistance of a sensor that is 50 Ω at the ice point falls to 25 Ω at the temperature of liquid helium rather than to 0.1 Ω as is the case with pure gold. Most of this additional resistance is temperature-independent, but the alloy still displays a positive slope of resistance against temperature from a value that is probably as low as 0.1°K. If this sensor fulfils its promise, it will provide a useful new instrument for both laboratory and industrial applications.

5. Gold Alloys for Thermocouples

5.1 Gold-palladium alloys (for higher temperatures)

The thermal emfs of Au-Pd alloys are negative to platinum in the thermoelectric series, vary with composition of the alloy and exhibit a maximum with the 60Au-40Pd alloy.[32]

These alloys find considerable application in thermocouples, for the measurement of high temperatures, but experience has shown that they generate higher emfs when they are combined with certain platinum alloys rather than with platinum itself. Couples which have been applied widely are[33] the following:

Negative leg	*Positive leg*
60Au-40Pd	90Pt-10Ir
60Au-40Pd	90Pt-10Rh
65Au-35Pd	14Pt-83Pd-3Au
65Au-35Pd	31Pt-55Pd-14Au

Major applications are in aircraft engines where thermocouples having high thermal emfs and good service lives at up to 1000°C are required.

High purity and freedom from physical strains are essential in the alloys used in such couples, the wire for which is normally supplied in an annealed condition. Nevertheless after the junction has been formed it is desirable to re-anneal under appropriate conditions and to avoid subsequent bending of the wires. Slow changes in the electrical properties of Au-Pd alloys which are associated with ordering[10] processes have already been mentioned.

5.2 Gold-cobalt and gold-iron alloys (for low temperatures)

The difficulties associated with the measurement of temperatures or temperature differences at temperatures approaching absolute zero are considerable, and have constituted a considerable obstacle both to research and development in the cryogenic field.[34]

The discovery that alloys of gold which contain small amounts of iron have very large and reproducible thermoelectric powers at very low temperatures has, however, led to their successful use as the active arms of sensitive and stable low temperature thermocouples. Prior to this development thermocouples had not been much employed at very low temperatures because the thermoelectric power of most thermocouple combinations decreases gradually to zero at absolute zero.

The anomalous behaviour of the Au-Fe alloys at very low temperatures is shared by a number of noble metal alloys containing small quantities of transition metals,[35] the effect of which is to enhance the thermopower very greatly, even to the extent that over certain ranges of temperature it may increase with decreasing temperature before finally tending to zero.

The first exploitation of this anomalous behaviour was the use of an Au-Co alloy containing 2.11 atomic % Co. For a number of years thermocouples in which one arm consisted of this alloy were used widely. This alloy, although it may be drawn in the homogeneous solid solution state, becomes supersaturated at lower temperatures with the result that cobalt phase may precipitate to a greater or lesser degree depending on the exact thermal or mechanical treatment of the wire. Calibrations were therefore difficult to reproduce. Apart from this the sensitivity becomes small at the temperature of liquid helium.

As a result of the studies by MacDonald and his collaborators[36,37] and others[38,39] of the thermal power of a wide range of dilute alloys, however, alloys containing 0.03 to 0.07 atomic % Fe in very pure gold have been developed[40-42] for use in various thermocouple combinations. In the United Kingdom the wire available commercially contains 0.03 atomic % Fe, whereas in the United States the standard material contains 0.07 atomic % Fe. Spectroscopically pure gold is used in their manufacture.

The large thermopower of the Au-Fe alloy is due to scattering of the conduction electrons in the gold by the magnetic moments of the iron. If the % Fe is "high", interactions between these magnetic moments tend to neutralise this effect, and the thermopower is reduced at low temperatures. If the % Fe is "low", however, the magnetic scattering at the higher temperatures is dominated by the normal scattering of electrons induced by lattice vibrations, and the anomalous thermopower is suppressed, and develops only as decreasing temperatures reduce the normal scattering induced by the lattice vibrations.[43] High purity gold is essential in order to ensure that electron scattering by "normal" impurities is small.

The effects of magnetic impurities in gold are such that they influence its electronic properties at concentrations at the ppb level, and their effects upon its thermoelectric power are sufficiently great for measurements of such power to be used for detection of such impurities.[44,45]

6. Gold and Gold Alloys in Conductor Applications

6.1 Gold and gold alloys for electrical contacts

Mated surfaces touch only at their high spots, which have a total area which is much less than that of the geometrical contact area. The total electrical resistance across their interface therefore has two components, firstly, a constriction resistancre component arising from the channelling of current flow through the limited area of the touching high spots and, secondly, a film resistance component arising from insulating oxide films if present on the mated surfaces. Low contact resistance is therefore associated with the absence of or easy penetration of tarnish or lubricant films, and the establishment of metallic contact across a high percentage of the area of the mated interface.

In the low voltage solid state devices which are so widely applied in computers, communication systems and instrumentation, high contact loadings cannot be applied to break tarnish films and to establish high areas of metallic contact across mated surfaces, so that tarnish and corrosion resistant metals must be used for the contacts. Amongst these, gold and its alloys are outstanding and are used widely, both with and without lubricants. They are used in wrought, clad and electrodeposited forms, and they have an advantage over metals such as platinum and palladium in that they do not[46] catalyse strongly the formation of contaminating polymeric material when contacts made of these alloys are operated in the presence of low concentrations of organic vapours. In fact, it is found advisable where palladium contacts are employed in communication systems, to coat one of the contact surfaces with a hard gold alloy. In most situations, however, where there is risk of exposure of high concentrations of organic vapours, gold alloys such as the 95Au-5Ni alloy may be preferable to platinum group metals for make-and-break contacts.

Electrodeposited gold alloys are used widely[47] for connectors, the hard wear resistant acid golds containing less than 1% Co or Ni being used the most, since pure soft gold wears rapidly in such situations, and pure gold surfaces have a tendency to cold-weld to one another. Use is made of barrel or jet[48] plating in preference to rack plating wherever possible. Electrodeposited alloys of lower gold content have not yet been used to any large extent.[49] The performance of plated golds is determined in large measure by their resistance to wear[50] and to tarnish. While the former can be controlled to a considerable extent by the type and structure of the plating applied and by use of lubricants, the latter is determined mainly by the integrity of the plating, through which substrate base metal may be exposed either as a result of pinholes or wear, or by diffusion through the gold coating. Diffusion effects are especially significant when copper is a substrate metal, because of the ease with which it diffuses through gold; they can be limited in such cases by the application of an undercoat of nickel as a diffusion barrier.[50,51] Much depends upon service conditions, such as temperature, however, and upon plating thicknesses, which can also be adjusted. A factor, the significance of which has not yet been fully defined, is the presence in hard acid gold platings of co-deposited polymer. With plated and other gold coatings the mechanical properties of the substrate may also affect contact behaviour.

Clad and inlay clad gold alloys, which are discussed later, are also used extensively for electrical contacts.[53,55] Cladding has an advantage in that a wide range of gold alloys can be applied. Performance data on Au-Cu, Au-Ag-Pt, Au-Ag-Ni, Au-Ag and Au-Ni clad coatings have been reported.[53]

A wide range of gold alloys finds application in contacts in other situations. Weldments on the ends of spring members of contacts are usually gold alloys, and relays may have contacts made of 75Au-25Ag or 69Au-25Ag-6Pt. Platings of silver hardened gold may also be used in place of the nickel and cobalt hardened acid golds. Also as mentioned above, an important advance was the use of solid palladium for one contact and a 22 carat hard gold alloy over palladium for the other. Apart from its relative freedom from the catalytic formation of polymer, this combination has been found to resist both wear and sliding damage. In dry reed relays for fast switching at low current, the ends of the reeds are usually plated with hard acid golds.[56] The arcing voltage of sealed reed contacts has been shown to be reduced when cobalt hardened acid cyanide gold plate is used, because of the presence of potassium in the gold plating.[57] For extremely sensitive relays and electrical measuring instruments, Russian workers have shown that a 97Au-3Re alloy is superior to a 60Au-30Pd-10Pt alloy previously used.[58]

Sliding contact materials, in addition to the requirement that they must be highly tarnish-resistant, which is essential for low contact resistance and freedom from noise, must also operate with minimum friction and wear. In this respect lubrication plays a role,[59] but the choice of alloys for the mating partners is all important.[60] The tribological properties of gold for electric contacts have been reivewed by Antler.[61]

Instrument slip rings are usually plated with 2.5-12.5 μm of low alloy acid golds hardened with nickel or cobalt, and pure gold may be used as substrate. Wrought alloys such as 90Au-10Cu, 75Au-22Ag-3Ni, and 69Au-25Ag-6Pt are also common ring materials. A widely used brush or wiper material is a 62.5Au-7.5Ag-30Cu alloy in which high strength and high hardness (275 Vickers) are developed by heat treatment to promote Au-Cu ordering. Other alloys used are a 72.5Au-4Ag-14Cu-8.5Pt-1Zn alloy and a 10Au-30Ag-14Cu-35Pd-1Zn alloy. An Au-Co-Cr-Zr alloy, in which the necessary hardness is developed by precipitation hardening, has also been described[62] as performing well in sliding contact tests. A 55Au-39Ag-3Cd-3In alloy, despite its high silver content, has been reported[63] to be resistent to sulphide tarnish. Schiff *et al.*[64] reported recently on a study of the composition and contact resistance of tarnish films on a variety of alloys.

A novel finding, namely, that of the very good performance in make-and-break contacts of gold which has been hardened by incorporation of cerium oxide as a disperse phase is discussed later in connections with gold composites.

6.2 Gold wire and gold coatings in electronic circuitry

Technically important in regard to the use of gold wires in electronic circuitry are the techniques employed to bond them in position, and the metallurgical considerations[65,66] which arise in this connection. These aspects are referred to later in discussion of gold wire.

Of interest in respect of the use of thin gold film conductors are diffusion or electromigration effects which may give rise to failures where the conductors are operated in integrated circuits at high current densities over extended period of time.[67,68] It has been demonstrated[69-72] that with gold films deposited on transition metal substrates such electromigration of metal occurs in the direction of electron flow, so that with time voids appear and grow at the cathodic ends of such conductors and rounded hillocks of metal build up at the anodic ends. On the other hand it has been found[73] that in annealed pure gold films deposited on glass electromigration occurs against the direction of electron flow. The precise mechanisms of these effects have not been established clearly.

Another point of interest is the extent to which the electrical properties of bulk gold are retained in gold films as their thickness is reduced. This will be referred to later in discussing the use of electrically heated gold coatings on glass. Suffice it to mention here that the minimum thickness at which gold films on glass begin to conduct is 0.3 μm.[74] The stage at which island formation on a gold surface coating begins depends, however, on the nature of the substrate, and ultra thin conducting films of gold, produced on mica substrates by special procedures, are highly conductive at thicknesses as small as 0.1 μm.[75]

6.3 Gold coatings as current collectors in solar cells

In solar cells sunlight striking the surface of a silicon solar cell generates a small current through an external circuit which connects silicon p- and n-regions, which are formed by doping each silicon wafer with boron and arsenic respectively. Large numbers of such cells have to be assembled to produce even moderate (e.g. 500 watts) power output, and it has been found that the current is collected most effectively from each cell by way of gold plated conductor grids deposited on the silicon cell surface. Not only is the gold highly conductive and resistant to corrosion, but the ease with which it can be soldered facilitates the assembly of power packs.[76]

6.4 Gold in microwave tubes

In microwave tubes, certain of the base metal structures are coated with gold to provide conductive paths for the ultra-high frequencies, which are so high that the currents are forced to the outermost layers of the device. Although silver coatings would provide high conductivity paths for these currents, gold is preferred because of its resistance to tarnish.

6.5 Gold in vacuum tubes

The leads to vacuum tubes are normally of Ni-Cr-Fe or Fe-Co-Ni alloys the thermal expansion characteristics of which match those of the glass. These alloys have electrical resistivities which are great enough, however, to cause local heating of glass to metal seals using these alloys. This is overcome and the current carrying capacity of such seals increased substantially by coating the leads with a layer of gold thick enough (25-30 μm) to ensure adequate conductivity of the lead over the joint area.

Gold is also used as a coating for grid wires in vacuum tubes incorporating oxide coated cathodes. Free barium can evaporate from such cathodes, and if it is deposited on other warm surfaces, these can then act as unwanted sources of electron emission. Gold plating on such surfaces, and especially the surfaces of control guides, readily takes up the barium, which diffuses into the gold, so that a surface of high work function and low emission is maintained.[77, 78]

6.6 Gold in semi-conductor devices

Gold is used in a variety of ways in the production of semi-conductor devices,[78] both in the form of protective coatings, and a number of the features of these applications are discussed later. Certain aspects of the behaviour of gold in contact with semi-conductors such as silicon and germanium merit particular mention however. When, for example, gold coated slices of silicon are heated, either deliberately or during the manufacture of silicon semi-conductor devices, diffusion occurs across the interface. The gold which diffuses into the silicon acts as a p-type impurity, and this may call for corrective action, either by way of suitable pre-doping of the metal so as to ensure ohmic contacts, or by the application of undercoatings of materials which will act as diffusion barriers between the gold and the silicon. Evaporated titanium either alone or preferably overcoated with platinum may be used for such undercoatings.

An example of pre-doping is a procedure followed in the early 1950s in the production of n-type transistors from germanium wafers doped with antimony. Because of difficulties in plating and soldering connections to such wafers, an Au-Sb alloy of controlled composition was plated on a Kovar disc or washer. The pure germanium wafer was then placed on the plated surface and heat fused at 500°C without flux. During this step both gold and antimony diffused into the wafer, and the alloy composition was adjusted so that the "n" properties of the antimony predominated.

Of more positive interest, however, is the effect which the gold diffused into a silicon semi-conductor can have on the performance of such semi-conductor materials when used in high speed switching devices. In such devices the signal is carried by extra injected minority carriers (electrons or holes according to type), and these diffuse down the concentration gradient set up in the equilibrium ratio of holes and electrons previously obtaining. To re-establish equilibrium, recombination must occur and this can be too slow to prevent degrading of high frequency pulse shapes. Very small concentrations (e.g. 10^{16} atoms per ml) of gold in the silicon semi-conductor however, facilitate such recombination and decrease the effective lifetime of the carriers very greatly. Since it is this effective life which determines the time required to switch a transistor, the storage time in computing assemblies can, in this way, be reduced by several orders of magnitude.

In practice, therefore, gold is evaporated on to the back of silicon chips used in some logic integrated circuits and in fast switching transistors for high frequency devices, and diffused into the chip by a subsequent heat treatment. Any adverse doping effects must of course be compensated for as indicated above.

Arising from its use as a deep impurity in silicon diodes and transistors to increase their switching speed as described above, the behaviour of gold in silicon has been studied extensively[79-80] and the thermal and optical emission and capture of holes and electrons at gold centres in bulk silicon examined.[81-83] This has led recently to the development of gold doped metal-oxide-silicon field-effect transistors (MOSFET'S) for use as infrared detectors.[81-86] Applications of devices of this type in large scale integrated imaging arrays are envisaged by Forbes and his co-workers.[86]

6.7 Superconductivity in gold alloys

Although no intermetallic compound of gold has yet been found to exhibit superconductivity at temperatures that would make it industrially useful, studies of gold intermetallics have increased our understanding of this effect. Such studies have been reviewed by Khan and Raub.[87]

7. Gold and Gold Alloys in Miscellaneous Instruments

7.1 Gold in an instrument for the detection of mercury vapour

The operation of an extremely sensitive, lightweight, field portable mercury detector[88] is based on the fact that a thin gold film undergoes a significant increase in resistance upon the absorption of mercury vapour.[89-91] The change in resistance is caused by an increase in scattering of conduction electrons at the gold film surface upon absorption of mercury atoms. Two gold films, one a sensor and the other a reference, are connected into opposite arms of a single dc bridge circuit and balanced. The reference film serves the function of compensating for resistance changes due to thermal fluctuations. The optimum thickness and optimum annealing temperature for the gold film have been studied by Chaurasia *et al.*[92]

The instrument has been used [93] extensively in recent years in prospecting, mercury vapour in soil gas having been found to be an indicator of the presence of a variety of underlying base and precious metal deposits from which it is released in traces as the deposits oxidise. It is also used in studies of air pollution by mercury.

It has been claimed[94] that mercury can be removed from waste water by contacting the water with gold.

7.2 Gold-chromium alloys for high pressure gauges

The Au-Cr alloys developed for resistor applications (see above) find considerable application in measurement of pressures of up to 6×10^5 atmospheres by the Bridgeman technique,[17] in which the fine resistance wire is exposed to the hydrostatic pressure of the working fluid. Since changes in resistance with pressure are small compared with those of temperature, the low temperature sensitivity of the Au-Cr alloy wires and their linear response to pressure (Figure 23) give them practical advantages over, for example, Manganin wires, even though these are more sensitive to pressure.[18]

7.3 A gold-silver-copper alloy in moving coil meter suspensions

The moving coil meter depends for its operation on the efficiency of its suspension, which in commercial practice has incorporated hardened steel pivots rotating in jewelled bearings with delicate hairsprings to provide the restraining force. One manufacturer has recently been reported[95] to have developed a commercial instrument which operates with two ligaments held taut, one each side of the moving coil, which possesses improved sensitivity and robustness. The ligaments are of a stabilised 62.5% Au alloy containing undisclosed percentages of silver and copper.

7.4 Gold as the basis of a high pressure calibration scale

The melting point of gold increases linearly with pressure from 1063°C at atmospheric pressure to 1404°C at 65 kbar. On the assumption that this linear relationship continues at still higher pressures, it can therefore be estimated that its melting point at 200 kbar will be 1988°C, and melting points at intermediate pressures can also be assessed.

Akella and Kennedy[96] have therefore proposed that the graph obtained by plotting and extrapolating the melting point of gold against pressure should be used as a primary standard for pressure measurement for geophysical equip-

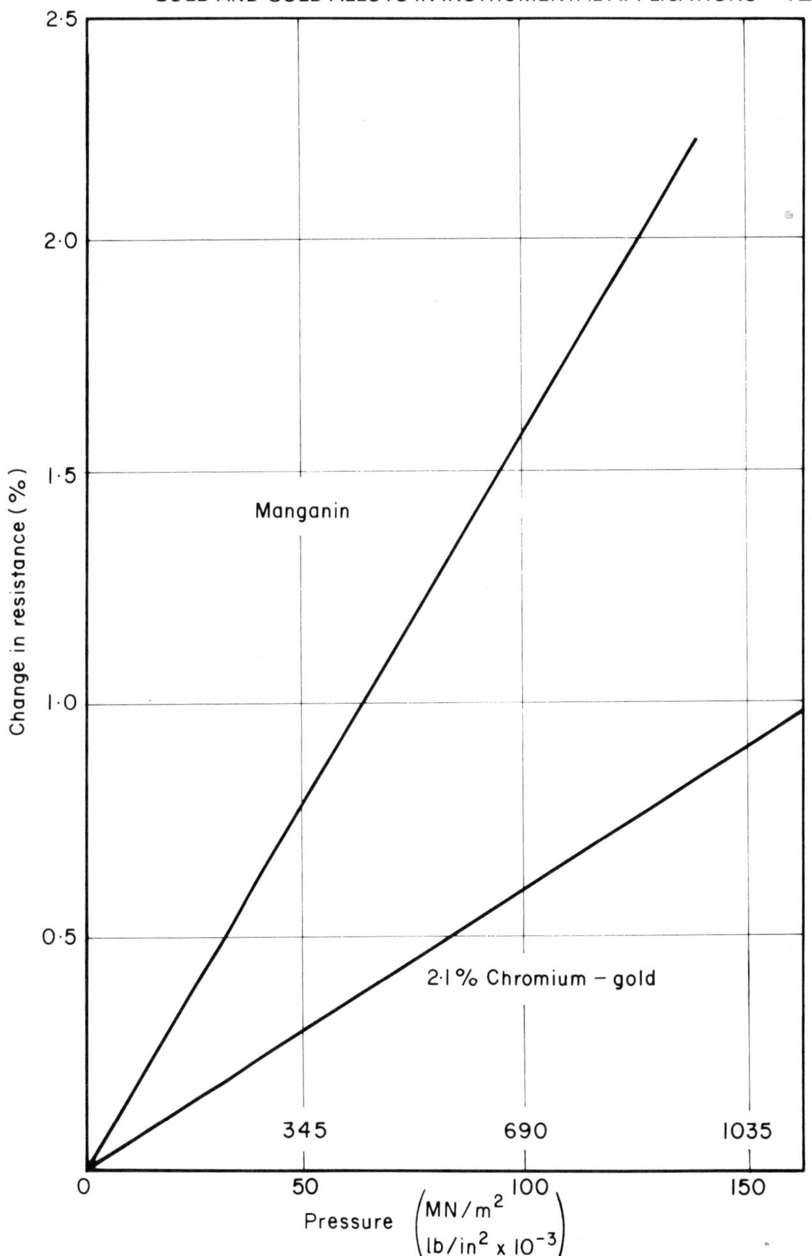

FIG. 23: Effect of pressure on the resistivity of the 97.9Au-2.1Cr alloy. Hydrostatic pressures of 100 × 10^3lb/in^2 (690 MN/m^2) increase the resistance of this material by about 0.5%, and the pressure resistance relationship is completely linear. (After Darling.[4])

ment at pressures in the region 100-200 kbar and over the temperature range 1600-2000°C.

7.5 Gold-containing membranes in ion-sensitive electrodes

Pellets made by washing, drying and pressing the material resulting from precipitation of silver salts (AgCl, AgBr, AgI and Ag_2S) on to finely divided gold of particle size less than 0.3 μm have been used as heterogeneous membranes in ion-sensitive electrodes and have been found[97] to show a selective ion-sensitive behaviour for halogens, silver and sulphide. Silver iodide on gold pellets is also sensitive for cyanide. The precise method by which these electrodes function is not known, but they are available commercially.

7.6 Gold coating of quartz crystals used for stabilising oscillatory circuits

A number of electronic devices are based upon the piezoelectric effect according to which when certain materials such as quartz or a titanate ceramic are compressed, their opposite faces exhibit a potential difference. Conversely, if a potential difference is applied across them, their dimensions change. A device of major importance based on this effect is the quartz crystal used for stabilisation of oscillatory circuits, the electrical frequency being stabilised by the frequency of vibration of the quartz. For such devices, a thin conducting film which does not interfere seriously with the fundamental mode of vibration of the crystal is required, and this is often provided by a sputtered or evaporated coating of gold.[98] The presence of the film does, in fact, change the frequency of vibration, and for accurate work it is necessary to carry out a second deposition to calibrate the crystal.

7.7 Gold coatings in a high sensitivity strain gauge

A strain gauge developed by the Electrical Research Association in the United Kingdom consists of a thin film of evaporated gold deposited between two thick planar electrodes on a glass sub-layer on a polished metal plate.[98] The gold film is of such a thickness that it has a discontinuous or "island" structure, so that the electrical resistance of the film arises mainly from the gaps between the gold "islands". Conduction across these gaps is by quantum mechanical tunnelling and is very sensitive to small changes in gap widths brought about by applied mechanical strains. In the case of ultra thin gold films evaporated on to plastic substrates, Stops and Azkan[99] have observed remarkable changes (10^3–10^4fold) in resistance as the plastic substrates are extended.

7.8 Gold grids in electronic humidistats

A new electronic humidistat[100] with fast and accurate response down to 1% relative humidity uses a sensing element consisting of a thin plastic wafer covered with two grid patterns of gold lines, and coated with lithium chloride which is very sensitive to moisture. The grids are connected to separate terminals. When the humidity increases, the area between the grid patterns becomes more conducting, so that greater current flows between them. This current is amplified and used to operate the humidity control equipment. A range of different sensing elements is available to measure a full range of relative humidities from 5–95%.

Gold was chosen as the grid material because of its high conductivity and resistance to corrosion.

7.9 Magnetic alloys of gold in switching and other applications

Although gold itself is essentially non-magnetic, gold alloys high in iron, cobalt or nickel are strongly ferromagnetic and some of these have found industrial applications.

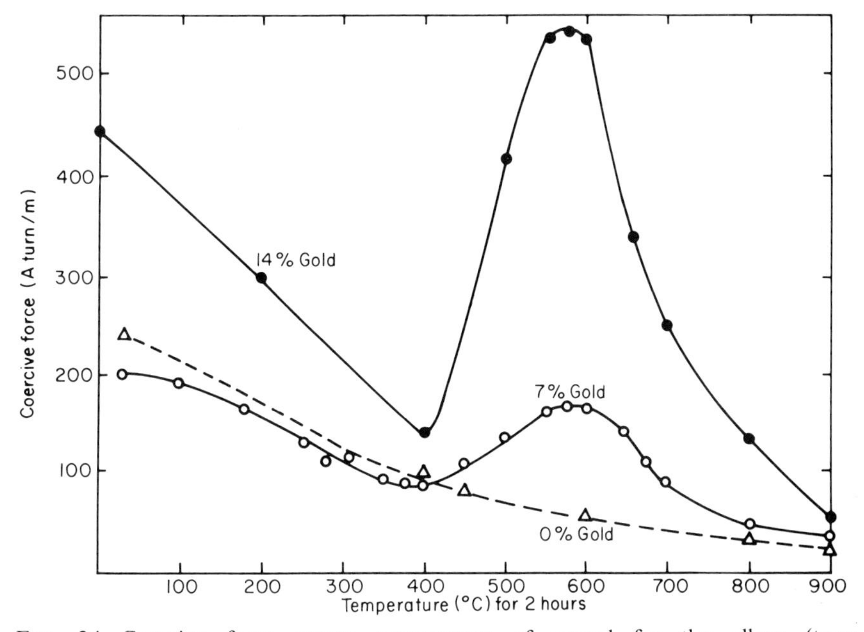

FIG. 24. Coercive force versus temperature of anneal for the alloys (tape 0.317 × 0.000317 cm) containing 0.7 and 14% Au by weight. Note the absence of an age hardening peak at 575°C for the 0% Au specimens. (After Nesbitt and Gyorgy.[102])

Thus the magnetic and other properties of Au-Ni-Fe alloys have been reported upon by Köster and Ulrich[101] and the effectiveness of additions of gold to Permalloy in reducing substantially the switching time in Permalloy type magnetic cores has been recorded by Nesbitt and Gyorgy.[102]

In the modified Permalloy alloys (71-78Ni, 15-20Fe, 2-14Au) the effect of the gold addition is to promote the precipitation of a second phase on heat treatment. In the presence of this phase the coercive force of the alloy is greatly increased (Figure 24).

Au-Co-Fe alloys have also found application in telephone switching systems[103] and permanent magnets of Au-Fe-Pt alloys have been described.[104]

7.10 Measurement of surface temperatures

Resistance thermometers consisting of vacuum deposited gold grids 4-6 μm thick and 0.1-0.25 mm wide have been formed[105] on glass and fused silica specimens and used for surface temperature measurements. Thinner films were not effective because of the formation of gold agglomerates during heating. It was found necessary to age the films by furnace soaking them at 807°C for 24 hours in order to ensure that they were stable. The temperature coefficients of resistance of the sensors were found to depend on the film thickness and were generally higher than that of bulk gold.

REFERENCES

1. Antler, M., Gold in electrical contacts, *Gold Bulletin*, **4** (1971), 42–46.
2. Wise, E. M., Editor, "Gold: Recovery, Properties and Applications". D. van Nostrand, Princeton, 1964, p. 301.
3. Sperner, F., A new gold resistance alloy: a study of the gold-palladium-vanadium system, *Gold Bulletin*, **6** (1973), 72–76.
4. Darling, A. S., Gold alloys for precision resistances, *Gold Bulletin*, **5** (1972), 74–81.
5. Matthiessen, A. and Vogt, C., *Trans. Roy. Soc.* **154** (1864), 167 ff.
6. Ford, P. J., Whall, T. E. and Loram, J. W., Electrical resistivity of the Au-V system, *J. Phys. F.* **4** (2), 225–231.
7. Whall, T. E., Ford, P. J. and Loram, J. W., Breakdown of Matthiessen's rule in gold alloys, *Phys. Rev. B.* **6** (1972), 3501–3509.
8. Rowland, T., Cusack, N. E. and Ross, R. G., Resistivity and thermoelectric power of the palladium–gold alloy system, *J. Phys. F.* **4** (1974), 2189–2202.
9. Köster, W. and Halpern, T., Leitfähigkeit und Hallkonstante. XXI. Gold-Palladium Legierungen, *Z. Metallk*, **52** (1961), 821–825.
10. Anon., Structural changes in gold–palladium alloys, *Gold Bulletin*, **6** (1973), 11.
11. Scheil, E., Specht, H. and Wachtel, E., Magnetic measurements in gold–iron alloys, *Z. Metallk.* **49** (1958), 590–600.
12. Schneider, J. F. and Sivil, C. S., "High Resistivity Alloy". U.S. Patent 2,780,543 (1957).
13. LINDE, J. O., Elecktrische Eigenschaften verdünnter Mischkristallegierungen.

III. Winderstand von Kupfer- und Goldlegierungen, *Ann. Phys. Leipzig,* **15** (5) (1932). 219–248.

14. Linde, J. O., "Electrische Widerstandeigenschaften der verdünnten Legierungen des Kupfers, Silbers und Goldes". Gleerupska Univ.–Bokhandeln, Lund, Sweden, 1939.
15. Thomas, J. L., Gold-chromium resistance alloys, *J. Res. U.S. Bur. Stand.* **13** (1934), 681–688 (R.P. 737).
16. Godfrey, T. B., Further data on gold–chromium resistance wire, *J. res. U. S. Bur. Stand.* **22** (1939), 565–571 (R.P. 1206).
17. Darling, H. E. and Newhall, D. H., A high-pressure wire gage using gold–chrome wire, *Trans. Am. Soc. Mech. Eng.* **75** (3) (1953), 311–314.
18. Lippmann, H. and Richard, M., Resistance manometers of Manganin, gold-chromium, nickel–chromium and iron–chromium alloys in the pressure range up to 6000 Kg/cm^2, *Wiss. Z., Karl-Marx-Univ., Leipzig, Math. Naturwissen. Reihe,* **21** (1972), 740–745 (*C.A.* **79** (1973), 56949q).
19. Solow, B. (Angstrom Precision Inc.), "Metal Film Resistor for Low Range and Linear Temperature Coefficient". U.S. Patent 3,356,982 (1967).
20. Schulze, A. and Eicke, H., Über Goldchrom—Normalwiderstände, *Z. Angew. Phys.* **4** (9) (1952), 321–325.
21. Schulze, A., Über die Gold—Chrom–Widerstandslegierung für Normalwiderstände, *Phys. Zeit.* **41** (6) (1940), 121–128.
22. Thomas, J. L., Gold-cobalt resistance alloys, *J. Res. U.S. Bur. Stand.* **14** (1935), 589–593 (R.P. 789).
23. Johnson Matthey and Company, "An Improved Electrical Resistance Alloy". British Patent 861,646 (1959).
24. Darling, A. S., Potentiometer slidewire materials: Metallurgical considerations involved in their selection and development, *Platinum Met. Rev.* **12** (1968), 54–61.
25. Linde, J. O., Resistivity—concentration dependence of alloys, *Helv. Phys. Acta.* **41** (6–7) (1968), 1007–1015.
26. Köster, W., Kehrer, H. and Rothenbacher, W., Leitfähigkeit und Hallkonstante. XXVIII. Nah- und Fernordnung in goldreichen Gold–Vanadium–Mischkristallen, *Z. Metallk,* **54** (1963), 682–688.
27. Köster, W., Rave, H. P. and Takeuchi, Y., Conductivity and Hall constants. XXIV. Gold–palladium–vanadium alloys, *Z. Metallk,* **53** (1962), 749–753.
28. Sperner, F. and Harmsen, N. (Heraeus W. C. G mb H), "Gold Alloys as Electrical Resistance Material, Especially for Potentiometers". German Patent 2,136,373 (1972) (*C. A.* **78** (1973), 90133 h).
29. Schmidt, D. W., Schmidt, W. and Wagner, W. J., A metal-film probe for the measurement of small and fast flow temperature fluctuations, *Wärme und Stoffübertragung,* **6** (1973), 221–227 (cf. *Gold Bulletin,* **7** (1974), 103).
30. Schmidt, D. W. and Wagner, W. J., Measurement of temperature fluctuations in turbulent wakes, *Z. Flugwiss.* **22** (1974), 10–14.
31. Anon., A low temperature resistance thermometer, *Gold Bulletin,* **6** (1973), 39.
32. Vines, R. F. and Wise, E. M., "The Platinum Metals and Their Alloys". International Nickel Co., New York, 1941 (cf. Ref. 2, p. 307).
33. Reference (2), pp. 307–312.
34. Berman, R., The measurement of low temperatures: the properties and applications of the gold–iron alloy thermocouple, *Gold Bulletin,* **6** (1973), 34–39.
35. Borelius, G., Keesom, W. H., Johannson, C. H. and Linde, J. O., Measurements

on thermoelectric forces down to temperatures obtainable with liquid or solid hydrogen, *Proc. Kon. Akad. Amsterdam*, **35** (1932), 15–24.

36. Pearson, W. B. and Templeton, I. M., Effect of dissolved iron on the thermoelectricity of silver at very low temperatures, *Can. J. Phys.* **39** (1961), 1084–1086.
37. MacDonald, D. K. C., Pearson, W. B. and Templeton, I. M., Thermoelectricity at low temperatures. IX. The transition metals as solute and solvent, *Proc. Roy. Soc., London*, **A266** (1962), 161–184.
38. Sparks, L. L. and Powell, R. L., Low temperature thermocouples. KP, normal silver and copper versus gold–0.02 at % iron and gold–0.07 at. % iron, *J. Res. U.S. Bur. Stand., Sect. A*, **76** (1972), 263–283 (*C.A.* **77** (1972), 22085y).
39. Knittel, T., Thermoelectric power dependence of a gold-iron alloy wire on the magnetic field orientation, *Cryogenics*, **13** (1973), 370–371 (*C.A.* **79** (1973), 71389a).
40. Berman, R. and Huntley, D. J., Dilute gold-iron alloys as thermocouple material for low temperature heat conductivity measurements, *Cryogenics*, **3** (1963), 70–75.
41. Berman, R., Brock, J. C. H. and Huntley, D. J., Properties of Au—0.03% Fe thermoelements between 1 and 300°K and behaviour in a magnetic field, *Cryogenics*, **4** (1964), 233–239.
42. Berman, R., Gold–iron alloys for low temperature thermocouples, *Temperature*, **4** (3) (1972), 1537–1542.
43. Berman, R. and Kopp, J., Thermoelectric power of dilute gold-iron alloys, *J. Phys. F.* **1** (1971), 451–468.
44. Kopp, J., Single impurity Kondo effect in gold: I. Thermopower, *J. Phys. F.* **5** (1975), 1211–1216.
45. Kopp, J., Low temperature resistivity and thermopower anomalies in pure gold, *Gold Bulletin*, **9** (1976), 55–57.
46. Hermance, H. W. and Egan, T. F., Organic deposits on precious metal contacts, *Bell Syst. Tech. J.* **38** (5) (1958), 739–776.
47. Antler, M., "Connectors", *in* "Gold Plating Technology", Ed. by F. H. Reid and W. Goldie, Electrochemical Publications, Ayr, Scotland, 1974.
48. Poll, G. H., Gold plating: spots, stripes and strips, *Prod. Finish. (Cincinnati)*, **39** (4) (1975), 46–54.
49. Mason, D. R., Blair, A. and Wilkinson, P., Alloy gold deposits—have they any industrial use?, *Trans. Inst. Met. Fin.* **52** (1974), 143–147 (cf. Foulke, D. G. and Duva, R., "Electroplating of 60–80% gold alloys for the contact industry". Plat. Electron. Ind. Symp., 4th, 1973, pp. 131–140).
50. Morabito, J. M., *et al.*, Material characterisation of Ti-Cu-Ni-Au (TCNA)—a new low cost thin film conductor system, *I.E.E.E. Trans.* **PHP–11** (1975), 257–262.
51. Anon., Diffusion barriers for gold plated copper: effectiveness of increasing thicknesses of nickel, *Gold Bulletin*, **9** (1976), 132–133.
52. Horn, G. L., and Merl, W. A., Friction and wear of electroplated hard gold deposits for connectors, *I.E.E.E. Trans.* **PHP–10** (1) (1974), 53–59.
53. Russell, R. J., "Properties of Inlay Clad Wrought Gold Alloys". Paper presented to the Precious Metals Alternatives and Fifth Plating in the Electronics Industry Symposium of the American Electroplaters' Society, New York, March 1975.
54. Tembe, G., Gold contact materials for connectors: factors affecting their selection and design, *Gold Bulletin*, **5** (1972), 14–18.

55. Lee, F. F. M., Clad metal inlays for connector springs. Paper to the 21st Annual Holm Seminar, Chicago, 1975 (cf. *Gold Bulletin*, **9** (1976), 23).
56. Houlston, J. F., "Reed Switch Contacts", *in* "Gold Plating Technology", Ed. by F. H. Reid and W. Goldie, Electrochemical Publications, Ayr, Scotland, 1974, pp. 533–541.
57. Augis, J. and Schubert, R., Influence of potassium on the arcing voltage of reed contacts. Paper to the 21st Annual Holm Seminar, Chicago, 1975 (cf. *Gold Bulletin*, **9** (1976), 23).
58. Anon., A rhenium–gold contact alloy for sensitive relays, *Gold Bulletin*, **9** (1976), 131.
59. Antler, M., *in* "Gold Plating Technology", Ed. by F. H. Reid and W. Goldie, Electrochemical Publications, Ayr, Scotland, 1974, pp. 259–276.
60. Steinmann, S., Flühmann, W. and Saxer, W., Verschleissverhalten und Struktur von galvanischen Edelmetallniederschlägen, *Metalloberfläche*, **29** (1975), 154–157 (cf. Anon., *Gold Bulletin*, **9** (1976), 11).
61. Antler, M., Tribological properties of gold for electric contacts, *I.E.E.E. Trans.* **PHP–9** (1973), 4–14 (*C.A.* **79** (1973), 24529m).
62. Brandes, E. A., A gold-chromium-cobalt alloy for sliding contacts, *Gold Bulletin* **8** (1975), 73–79.
63. Abbot, W. H., "Low Energy Electrical Contacts of Gold–base Alloys". U.S. Patent 3,661,569 (1969).
64. Schiff, K. L., Harmsen, N. and Schnabl, R., The tarnishing behaviour of gold–based alloys in corrosive atmospheres". Paper presented to the 21st Annual Holm Seminar, Chicago, 1975 (cf. *Gold Bulletin*, **9** (1976), 22–23).
65. Reid, F. H. and Goldie, W., "Gold Plating Technology". Electrochemical Publications Ltd., Ayr, Scotland, 1974.
66. Harper, C. A. (Ed.), "Handbook of Materials and Processes for Electronics". McGraw Hill Book Company, 1970.
67. Anon., Electromigration in thin gold films. Diffusion problems in integrated circuits, *Gold Bulletin*, **7** (1974), 20–21.
68. Anderson, J. C., Applications of thin films in microelectronics, *Thin Solid Films*, **12** (1) (1972), 1–15.
69. Hummel, R. E. and Breitling, R. M., Electromigration in thin silver, copper, gold, indium, tin, lead and magnesium films, *J. Phys. Chem. Solids*, **53** (1972), 845–852 (cf. *Appl. Phys. Lett.* **18** (1971), 373).
70. Klein, B. J., Electromigration in thin gold films, *J. Phys. F.* **3** (1973), 691–696.
71. Blech, I. A. and Rosenberg, R., Direction of electromigration in gold thin films, *J. Appl. Phys.* **46** (1975), 579–583.
72. Tai, K. L., Sun, P. H. and Ohring, M., Lateral self-diffusion and electromigration in thin metal films, *Thin Solid Films*, **25** (1975), 343–352.
73. Hummel, R. E. and Dehoff, R. T., On the controversy about the direction of electrotransport in thin gold films, *Appl. Phys. Lett.* **27** (1975), 64–66.
74. Cornely, R. and Fuschillo, N., Electrical and optical properties of radio frequency triode-sputtered gold films, *J. Vac. Sci. Technol.* **11** (1974), 163–166 (cf. *Bull. Am. Phys. Soc.* **18** (1973), 16.
75. Chaurasia, H. K. and Voss, W. A. G., Ultra thin conducting films of gold on platinum nucleating layers, *Nature*, **249** 3rd May, (1974), 28–29.
76. Anon., Gold in silicon solar cells: power supply systems for satellites, *Gold Bulletin*, **5** (1972), 31–33.

77. Kohl, W. H., "Handbook of Materials and Techniques for Vacuum Devices". Reinhold Publishing Corporation, 1967, pp. 534–535.
78. Davis, G. L., Gold in semiconductor technology—some unconventional metallurgical concepts, *Solid State Technol.* (1976), February, 49–53.
79. Collins, C. B., Carlson, R. O. and Gallagher, C. J., Properties of gold doped silicon, *Phys. Rev.* **105** (1957), 1168–1173.
80. Bullis, W. M., Properties of gold in silicon, *Solid-State Electron.* **9** (1957), 143–168.
81. Forbes, L., "Thermal and Optical Emission and Thermal Capture of Electrons—Holes at Gold Centers in Silicon". Ph.D. Thesis, University of Illinois, Urbana, 1970.
82. Hennig, F., "Emission and Capture of Electrons—Holes at Gold Centers in Silicon at Thermal Equilibrium". Ph.D. Thesis, University of Illinois, Urbana, 1974.
83. Sah, C. T., Forbes, L., Rosier, L. L., Tasch, A. F. and Tole, A. B., Thermal emission rates of carriers at gold centers in silicon, *Appl. Phys. Lett.* **15** (1969), 145–148.
84. Forbes, L. and Yeargan, J. R., Design for silicon infra-red sensing MOSFET, *I.E.E.E. Trans.* **ED–21** (1974), 459–462.
85. Parker, W. C., Wittmer, L. L., Yeargan, J. R. and Forbes, L., "Experimental Characterization of the Infra-red Response of Gold-doped Silicon MOSFETS (IRFETS)". Late News Paper 4.8 presented at the Int. Electron Devices Meeting, Washington, D.C., December 1974.
86. Parker, W. C. and Forbes, L., Experimental characterization of gold-doped infra-red sensing MOSFETS, *I.E.E.E. Trans.* **ED–22** (1975), 916–924.
87. Khan, H. R. and Raub, Ch. J., The superconductivity of gold alloys, *Gold Bulletin*, **8** (1975), 114–118.
88. McNerney, J. J. and Buseck, P. R., The detection of mercury vapour: a new use for thin gold films, *Gold Bulletin*, **6** (1973), 106–107.
89. Lythgoe, S., Robinson, P. J. and Sedgwick, R. D., Removal of mercury by gold surfaces and the prevention of mercury photosensitization, *J. Sci. Instrum.* **3** (1970), 401–402 (*C.A.* **72** (1970), 127243y).
90. McNerney, J. J., Buseck, P. R. and Hanson, R. C., Mercury detection by means of thin gold films, *Science* **178** (1972), 611.
91. McNerney, J. J., "Method and Apparatus for the Detection of Selected Components in Fluids". U.S. Patent 3,714,562 (1973).
92. Chaurasia, H. K., Huizinga, A. and Voss, W. A. G., Optimum gold films for mercury detection, *J. Phys. D.* **8** (1975), 214–218 (*C.A.* **82** (1975), 132524d).
93. McNerney, J. J. and Buseck, P. R., Geochemical exploration using mercury vapour, *Econ. Geol.* **68** (1973), 1313–1320.
94. Akabori, S., Wakamatsu, A., Kakuno, N. and Sakata, A., Removal of mercury from waste water by contacting with gold, *Jap. Kokai*, 75 67275 (October 1973).
95. Anon., New moving coil meter suspensions: a high strength gold alloy ligament, *Gold Bulletin*, **4** (1971), 71.
96. Akella, J. and Kennedy, G. C., Melting of gold, silver and copper—proposal for a new high-pressure calibration scale, *J. Geophys. Res.* **76** (1971), 4969–4977.
97. Van Osch, G. W. S. and Griepink, E., The use of a gold–containing membrane for ion-sensitive electrodes and their application in analysing systems. I. Preparation and some characteristics of the ion-sensitive electrodes, *Z. Anal. Chem.* **273** (1975), 271–274.

98. Anon., The vacuum deposition of gold: developments coating techniques and applications, *Gold Bulletin*, **4** (1971), 30–32.
99. Stops, D. W. and Azkan, D., Remarkable change in resistance, *Thin Solid Films*, **25** (1975), S7–S9.
100. Anon., Humidity control equipment: Gold grids in electronic humidistats, *Gold Bulletin*, **6** (1973), 98.
101. Köster, W. and Ulrich, W., Das Dreistoffsystem Eisen-Nickel-Gold, *Z. Metallk*, **52** (1961), 383–391.
102. Nesbitt, E. A. and Gyorgy, E. M., Two-phase permalloy for high-speed switching, *J. App. Phys.* **32** (1961), 1305–1308.
103. Moak, D. P. *et al.*, Improved melting and casting procedures for cobalt base magnetic alloys, *J. Vac. Sci. Technol.* **9** (1972), 1356–1359.
104. Shimizu, A. and Hashimoto, H., "Permanent Magnet Alloys of Pt-Au-Fe". Jap. Patent 70 31 056 (1970) (*C.A.* **74** (1971), 102347p).
105. Moeller, C. E., Gold film resistance thermometers for surface temperature measurements, *Temp.: Its Meas. Control Sci. Ind.* **4** (1972), 1049–1056.

6 Gold and Gold Alloys in Engineering Applications

1. Industrial Gold Solders and Brazes

1.1 Introduction

Gold solders for jewellery fabrication designed for working temperatures ranging from about 600 to 800°C have limited usefulness in non-jewellery applications where they have to compete with much less expensive silver solders and base metal brazing alloys. The growth of the electronics, nuclear power, aero-engine and spacecraft industries, however, brought with it the use of new structural materials, the mechanical properties and resistance of which to various environmental influences have had to be matched by the development of suitable joining materials and joining techniques. While the new base metal brazing alloys were found in some instances to be applicable, in others, designers and metal fabrication engineers were forced to have recourse to brazing alloys based upon noble metals such as gold.[1]

Table XXXVIII reproduced from Sloboda,[1] lists some commercially available brazing alloys in which gold is the major constituent, as well as gold itself, which is sometimes used for low-temperature brazing of diffusion seals and for joining tungsten. It will be noted that, in contrast to the position regarding carat gold solders, full details of compositions of the alloys are available, and in a number of instances standard specifications have been drawn up for the alloys.

They are in general simpler in composition than carat gold solders, and fall into well-defined groups, the basic features of which will be summarised briefly. In all cases the difference between the solidus and liquidus temperatures of brazing alloys is a significant feature, and should be as small as possible.

1.2 Gold-copper alloys

In these alloys the unusual character is exploited of the Au-Cu system, as revealed by its equilibrium diagram (Figure 6). Although the alloys form a continuous series of solid solutions, the liquidus and solidus curves converge near the 80Au-20Cu composition, at a temperature of 889°C. At no point are these curves widely separated, so that the melting ranges of all alloys are relatively small, and decrease to zero at the 80Au-20Cu composition, which is characterised by high fluidity and ability to penetrate narrow joint gaps and to form small radius fillets. The addition of small amounts of other metals such as iron, to retard Au-Cu ordering should be noted (table XXXVIII, alloy No. 5).[2] Ordering, if it occurs, is accompanied by volume changes which may affect the properties of joints.

Useful properties of these alloys as listed by Sloboda[1] include:

(i) Ability to wet copper, nickel, iron, cobalt molybdenum, tantalum, niobium, tungsten, and their alloys.

(ii) Ability to produce ductile joints without excessive inter-alloying. The latter factor is important in cases where excessive erosion of the work piece by molten brazing alloy could affect the dimensional accuracy of brazed parts or unduly weaken thin-walled structures.

(iii) Freedom from volatiles and from constituents that form refractory oxides. The former is important when brazing is used to fabricate parts that operate in vacuum at elevated temperatures; freedom from elements that form refractory oxides makes for easy brazing without a flux in reducing and neutral atmospheres or in vacuum.

(iv) Exceptional resistance to corrosion, especially in the case of gold-rich alloys.

(v) Ability to be produced in wrought forms such as foil, strip and wire. This is an important consideration in furnace brazing applications in which the brazing alloy must be placed in or near the joint area before brazing; shims and wire preforms often provide the most convenient means for doing it.

Au-Cu brazing alloys find extensive application in the electronics industry where they are used in the fabrication of wave guides, electron tubes, valves, radar equipment and vacuum devices.

1.3 Gold-silver and gold-silver-copper alloys

Au-Ag alloys of a wide range of compositions can be used as brazing alloys. As the Au-Ag phase diagram[3] shows, there is no point in this system at which there is any significant separation between the liquidus and solidus lines.

The use of Au-Ag-Cu alloys as carat gold solders, has already been discussed and needs little further mention here. The 60Au-20Ag-20Cu alloy is particularly useful because of its short melting range. A recent publication[4] indicates that by incorporation of 12–30% Au in Ag-Cu- Cd-Zn silver brazing

Table XXXVIII. Industrial Gold Brazing Alloys

	Composition, weight per cent						Melting range, °C		Specifications	
No.	Au	Cu	Ni	Ag	Pd	Other	Solidus	Liquidus	B.S.1845 (UK)	AWS-ASTM (USA)
1	100						1063	1063		
2	94	6					965	970		
3	81.5	15.5	3					910		
4	80	20					910	910		BAu-2
5	80	19				1.0 Fe	905	910	Au 1	
6	75	25					910	914		
7	62.5	37.3					930	940	AU 2	BAu-1
8	50	50					955	970		
9	40	60					975	995		
10	37.5	62.5					980	1000	Au 3	
11	35	62	3				973	1029		BAu-3
12	35	65					970	1005		
13	30	70					995	1020	Au 4	

Table XXXVIII *(Continued)*

	Composition, weight per cent						Melting range, °C		Specifications	
No.	Au	Cu	Ni	Ag	Pd	Other	Solidus	Liquidus	B.S.1845 (UK)	AWS-ASTM (USA)
14	20	80					1018	1040		
15	70			30			1030	1040		
16	60	20		20			835	845		
17	82.5		17.5				950	950	Au 5	BAu-4
18	68	22	8.9			1.0 Cr 0.1 B	960	980		
19	75		25				950	990	Au 6	
20	72		22			6.0 Cr	975	1038		
21	65		35				950	1070		
22	92				8		1180	1230		
23	87				13		1260	1300		
24	75				25		1375	1400		
25	70		22		8			1045		
26	50		25		25			1121		

After Sloboda[1]

alloys, alloys with low working temperatures can be obtained which, because of their reduced content of Cd and Zn are sufficiently ductile to be cold worked without annealing.

1.4 Gold-nickel alloys

As will be seen from Figure 11, the Au-Ni system is metallurgically similar to the Au-Cu system, with the liquidus and solidus curves converging at a composition of 82.5Au-17.5Ni. A significant difference, however, is that the liquidus and solidus curves diverge considerably with increase of the nickel content above 17.5%, to such an extent that the melting ranges and viscosities of alloys containing more than 35% Ni make them unsuitable for brazing. The 82Au-18Ni alloy is used extensively.[5,6]

Although the commercially available Au-Ni brazing alloys have melting ranges not greatly different from those of Au-Cu brazing alloys, they have much greater mechanical strength and much greater resistance to oxidation at high temperatures than the latter alloys.[7-9]

1.5 Gold-nickel-copper, gold-nickel-chromium, gold-nickel-molybdenum and gold-nickel-tantalum alloys

The various brazing alloys of these types can be treated conveniently as Au-Ni alloys whose properties and costs have been modified by incorporation of other alloying metals.

As regards incorporation of copper, it will be noted from the Au-Ni-Cu diagram (Figure 11) that alloys Nos. 3, 11 and 18 of Table XXXVIII are all based on compositions falling in or close to a valley of low liquidus temperatures. The 35Au-62Cu-3Ni alloy (No. 11) has better flow properties on iron alloys, nickel and molybdenum than the binary Au-Cu alloys, and does not penetrate the grain boundaries of glass sealing alloys such as Kovar and Rodar as readily. It is therefore used widely for brazing these materials. Its grain size is smaller, and it is less subject to hydrogen embrittlement, so that the possibility of slow leaks through grain boundaries is reduced.[5] The 81.5Au-15.5Cu-3Ni alloy (No. 3) was developed specifically to meet the need for an alloy more ductile than the 80Au-20Cu binary alloy, which it can replace. The 68Au-22Cu-8.9Ni-1Cr-0.1B alloy (No. 18) was developed[10] as a less expensive equivalent of alloy No. 17 with comparable wetting and free flowing properties, high temperature strength and oxidation resistance. Indium has also been incorporated successfully in Au-Ni-Cu brazing alloys.[11]

Chromium, molybdenum or tantalum confer on Au-Ni alloys increased resistance to oxidation as well as ability to wet graphite, diamonds and carbon.

Thus the 72Au-22Ni-6Cr alloy (No. 20) is particularly suited[12] to joining heat-resistant base metals where service at high temperatures is involved. It wets and brazes a wide variety of heat-resistant materials such as stainless steels, Hastalloys and René 41 as well as graphite and carbon.[5]

For the brazing of joints with graphite, however, Au-Ni-Mo and Au-Ni-Ta alloys are the most frequently used.[13] For example 35Au-35Ni-30Mo alloy[14] and 60Au-10Ni-30Ta alloy are used for the brazing of graphite-to-graphite and graphite-to-metal joints. The former alloy has, in addition, good corrosion resistance, even to molten fluorides.

A brazing alloy with a liquidus temperature of less than 1040°C which contains 35-45Au, 1-3Si, 0.5-2.0B and the remainder Ni has also been described.[15] Another low-gold alloy which has been described[16] shows a small temperature difference of about 60°C between liquidus and solidus, and is useful at 1105–1214°C. It is reported as containing 14-25Au, 1-5Si, 1-4B, 2-12Cu, 1-5Fe and up to 17Pd plus Ni. Up to 50% of the nickel can be replaced by cobalt. The chromium and iron are claimed to reduce the liquidus-solidus range, and the chromium to improve the resistance to oxidation.

1.6 Gold-palladium and gold-palladium-nickel alloys

The binary Au-Pd, Au-Ni and Ni-Pd systems are all of the fully solid solution type, and in the liquidus plot for the ternary Au-Ni-Pd system Figure 25) a valley of low temperatures extends across the systems connecting the minima of the Au-Ni and Ni-Pd binaries. The spread between liquidus and solidus temperatures in this region is negligible, which makes alloys along this area useful where a sharp melting and strong brazing alloy is required. Indeed this system is to be compared with the Au-Cu-Ni system to which reference has been made above. The two systems are similar in type, that based on Au-Cu-Ni exhibiting higher liquidus temperatures. Like the Au-Cu-Ni alloys, alloys of the Au-Ni-Pd system, while single-phase just below the solidus, separate into two phases over a wide area of compositions containing up to 20% Pd.[17]

When low vapour pressure and complete resistance to oxidation at high temperature in a braze are needed, Au-Pd alloys can be used. The 92Au-8Pd alloy is useful, for example, in the brazing of cathode structures. It also wets and flows easily on tungsten, molybdenum, nickel and stainless steel. In many instances, however, the cheaper Au-Pd-Ni alloys offer a satisfactory and cheaper alternative over a considerable range of temperatures. As illustrated in Table XXXVIII and in various specifications,[18] they have found extensive application in the aerospace industry. The 50Au-25Pd-25Ni alloy, like the 92Au-8Pd and 72Au-22Ni-6Cr alloys, wets and flows on tungsten, molybdenum, nickel and stainless steel.

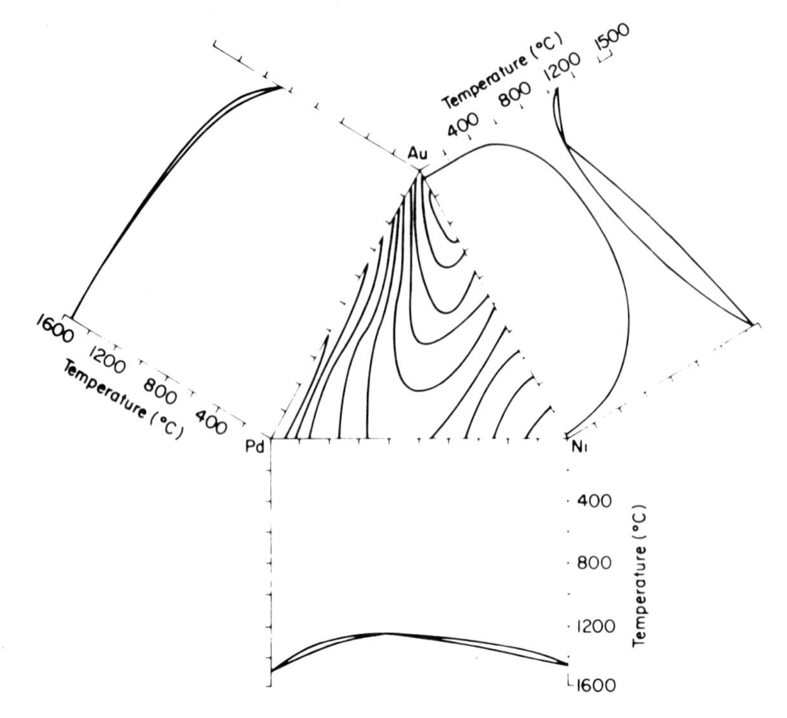

FIG. 25. Liquidus isotherms of the Au-Ni-Pd system in relation to the phase diagrams of the binary Au-Ni, Au-Pd and Ni-Pd edge systems. (Compositions in weight per cent.) (After Hansen,[3] Wise[9] and Gregor'ev *et al.* [17])

1.7 Other high-melting alloy systems

Although not represented in Table XXXVIII, there are other alloy systems, such as the Au-Cu-Pd and Au-Pd-Pt systems which also exhibit areas of composition over which there is either only a small or a moderate difference between the liquidus and solidus temperatures.

Certain Au-Cu-Pd alloys have already found application as industrial brazes, and certain Au-Pd-Pt brazes have been used in the fabrication of dental appliances.

In general, therefore, the potentialities for high temperature brazing alloys to meet specific applications as they arise are considerable.

1.8 Gold-silicon and gold-germanium alloys

The characteristics of systems based on gold, silver, silicon and germanium will be apparent from the fact that both gold and silver form simple eutectics with silicon and germanium, with melting points and compositions as follows:[3]

Eutectic	*Melting point*	*% Au or Ag*
Au-Si	370°C	94
Au-Ge	356°C	88
Ag-Si	830°C	95.5
Ag-Ge	651°C	89

As the Au-Ag system consists of a continuous solid solution, no eutectic four-phase equilibrium occurs in the ternary systems Au-Ag-Si and Au-Ag-Ge. A binary eutectic curve runs through each of these ternary systems, originating from the marginal Ag-Ge and Ag-Si systems and ending in the marginal Au-Ge and Au-Si systems. This means that in the Au-Ag-Ge system,[19] for example, (see Figure 26) the liquidus temperatures on the eutectic line are found between 651°C and 356°C, and in the Au-Ag-Si system[20] between 830°C and 360°C.

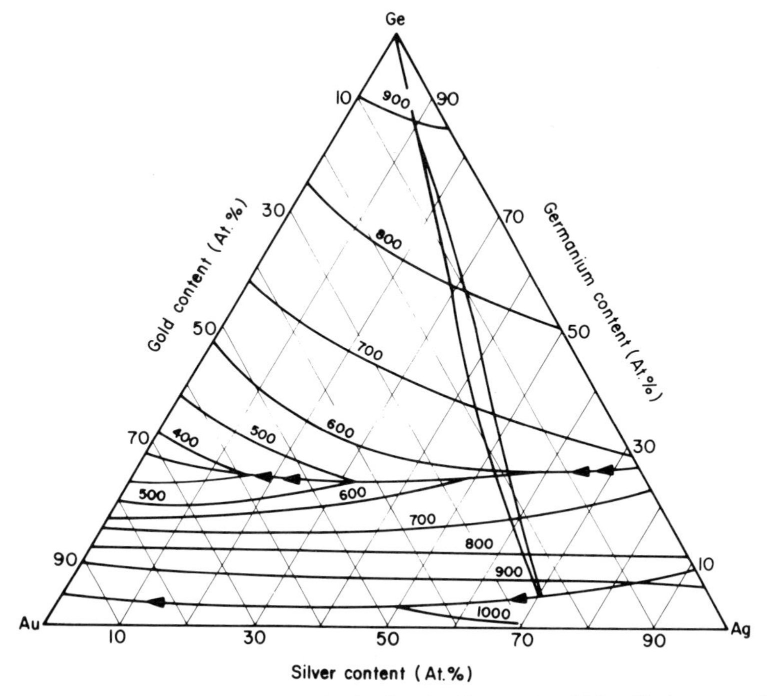

FIG. 26. Liquidus isotherms in the Au-Ag-Ge system. (After Zwingmann.[19])

There may be possibilities for the use of solders based on these ternary systems, and upon the Au-Ag-Ge-Si quaternary system, but the formation of silicides with metals such as copper limits their field of application severely.

Both the Au-Si eutectic and the Au-Ge eutectic are involved, however, in semiconductor technology in the eutectic die bonding of silicon and germanium devices, and are used as solders in situations where low melting points and high resistance to corrosion are required.[21,22]

A North America Rockwell Corporation patent[23] describes the bonding of microcircuit chips to substrates, while forming electrical connections and a hermetic seal, by making interconnecting islands surrounded by a sealing ridge. The islands and ridges comprise silicon coated with a metal such as gold which forms a low-temperature eutectic with silicon, and bonding is effected by heating to the eutectic temperature. The use of Au-Ge alloy to bond a semiconductor device to the metallic ring seal of a glass envelope has also been patented,[24] and International Business Machines Corporation[25] has patented the bonding of substrate to a circuit chip by frictional movement across a fusible metal layer such as gold placed between them. International Business Machines Corporation and Motorola Inc.[26] have also patented a semiconductor circuit which includes a silicon chip and a molybdenum plate both of which are gold plated, so that in operation they are bonded by an Au-Si eutectic. As an example of a somewhat different type, the English Electric Co. Ltd.[27] has patented articles of self-bonded SiC containing silicon bonded together by brazing with an alloy of silicon and gold or certain other metals. Many other examples of the use of Au-Si and Au-Ge alloys for bonding purposes are recorded in the literature.

The Au-Si and Au-Ge eutectics display certain interesting properties; for example when they are solidified rapidly they are white and very brittle. On being warmed to 200°C, however, the Au-Si eutectic becomes yellow and very tough within a few minutes, as a result of precipitation of the silicon in an essentially pure gold matrix. The Au-Ge eutectic behaves similarly. The effects of varying the rate of cooling of the Au-Si eutectic from the liquid state on the microstructure of the solid eutectic have been reported upon by Philofsky *et al.*[28] Very high velocity or splat-cooling of liquid Au-Si, Au-Ge and other gold alloys to yield glassy gold alloys containing metastable phases has been reviewed by Rivlin *et al.*[28a] These authors have also drawn attention to the differences between the structures of liquid and solid Au-Si and Au-Ge eutectics and to the significant extent to which these eutectics expand on solidification because of these structural differences.

The terminal solubilities of silicon and germanium in gold are extremely low, so that when gold is melted on a silicon chip, almost pure gold and silicon separate out on cooling. The gold deposited is therefore highly conductive and use can be made of the depositing semi-conductor, if it grows epitaxially on the parent wafer, to generate p-n junctions. In making so-called gold bonded diodes, therefore, a pointed gold wire is pressed into a germanium slice and given an energy pulse sufficient to cause melting at the contact point. Germanium is dissolved in the gold and re-precipitates, at least partly epitaxially, on the unmelted slice. If the gold wire contains a small concentration of a third element the reprecipitated Ge zone can be of such a type as to give a sharp p-n junction. Gallium-doped gold is used in bonding to Ge chips in this way.

1.9 Gold-tin-germanium alloys

An Au-Ge-Sn alloy has been claimed as a "non-migrating" solder suitable for use in conjunction with non-migrating conductor compositions in the fabrication of circuits of high reliability. E.I. du Pont de Nemours & Co.[29] have patented a solder which contains 80-85%Au, 1–10% Ge and 5–19% Sn and which melts below 325°C, and which has these characteristics.

1.10 Gold-tin alloys

The Au-Sn system (Figure 27) is of interest from two points of view so far as gold usage is concerned. Not only do Au-Sn alloy solders (e.g. an 80Au-20Sn alloy) find application in the electronics industry[30,31] but problems exist in relation to the soldering of electrical connections to gold-plated surfaces with Sn-Pb solders which involve the Au-Sn and the Au-Pb-Sn systems. These problems have been the subject of recent reviews, [32,35] and are discussed in a later section.

The addition of 0.5–0.6% Bi to the 80Au-20Sn alloy is said[36] to improve wettability when the alloy is used for joining gold-plated products.

1.11 Gold-indium and gold-indium-bismuth alloys

The Au-In system (Figure 28), although somewhat complex like the Au-Sn system, shows a eutectic at a gold content of 73% with a melting point of 451°C which is referred to by Wise[3] as being a useful solder. An alloy of composition 95Au-5In, with a melting range of 647–830°C and a working temperature of 830°C is also used for vacuum work. Moreover International Business Machines Corp.[37] has patented a solder melting at 71.5–74.0°C and which consists of 69–71% In, 27–29% Bi and 1.5–2.5% Au. This solder is claimed to have excellent ductility and corrosion resistance, and a relatively low electrical resistivity of about 28 μohm cm.

1.12 Gold-antimony and gold-arsenic alloys

These have found application in semiconductor technology. Thus A. G. Brown, Boveri et Cie[38] have claimed that a ferrous metal and an n-type semiconductor such as lead telluride can be joined with an alloy of gold con-

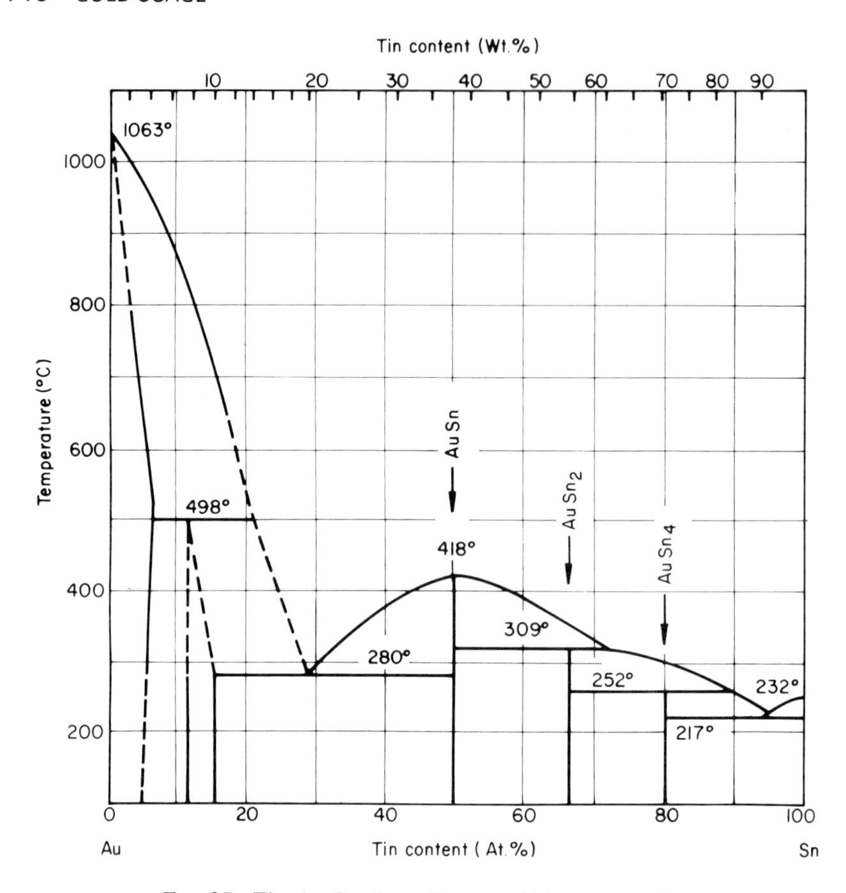

FIG. 27. The Au-Sn phase diagram. (After Hansen.[3])

taining, for example, 0.5–0.1% As or Sb. Pyatyshev *et al.* have claimed[39] that 20–25% Sb in an Au-Ge-Sb alloy containing 10–15% Ge improves the wetting action of the alloy.

1.13 Gold-tantalum and gold-niobium alloys

Mention was made earlier[13] of the fact that incorporation of molybdenum and tantalum in Au-Ni alloys confers on such alloys the ability to wet graphite, so that they can be used for brazing of metals to graphite in chemical engineering construction. Du Pont de Nemours & Co.[40] have recently patented the use of Au-Ta alloys containing 1–25% Au, or Au-Nb alloys[41] containing 1–10% Au, for attaching diamonds to metal mounts.

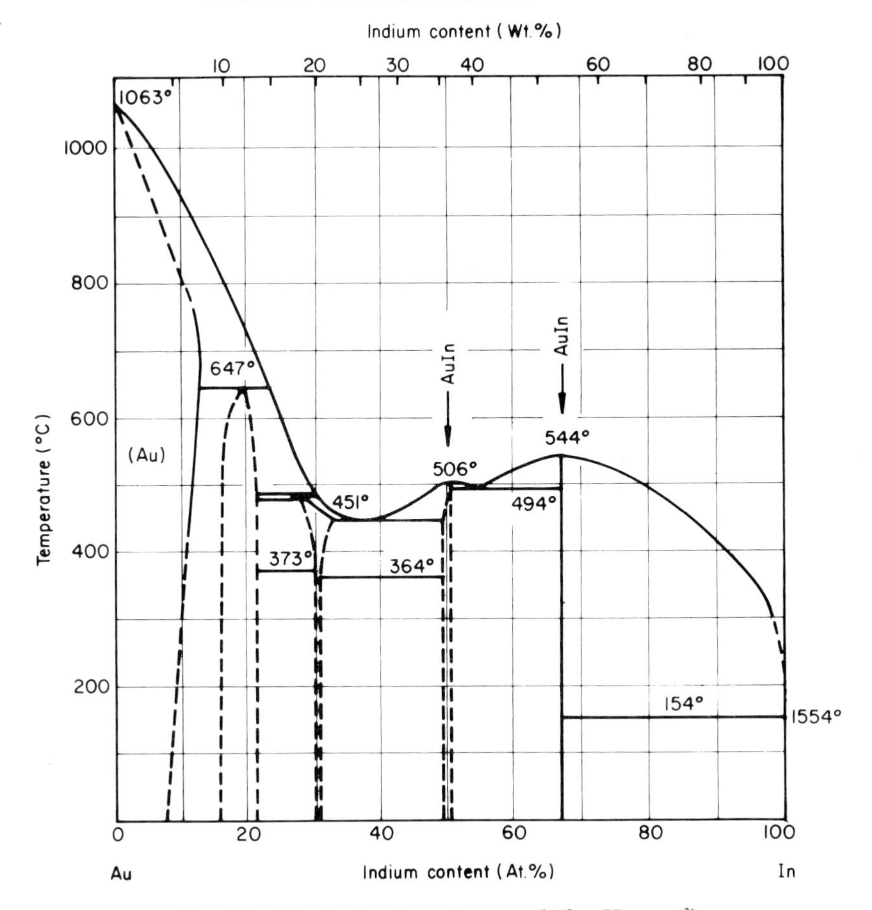

FIG. 28: The Au-In phase diagram. (After Hansen.[3])

2. High-strength, Corrosion-resistant Gold-Platinum-Rhodium Alloys for Spinnerets and other Applications

Bushings or spinnerets, the faces of which are pierced by as many as 15,000 holes ranging from 40–120 μm in diameter, play an important role in the production of viscose rayons and other man-made fibres. Au-Pt alloys have been found to be particularly suited for the making of such spinnerets, because of their resistance to corrosion and because of their hardness. Since the spinnerets are used in large numbers they provide a significant outlet for gold alloys.

The development of spinneret alloys has been described by Funk and Reinacher.[42] The first metal spinnerets, which were produced[43] in 1908 were made of pure platinum. They were of inadequate hardness (40 Brinell), as was the case with Pt-Ir,[44,45] Pt-Ru,[46] Pt-Au[47] and Pd-Au[46] alloys, all of which were tested and/or applied during the following twenty to twenty-five years.

A big advance resulted, however, in 1930 as a result of successful precipitation-hardening of the gold alloys. Initially base metals such as iron, cobalt, nickel,[48-50] zinc,[51] tin [52] and magnesium[53] were used as additives, and although they enabled Brinell hardnesses of up to 130 to be achieved, the hardened alloys were insufficiently corrosion-resistant, because of the formation of relatively base phases during hardening. Attempts to use them therefore ceased when it was found that corrosion-resistant gold-platinum alloys, without any base metal additives, could be precipitation hardened if the platinum content was raised to over 20%. The background to this development will be clear from the miscibility gap in the solid state revealed by the phase diagram for the Au-Pt system, which was reported upon by Grigorjew[54-55] in 1928 and 1929, by Johansson and Linde[56] in 1930, and then later by Wictorin[57] and by Darling, Mintern and Chaston.[58] The wide spread between liquidus and solidus caused difficulties in determining the Au-Pt phase diagram (see Figure 29).

Nowack[49] reported in 1929 and 20% and 25% Pt-Au alloys could be precipitation hardened, and after 1933 a 30% Pt-Au alloy was introduced which could be hardened up to 200 Brinell. It made possible the production of thinner walled spinnerets, and improved polishing of the bore walls and outer spinneret face with consequent improvements in rayon quality. Problems arising from segregation during hardening which led to variation in hardness, and from brittleness which developed with maximum hardening, were overcome after 1935 by the incorporation[59] of 0.1–0.6% Rh in the alloys.

Even harder spinneret alloys were soon called for, however, by rayon manufacturers, because of their desire to spin threads containing titanium dioxide, which acted as a scouring agent on the spinneret holes. Such harder alloys became available in 1936 when a method was found[46] to work a 50/50 platinum-gold alloy, capable of being precipitation hardened up to about 350 Brinell. Rhodium (1%)[60] was incorporated in these alloys, and rhenium (0.1%)[61] and iridium[62] were found to be similarly effective in achieving hardnesses of about 300 in 40/60 platinum-gold alloys. Spinneret alloys of this type containing up to 85% gold have been claimed recently.[63]

The mechanism by which rhodium enhances precipitation hardening of these platinum-gold alloys has been elucidated by Schmid,[64] and the Au-Pt-Rh system reported upon more generally by Raub and Falkenburg.[65]

Au-Pt-Rh alloys have also been applied in other areas. Gold is not wetted by molten glass, so that the behaviour of high melting alloy compositions, which are essentially Pt-Rh alloys containing small quantities of gold have been examined for applications in the glass industry.[66-68] Their usefulness has been limited, however, by the relative ease of evaporation of the gold from the alloys

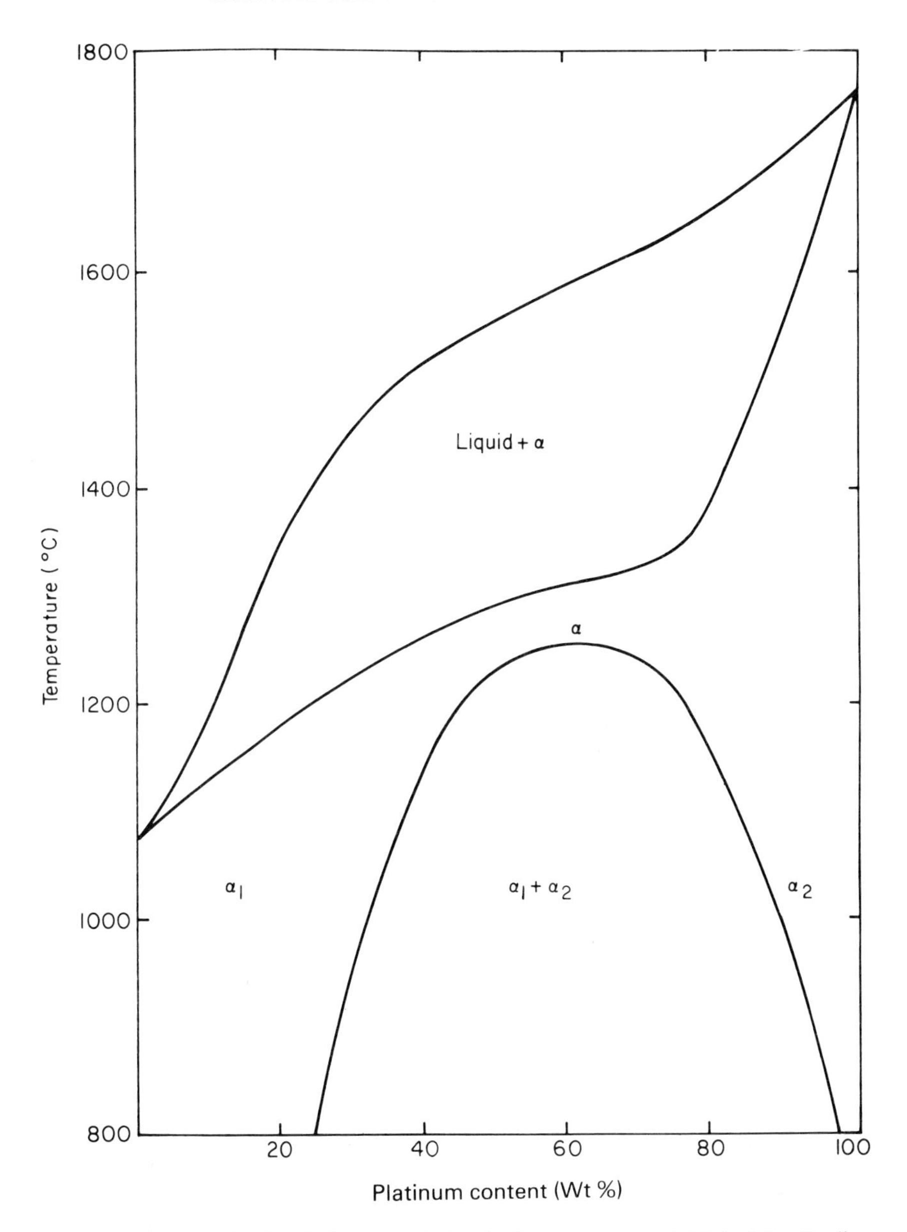

FIG. 29: The constitutional diagram of the Au-Pt system as established by Darling, Mintern and Chaston.[58] Alloys containing 50–80% Au, quenched from temperatures in the region of 1100 to 1200°C, can be precipitation hardened after fabrication into spinnerets to give very high hardness values. (After Funk and Reinacher.[42])

at high temperatures, which can cause them to crack. Crucibles and other containers of an Au-Pt-Rh alloy of composition 84:15:1 have been used successfully, however, for the handling of a variety of molten salts at temperatures of up to 800°C.[69]

3. Applications Based upon the Optical Properties of Gold Alloys

3.1 Introduction

Applications of the optical properties of gold fall into two main classes. The first is the wide use in jewellery and the arts of gold and its variously coloured alloys in both "massive" form and in decorative coatings and finishes. In these applications the basic considerations are the factors governing the absorption and reflection of *visible* radiation from the metal surface. The second comprises a variety of applications involving the control or the use of radiant energy, usually with the aid of surfaces coated with gold or gold alloys. Here the basic considerations are absorption, transmission and reflection of primarily infrared or near-infrared radiation from such surfaces. Additionally, there are a few applications in which the behaviour of gold surfaces exposed to U.V. radiation or X-rays is of significance. In some of these applications the properties of gold and its alloys which are exploited are unique, while in others the properties exploited are shared by other metals which are not used because of their susceptibility to tarnish.

3.2 Absorption, reflection and transmission of radiation by gold and its alloys

The optical properties of gold and its alloys were reviewed in 1972 by Loebich[70] who discussed, firstly, the properties of opaque gold layers, secondly, those of semitransparent gold deposits and, finally, the effects on these of the incorporation of alloying metals in the gold. For the sake of convenience this treatment will be followed here. The factors affecting the colour of gold alloys have also been reviewed by Saeger and Rodies.[70b]

Opaque gold films and massive gold

Figures 30 and 31 illustrate the combined findings of various authors on the reflection and absorption of radiation by opaque layers of solid fine gold and vacuum-deposited films.

Figure 30 shows that in the i.r. and longer wave-length ranges of visible light, gold reflects more than 95% of incident light. Below $\lambda = 650$ nm, the reflectively falls off rapidly with diminishing wave-length. The results obtained by different investigators are shown to cluster along two curves with an absorption edge at or near $\lambda = 500$ nm. Loebich expresses the view that those observations which indicate a less gradual fall off have been derived from samples with lattice imperfections and surface defects which were avoided or corrected for by later investigators. The attractive yellow colour of gold results from its low reflectivity for wavelengths below about 500 nm.

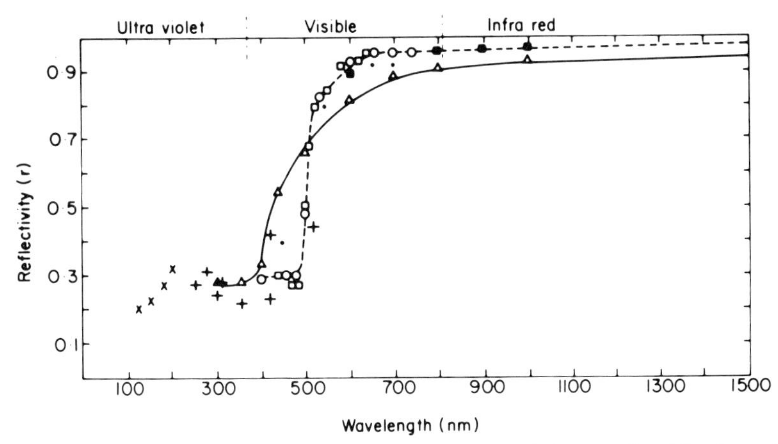

FIG. 30: Reflectivity r of gold in terms of wavelength λ in nm. Combined results of several studies.

■	Dickson and Jones[70a]	+	Pfestorf[71]
○	Knosp[72]	×	Robin[73]
□	Otter[74]	•	Schultz[75]
+	Philip[76]	△	Stabe[77]

Figure 31, on the other hand, shows how, in agreement with the reflectivity data, the absorption factor for gold, which is low at longer wavelengths, increases rapidly at about 500 nm and shows maxima at $\lambda = 400$ nm and 320 nm. Again there are discrepancies between recorded data probably because of the presence of surface defects on some of the samples studied.

Both the reflectivity and the absorption or emissivity (Kirchoff) are temperature-dependent,[79a] and the general trends of the effects are illustrated in Figures 32 and 33. More short wavelength radiation with $\lambda < 500$ nm is reflected by a hot gold surface than by a cold one, whilst with $\lambda > 500$ nm the reverse is the case. At $\lambda = 500$ nm, however, the reflectivity of gold is unaffected by temperature. The observed data for emissivity ($e = 1 - r$ for opaque films) correspond with those for reflectivity with e being practically unaffected by temperature in the region around $\lambda = 500$ nm. At longer wavelengths, however, the emissivity increases with temperature, corresponding to decreasing reflectivity.

Semitransparent gold films

Gold films and coatings which are thin enough transmit radiation. Gold leaf about 100 nm thick transmits a small amount of green light, and gold deposits 5-40 nm thick transmit considerable amounts of such light. The dependence of such transmission on wavelength is illustrated in Figure 34, and the relation between reflectivity and transmission for gold films of different thicknesses for $\lambda = 492$ nm is illustrated in Figure 35.

From these it will be seen that thin gold deposits largely reflect red light and i.r. radiation, while transmitting, to an extent dependent on their thicknesses, light of shorter wavelengths.

Gold deposits of thickness less than 4 nm transmit light which may be grey, yellow, red or violet according to the process used for deposition. As in the

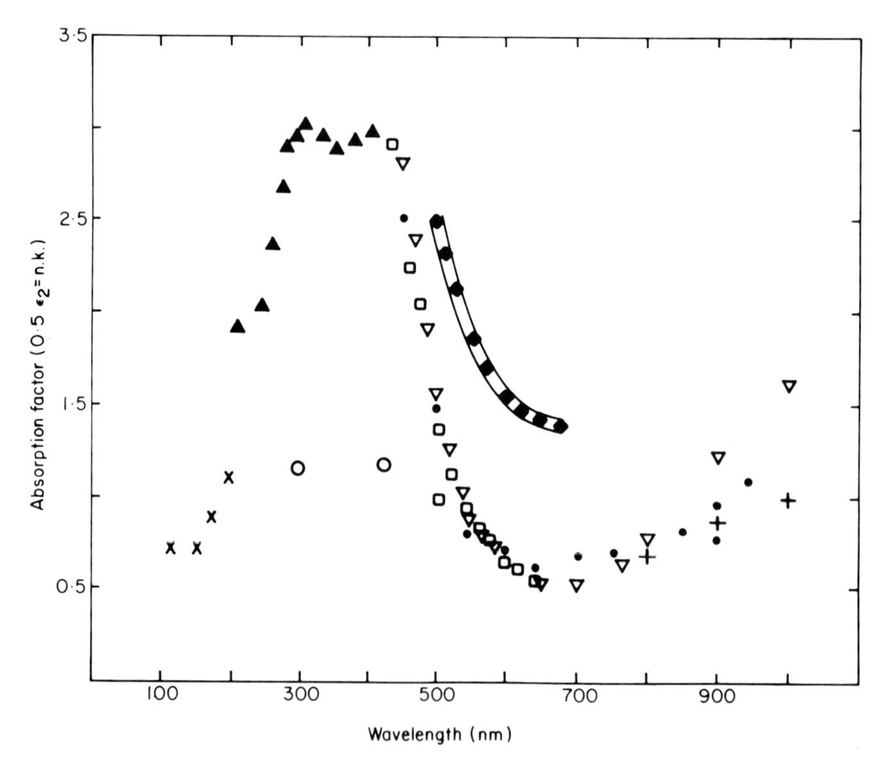

FIG. 31: Absorption factor $\frac{1}{2}\varepsilon_2 = nK$ of gold in terms of the wavelength λ of light in nm.

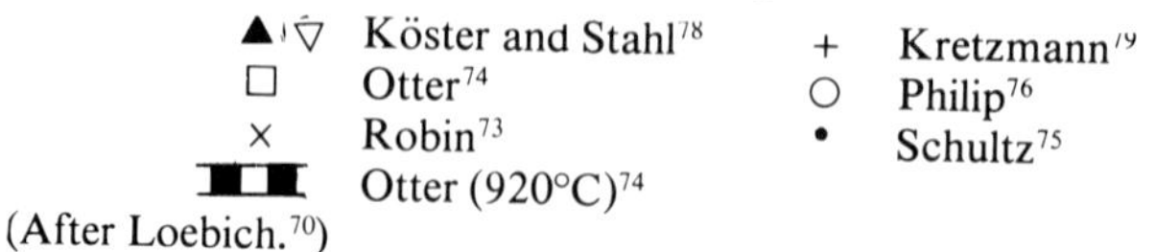

(After Loebich.[70])

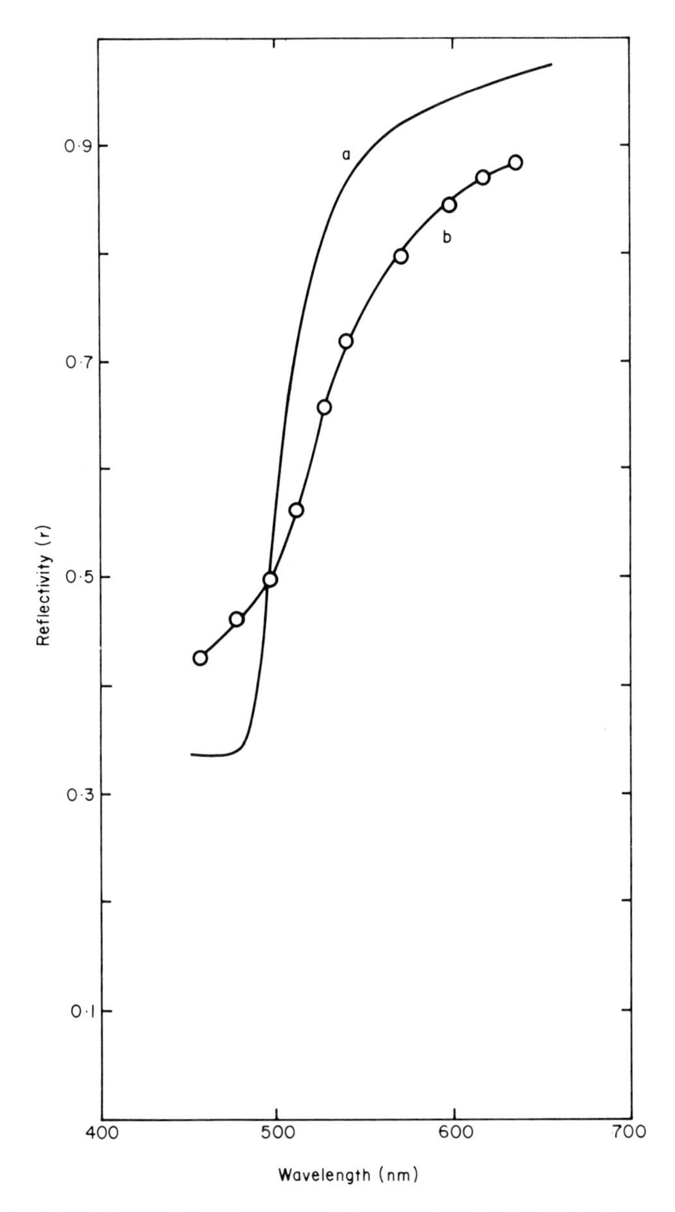

FIG. 32. Reflectivity r of gold in terms of the wavelength λ of light in nm, at 10 and 920°C.[74]

a, 10°C
b, 920°C

(After Loebich.[70])

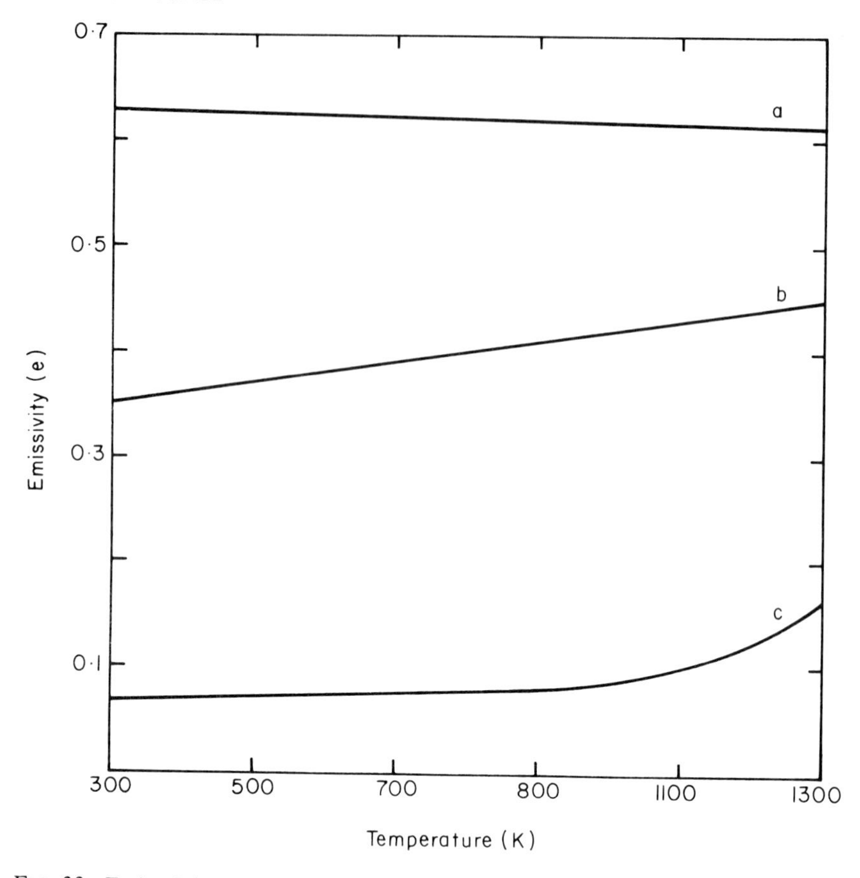

FIG. 33. Emiossivity e of gold as a function of temperature between 300 and 1280 K, according to Worthing.[80]

a, λ = 460 nm
b, λ = 535 nm
c, λ = 665 nm

(After Loebich.[70])

case of colloidal aqueous gold solutions this is explained as being due to a colloid-like grain structure of the deposit.

The effects of alloying metals

The effects of alloying silver with gold upon the optical properties discussed above have been studied by Köster[79] and Pepperhoff.[81] Figure 36 shows that with increasing silver content, the reflectivity curve shifts steadily from the visible into the ultra violet range. The absorption data of Figure 37 are in reasonable agreement with the reflectivity data.

Gold-copper alloys, the optical properties of which are affected by ordering

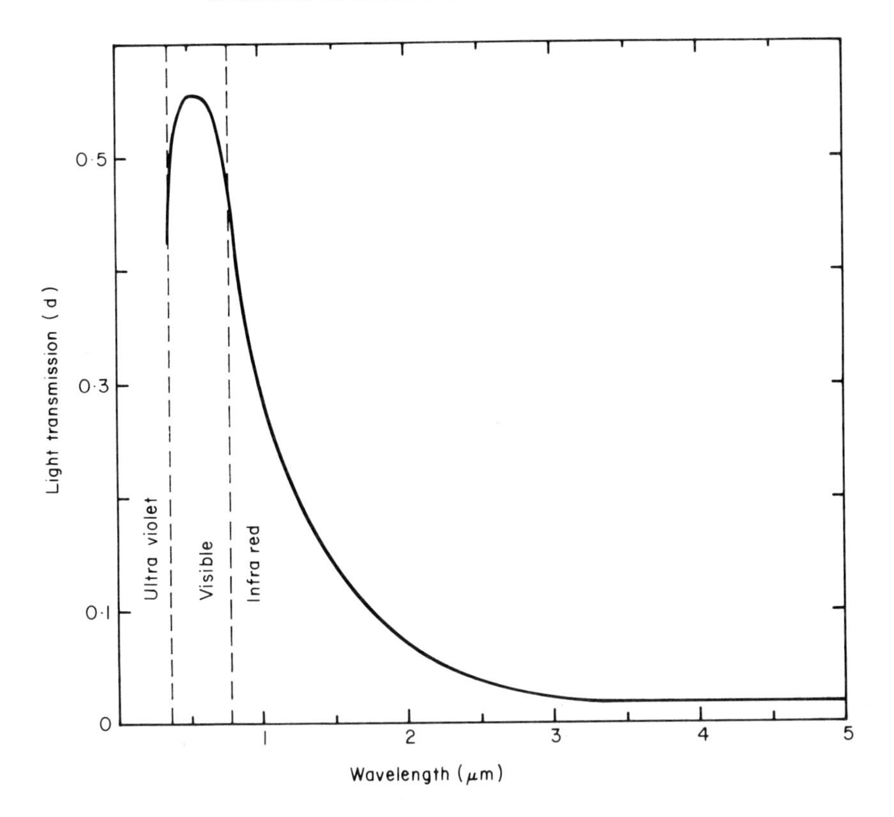

FIG. 34. Light transmission d of vacuum deposited gold on commercial glass in the wavelength range from 0.4 to 5 μm (reference 81, and Prospektangaben der Fa. DETAG A. G., Witten/Ruhr). (After Loebich.[70])

and disordering, have been discussed also by Köster and Stahl,[79] by Stahl *et al*[82] and by Rivory,[83] and dilute gold-zinc alloys by Weiss and Muldawer.[84]

An understanding of the relationships between the colour and the constitution of gold and its alloys in terms of electrical properties and the state of the electrons in the metal lattice is important both for the development of industrial applications for gold alloys in which the optical properties of the alloys are exploited, and for the development of coloured carat golds for jewellery.[70a]

In general terms it may be stated that the high reflectivity of gold at long wavelengths is associated with the high electrical conductivity of the metal and the number and mobility of its free s-electrons. At shorter wavelengths, reflection and absorption behaviour are more complicated, first in a narrow energy band by plasma resonance caused by interaction of free and bound electrons, and then by inter-band transference of d-electrons. In the case of gold these latter phenomena occur at wavelengths at the middle of the visible spectrum and are responsible for the yellow colour of the gold.

A recent publication[70b] reviews the mechanisms underlying the variations in the optical properties and therefore the colours of gold alloys, and incorporates consideration of reflectivity curves for the Au-Ni and Au-Pd alloy systems, and for such intermetallic compounds of gold as $AuAl_2$, $AuIn_2$ and $AuGa_2$.

3.3 Gold and gold alloys for solar energy collectors and concentrators

The fact that gold films which are thin enough to be virtually transparent to solar wavelengths nevertheless retain the low emissivity at longer wavelengths which is characteristic of gold in more massive form, is likely to ensure gold an important role in the utilization of solar energy.[85,86] It implies that when a solar energy collector is coated with a gold film of appropriate thinness, the absorption of energy is determined by the absorptivity of the substrate material of the absorber, while its emissivity is controlled by the gold film on its surface, so that high values for the ratio R of solar absorptivity to long wavelength emissivity can be achieved.

Whether the solar energy is used to heat water, or to operate a heat engine the effects of such high values are very benficial. Figure 38, for example, illustrates how the output of thermal energy from a flat plate collector in a water heater is increased with increase in R.

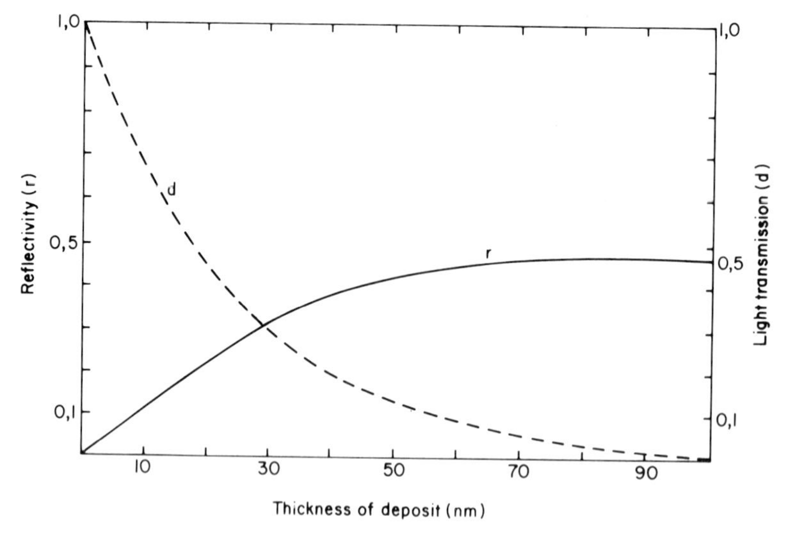

FIG. 35. Reflectivity r and light transmission d of thin gold films in terms of their thickness in nm for light of wavelength $\lambda = 492$ nm. (After Loebich.[70])

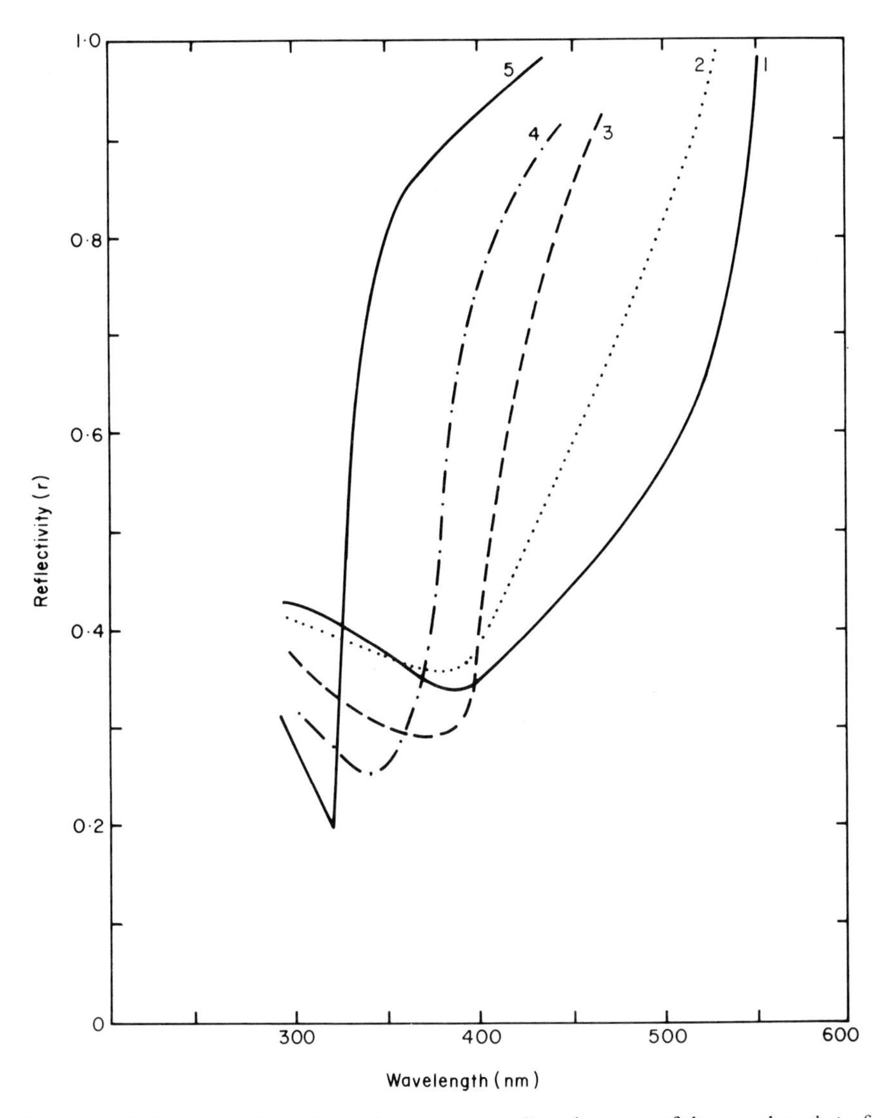

FIG. 36. Reflectivity of Au, Ag and three Au-Ag alloys in terms of the wavelength λ of light in nm, according to Pepperhoff.[81]

Curve 1	Au
Curve 2	85Au-15Ag
Curve 3	50Au-50Ag
Curve 4	30Au-70Ag
Curve 5	Ag

(After Loebich.[70])

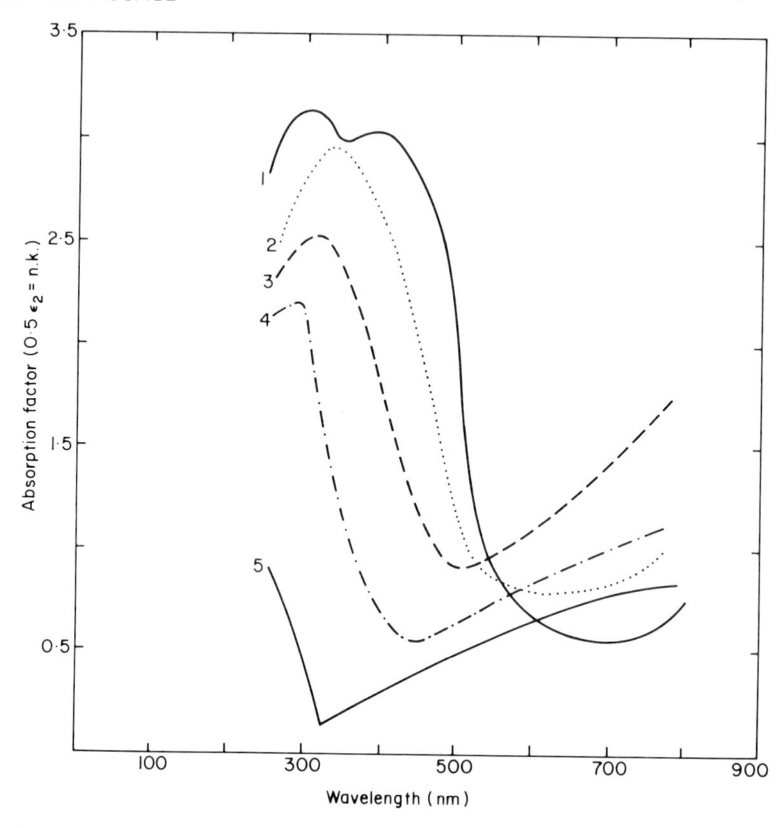

FIG. 37. Absorption factor $\frac{1}{2}\varepsilon_2 = nK$ of Au, Ag and three Au-Ag alloys in terms of the wavelength λ of light in nm, according to Köster and Stahl.[79]

Curve 1	Au
Curve 2	80Au-20Ag
Curve 3	50Au-50Ag
Curve 4	20Au-80Ag
Curve 5	Ag

(After Loebich.[70])

Figure 39, on the other hand, illustrates the marked effect of an increase in *R* on the efficiency of a solar-powered heat engine. In this case the efficiency is a function of the difference between the maximum and minimum temperatures of the cycle. As will be noted from Figure 38, the energy output from a collector falls as the output temperature increases. On the other hand increases in collector output temperature improve engine efficiency, so that the overall efficiency passes through a maximum. The value of this maximum increases as *R* is increased.

Collector output temperatures can be increased by concentrating the solar energy from a large area on to a smaller absorber area by means of reflecting devices. The chemical stability and high reflectivity of certain gold alloy coatings could well give these coatings a competitive advantage over other materials in such reflecting devices. Transmission and reflectance measurements on sputtered Au-Cr films of both thick (150 nm) and thin (10-40 nm) types have been described recently in this connection.[87]

Gold coatings have also been developed for use in thermoelectric solar cells.[88,90] For these it was found that the low emissivity of gold and high absorptivity for peak solar radiation were exhibited by composite gold coatings applied to Inconel by liquid gold techniques. The liquid golds used were such as to generate on the Inconel films containing 89.5% gold, together with the 10.5% of a mixture of oxides of rhodium, bismith, barium, chromium and

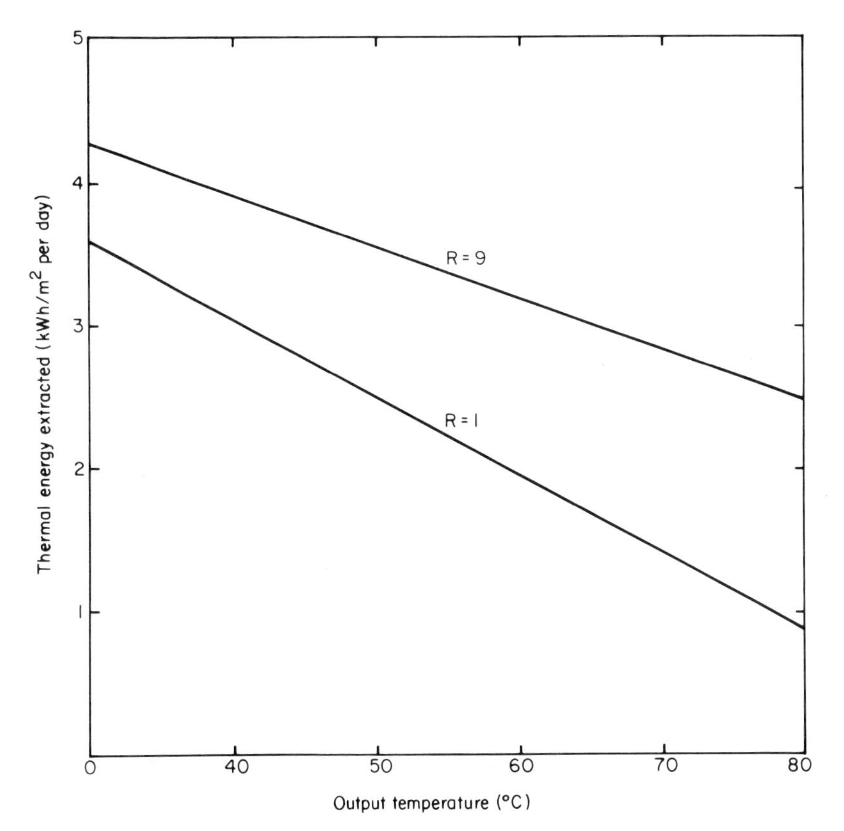

FIG. 38. The output of thermal energy from a flat-plate collector in a water heater operating in latitude 23.5°C. Treatment of the collector plate to give a high ratio R of solar absorptivity to long-wave emissivity increases the output at a given temperature. (After Brinkworth.[85])

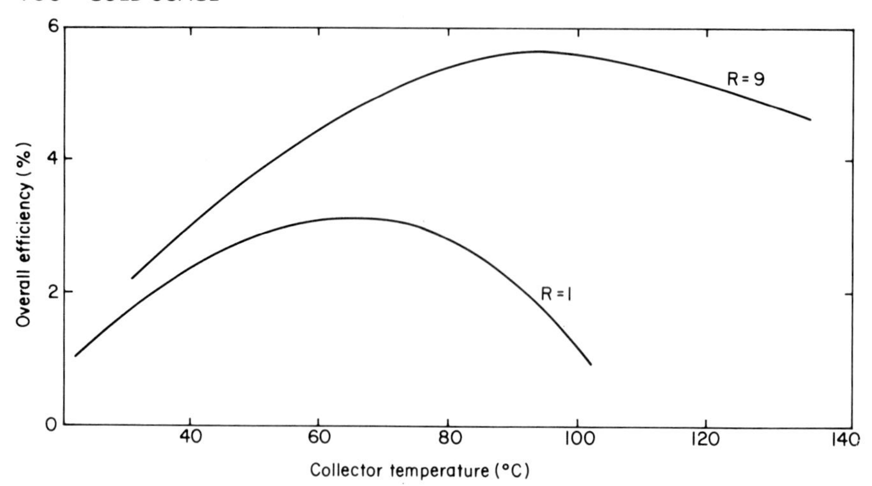

FIG. 39. The efficiency of a solar-powered heat engine showing the marked effect of treating the collector plate to give a high ratio R of solar absorptivity to emissivity. (After Brinkworth.[85])

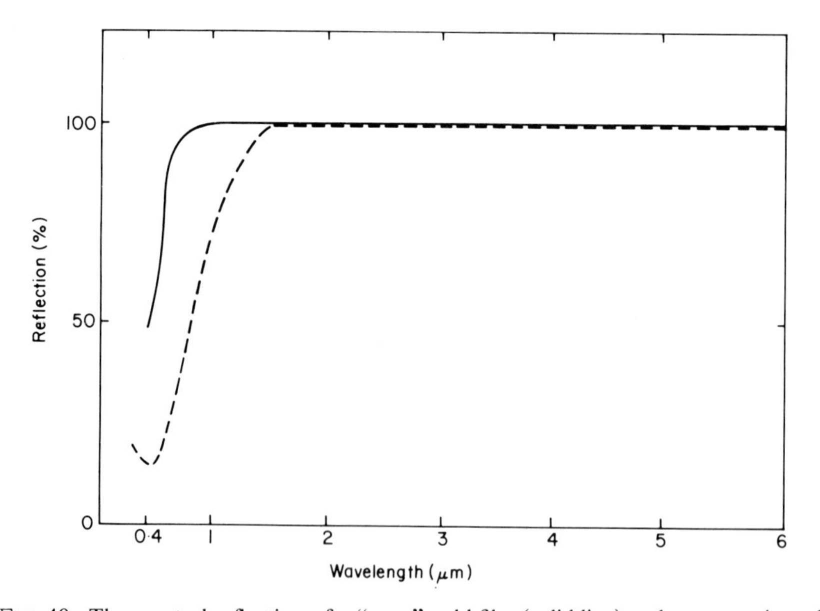

FIG. 40. The spectral reflection of a "pure" gold film (solid line) and a composite gold film developed as a selective solar absorber. The composite film has a minimum reflection at a wavelength of 0.5 μm, the wavelength of maximum solar radiation, while in the near infrared it has low emissivity comparable with that of pure gold. (After Langley.[90])

silicon. Such films absorb a high proportion of radiation of wavelength > 700 nm, in contrast to pure gold films which reflect more than 95% of radiation in this range.

3.4 Gold and gold alloys for heat reflectors in heating installations

Gold-coated reflecting surfaces find application in a variety of situations where control of radiant heat is required in heating equipment. Domestic cookers may, for example, be fitted with high-speed heaters emitting i.r. radiation of wavelength 750–3000 nm, which are placed at the focal point of a gold-coated parabolic metal reflector. The thickness of the gold coating is about 200 nm and a barrier layer of oxide may be placed between the substrate metal of the reflector and the gold layer to prevent diffusion of the gold into the reflector surface.[91] In the same manner radiation of wavelength 5000–15000 nm may be controlled in the drying of paper after printing, and Tugwell[89] has referred to the use of gold coatings for infrared reflection behind the heating bars in an installation used for annealing the moulded steel domes of a space capsule at 650°C.

In some applications, the heating elements may be enclosed in quartz glass tubes which are transparent to infrared radiation and which are coated on the reverse side with gold.[92,93] Such devices have the advantage that cleanliness of the reflecting gold surface is ensured.

The high reflectivity of gold in the infrared is also of interest in the design of high-power-density lasers. For these, reflective surfaces of high purity metals are required, which are free from even minute defects, such as inclusions or voids which might lead to damage. Gold, plated over suitably prepared surfaces, appears to have the reflectivity and other properties which are called for.[94]

3.5 Gold-coated glasses for windows in buildings and in transport vehicles

Ordinary window glass is almost completely transparent to solar radiation in the range extending from the ultra violet to the infrared (wavelength 0.3-2.1 μ m), the energy of about half of which les in the infrared outside the visible range. The large window areas of many modern buildings therefore cause overheating of offices and increase the load on air conditioning installations. These problems can be reduced by the use of gold-coated window glasses.

In the design of such glasses, the screening-off of infrared radiation must take place as far as possible by reflection, since absorbed solar energy heats the glass, which in its heated condition can transfer a considerable amount of this energy into a room by convection and by long-wave secondary radiation.

Moreover, in practice the amount of visible light which is transmitted by the glass must be controlled both by the thickness of the gold film and by the nature of the glass.

Figure 41, taken from an article by Groth and Reichelt[95] gives a schematic representation of a double glazed solar insulation unit in which the heat-reflecting gold film is applied to the inner surface of the outer glass pane, where it is protected from mechanical damage. From this it will be seen that the total energy entering the room is made up partly of energy transmitted *through* the glass and partly by energy transferred *from* the glass by secondary emission and by convection.

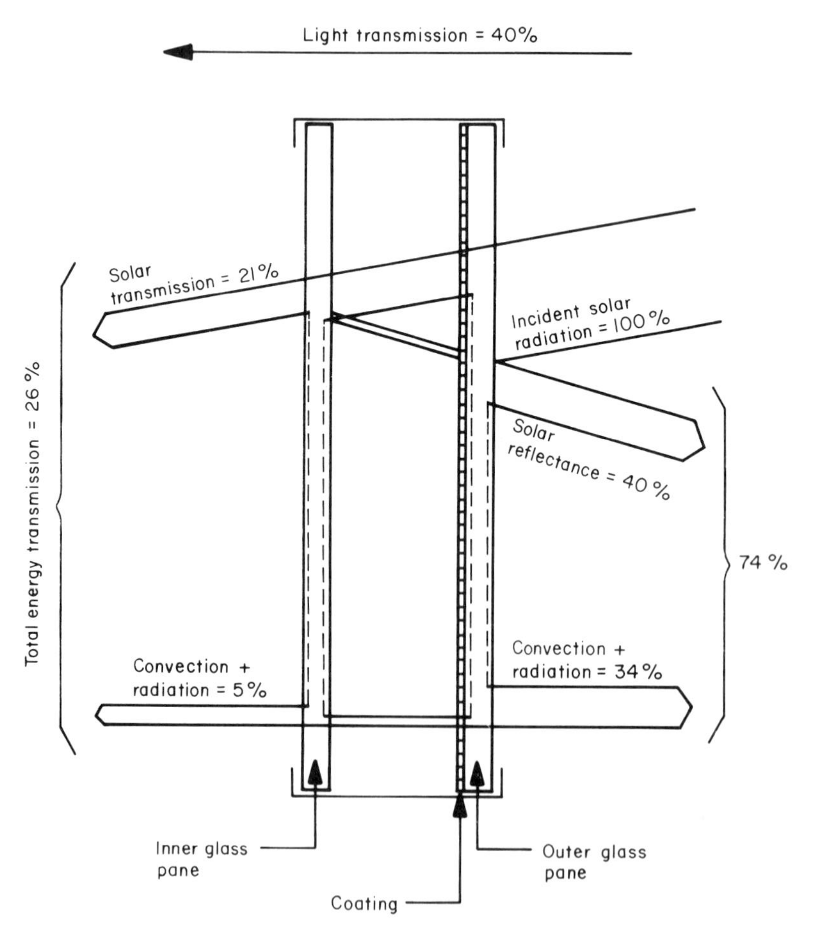

FIG. 41. Schematic representation of a gold-coated solar insulation double glazed unit with the corresponding distribution of solar energy. (After Groth and Reichelt.[95])

Two types of gold film are used to produce these effects, namely, simple transparent gold films, and gold interference layer coatings.

The visible light transmissions which are practical and feasible with simple gold coatings on glass lie in the range of 20–40%. Transmissions greater than 40% can be achieved only by reducing the film thickness to the point at which it loses some of its high reflectivity in the infrared.[96] At transmissions of less than 20%, on the other hand, the intensity of the green colouring of the glass becomes too great.

A problem which has had to be solved has therefore been that of designing gold films through which light transmission is increased without development of colour bias, while high heat reflectivity is retained. This has been achieved in present-day practice with what are called gold interference layer systems. These are made by a method closely related to that used for reducing the reflectivity of glass surfaces used in photographic apparatus and in the blooming of instrument lenses. By the use of thin dielectric films with a suitable refractive index and thickness, the reflected light rays at the surface and interface of the extra film can be obtained with almost equal amplitudes but with a shift of phase angle of 180°, so that they almost cancel each other out. As a result, since absorption is minimal, the non-reflected light appears as enhanced transmission.[95] The effect is illustrated in Figure 42.
It will be noted that reflectivity in the blue-green spectral range is low and transmission increased to about 70%. In transmitted light these films are slightly yellowish and give clearer vision and contrast.

In practice the gold film is most effectively sandwiched between two dielectric layers. This gives satisfactory bonding of the coatings to one another and to the glass substrate. Methods have been developed[97] for calculating film thicknesses, and practical tests can conveniently be carried out by evaporating a wedge-shaped film of one dielectric material on to a glass substrate, applying to this a uniform transparent film of gold, and evaporating on to this a film of a second dielectric material, which is also wedge-shaped, but so that the two wedges are at 90° to one another. In this way an impression can be gained of the range of effects possible with different thickness combinations of the two dielectric layers and a required combination selected. In this way glasses can be produced of various colours and with light transmissions ranging from 66–22% and with the corresponding energy transmissions between 44–15%.[98]

Double glazing, with gold coating, is also effective in reducing heat loss to the outside of buildings during winter. Gold-coated solar insulating glass is also being used increasingly for the windows of air conditioned trains and cars. Gold alloys may replace gold in these applications in some instances.

The heads of missiles may also be constructed of plastic-glass domes coated on their concave surface by a gold film which is designed not only so that an

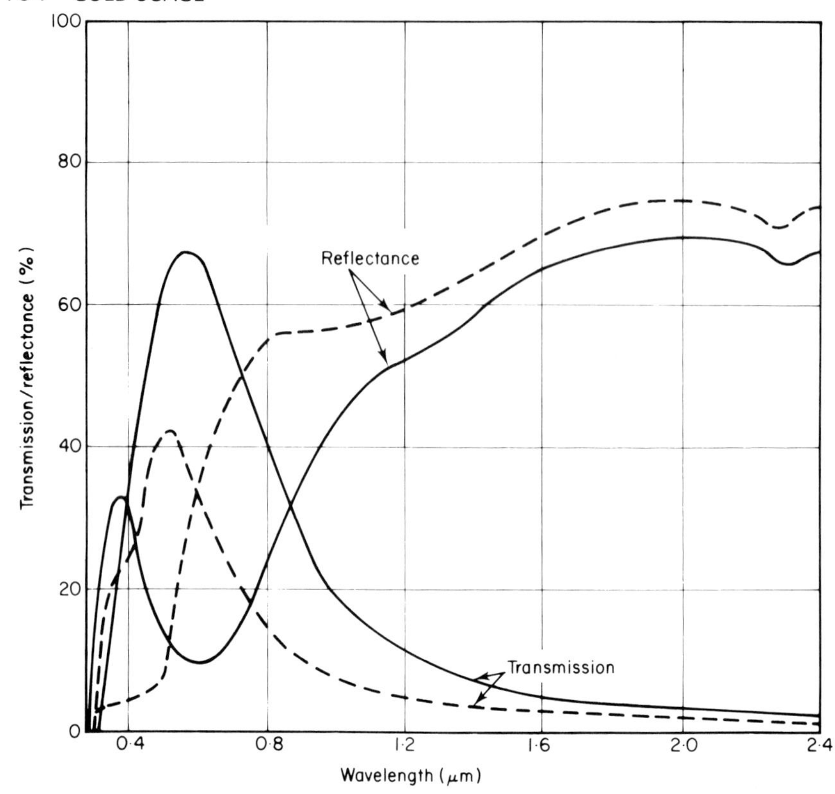

FIG. 42. Spectral transmission and reflectance of solar insulated double glazing for single gold film (dotted lines) and a gold interference layer system (solid lines).
With the interference layer system the transmission of visible light rays is greatly increased by comparison with the single gold coating, while the reflectivity in the infrared range is largely retained. (After Groth and Reichelt.[95])

optimum amount of light is transmitted into the dome for photographic recordings, but also so that the gold film will be sufficiently conducting to attenuate RF radiation to the point at which it does not interfere with the electronic systems inside the dome.[99]

Vacuum coating techniques are used exclusively in producing these gold-coated glasses, since they are the only techniques which give sufficiently continuous films for high i.r. reflectively.

3.6 Electrically-heated transparent gold coatings on glass

Of special interest is the fact that if glass surfaces are precoated by evaporation with Bi_2O_3, then even very thin (0.3-0.6 nm) gold films which may be applied, whether by sputtering,[100] or evaporation,[101] are so even and uniform that they can be heated electrically, and the coated glass therefore used in windscreens to control icing and misting conditions.[1] The films retain many of the electrical properties of gold at thicknesses down to less than 10 nm. Their conductivity and transparency can be increased by heat treatment at 200°C and an outer film of oxide may also be deposited on the gold.

Where gold coatings on glass are used as conductors, their temperature dependent degradation and their susceptibility to deterioration are also reduced by the application of outer films of a variety of metal oxides.[102]

3.7 Gold and gold alloy coatings for protection against radiant energy

The extensive use of gold coatings for temperature control in space exploration has received considerable publicity[88,102] and calls for no further discussion here.

Less well known, perhaps, is the use of gold films in cryogenic engineering, for example on the inside of Dewar flasks. With the rapid growth of applications of low temperature technology, it seems likely that use of gold in this area will increase accordingly.[90]

3.8 Gold blacks and their special properties

Whereas the condensation of gold and a number of other metals from the vapour phase under high vacuum leads to the formation of bright surface coatings, condensation of the metals at relatively high (1–10 mm) pressures gives rise to the deposition of metallic "blacks". This latter phenomenon was first reported by Pfund,[104] who also drew attention[105] to the high *transparency* of gold blacks to infrared radiation, and to their potential as coatings for the receiving areas of thermopiles and radiometers.

Because of their particular merits from this point of view, and because of the significance from other points of view of an understanding of the transmission of radiation through a scattering medium accompanied by absorption, the production and properties of gold blacks have been studied by a number of investigators,[106-113] and reviewed by Ito.[114]

The basic building blocks of gold black structures are gold crystallites about 10 nm in diameter, which are formed in the vapour phase, coalesce into chains and aggregates and build up low-density particles about 200–800 nm in diameter, which settle out on the surface of the substrate.[106-110, 115-117] Their properties, and in particular their infrared transmission are critically dependent[113] on the conditions used for their preparation, and especially upon the pressure of inert gas in the vessel, and the extent, if any, of its contamination with oxygen. The latter, in the presence of the tungsten filament used for vaporising the gold, gives rise to deposits in which tungsten oxide may be detected.[118-119] The condition of the filament surface also affects the degree of contamination with tungsten oxide. A nearly constant transmission can be realized, however, in the 2–15 μm wavelength region.[113] The transmittance does not fall off exponentially with thickness and this conforms to the view that radiative transfer takes place with both absorption and scattering.[120,121]

Coatings of gold black backed by substrates of low emissivity could have application in the design of solar energy absorbers,[85] since they should provide high solar absorptivity, at the wavelength (up to 2000 nm) over which solar energy is mostly transmitted, combined with low emissivity at the longer wavelengths over which energy absorbed by the collector might otherwise be lost.

4. Gold and Gold Alloys in Miscellaneous Engineering Applications

4.1 Gold-palladium alloys in spark plug electrodes

A range of spark plugs has been introduced[122] in which the centre electrode is tipped with a 60Au-40Pd alloy and which offer the advantages of easier starting, specially in adverse weather conditions, and less gap erosion. The corrosion resistance of the alloy has also made it possible to design these plugs with a smaller diameter insulator tip, which gives greater clearance between the insulator and the shell at the firing end. This results in better scavenging of fuel deposits and better protection against fouling.

4.2 Gold and gold-plated condenser surfaces for dropwise condensation of steam

Surfaces upon which dropwise, rather than continuous film, condensation of steam occurs are of interest because of the improved heat transfer which results from their use. Dropwise condensation of steam occurs[123-125] on ordinary gold and gold-plated surfaces and considerable effort has been devoted to attempts to exploit this.[126]

Pure gold surfaces, however, are inherently hydrophylic in character,[127] and hydrophobic or non-wetting behaviour of gold surfaces is to be attributed to the presence of hydrophobic contaminants. Thus using solid gold tubes, dropwise condensation cannot be maintained unless minute amounts of substances such as long-chain mercaptans are introduced, which form hydrophobic films on the surface of the gold.[128] Dropwise condensation is more easily maintained on gold-plated surfaces, probably because of the presence of absorbed hydrophobic substances on these surfaces which are derived from the brightening agents used in the plating baths.

In trials of gold-plated condenser surfaces, however, in which sea water steam was used in distillation units, loss of ability to promote dropwise condensation has been found to result from other causes as well, such as contamination of the gold surfaces with iron oxide and other particulate matter, and diffusion of the substrate metal into the gold plating.[126]

Industrial application of gold surfaces bearing hydrophobic contaminants in steam condensers therefore awaits further development.

4.3 Gold-palladium alloys as thermal fuses for electric furnaces

A characteristic feature of the Au-Pd system, which forms a continuous series of solid solutions, is the small separation between solidus and liquidus curves. A series of alloys melting each within a range of about 10°C over the range 1063–1500°C is thus available for use as thermal fuses for protecting electric furnaces from overheating.[129] The alloys are ductile and the fuse elements which are available commercially are in the form of 1 mm-diameter wire which is enclosed in a ceramic sheath and inserted in the furnace. They are appropriately connected into the control circuit so that a power cut-off is initiated if the wire melts. Because of their oxidation resistance, these fuses are very stable in use.

Gold-silver alloys apparently find similar application,[130] and a patent has been granted for the use of gold as the fuse material in a high-temperature fuse device for internal combustion engine catalytic converts.[131]

4.4 Gold bursting discs and gold containers

Bursting discs for the protection of chemical plant operating under relatively low pressures can, with advantage in many instances, be made of pure gold because of its ductility and its resistance to corrosion.[132,133] Gold discs fitted to liquid ammonia tanks, for example, show long lives as compared with those of discs of other metals.

Likewise, containers for a number of corrosive gases such as SO_2 and H_2S at high temperatures may be of gold or of gold-platinum alloy.[130] Gold vessels also find application in a variety of processes where its resistance to corrosion gives it advantages over other metals. Gold containers have been used,[134] for example, for the growth of single crystals of lead germanate for infrared detectors, because of their chemical compatibility and the ease with which containers could be produced in the shapes required by electroforming. An expensive alloy of gold (75Au-25Pt) with m.p. 1370°C has also found limited use for crucibles, where it has been found resistant to even alkali carbonate and bisulphate fusions.

As already mentioned, the 84Au-15Pt-1Rh alloy developed for spinnerets can also be considered for high temperature applications where corrosion resistance is required. The interior gold coating of autoclaves by an explosive plating technique has been described.[135]

4.5 Gold coatings as diffusion barriers.

The development of chromium alloys for high-temperature applications has been hindered by the lack of an effective treatment to prevent their embrittlement by nitrogen absorbed during exposure to hot gases. Nitrogen has very low solubility in gold,[136-137] and it has been demonstrated[138] that a protective coating of gold can prevent absorption of nitrogen by chromium. Moreover, the temperature range over which such a coating can be used can be extended by alloying the gold with palladium in order to raise its melting point.[138-139] The potential for the application of such coatings to chromium alloys and other materials used for both turbine blades and nozzle guide vanes has still to be assessed.

Gold coatings have also been explored[140-141] and found effective as diffusion barriers in electroplating to prevent ingress of hydrogen into the substrate metal, and its consequent hydrogen embrittlement.

4.6 Gold in fuel cell construction

Gold and gold alloys are mentioned frequently in literature relating to the design and development of fuel cells, in which their use as corrosion-resistant materials of construction, in hydrogen-permeable membrane electrodes and in catalytic materials is described.

Alloys consisting of 25–35% Au+Ag, 0.2–5.0% of either Pt, Rh or Ru, and the rest Pd have been described[142] as having hydrogen permeabilities approximately the same as, or higher than, those of palladium or platinum, and as being strong and corrosion resistant.

The use of other gold alloys in fuel cell electrodes has been reported upon by Shirogami[143] and Grevstad[144] and a number of others. The extent to which they have found practical application, however, is not clear.

4.7 Gold coatings to assist heat dissipation

Many electronic components during operation generate heat which must be removed from what is frequently a very localised zone. Insulating materials with high thermal conductivity are therefore desirable in such circumstances and beryllium oxide with a resistivity of 10^4 ohm cm and a thermal conductivity exceeding that of aluminium has found application in such components as high-frequency transistors, power transistors, klystrons, thyristors and integrated circuits to help transfer heat to the heat sink.[145] Thermal barriers which exist at interfaces with the beryllium oxide must be overcome, however, and to this end beryllia components may be metal coated. In at least one company gold coatings applied by sputtering have been found to be most effective for this purpose.

Gold-diamond composites have also been claimed as heat-sink mounting materials for semiconductor devices, because of their high thermal conductivity.[146-147]

4.8 Gold plating in submarine telephone cable repeaters

Signal strengths are maintained in submarine telephone cables by repeaters every few miles along their length. Such repeaters are, in effect, two amplifiers back to back so that signals in each direction are amplified, with the lower frequencies being used in one direction and the higher frequencies in the other. One amplifier is fed through low-pass filters and the other through high-pass filters. The power to operate the repeaters is supplied from the ends of the cable; thus the first filters in the repeater are the power separating filters, which separate the signal from the power. The capacitors in these filters have to withstand the full voltage of the system and are probably the most highly stressed components in the repeater.

Since the repeaters must be sturdy and operate over long periods without deterioration or failure, gold plating is used extensively in their construction. Apart from its inertness, gold does not, so far as is known, grow metal whiskers which are a problem with other finishes such as tin, and gold coatings are therefore used where the formation of such whiskers must be avoided.[148] Other details of its use in repeater construction have been given by Girling[149] from whose paper the above information has been obtained.

4.9 Organic gold compounds as possible lubricating oil additives

The use of additives in lubricating oils to inhibit corrosion and reduce wear is well established and a number of such additives are zinc 0,0- dialkyl-phosphoro-dithioates, which possess[150] structures of the type:

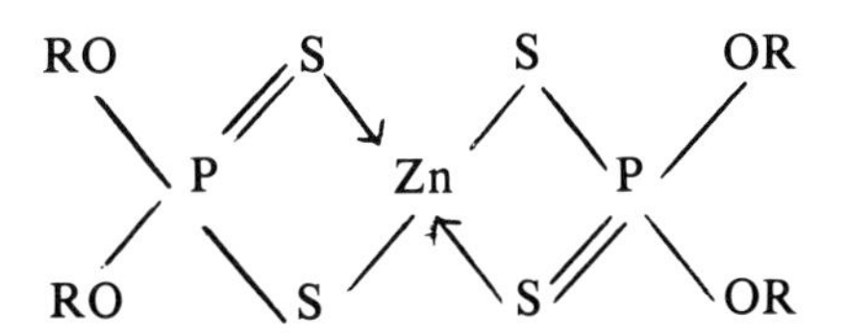

It has been established that the effectiveness of these compounds depends upon their rate of thermal decomposition, which apparently leads to the formation of a corrosion-resistant film by reaction with the metal bearing surfaces.

This reaction gives rise, however, to a certain amount of chemical attack, and in an attempt to overcome this, the behaviour of the corresponding gold (I) 0,0- dialkyl-phosphoro-dithioates in lubricating oils has been explored. These were found[151-152] to undergo decomposition at the rubbing surfaces with deposition of gold. Using the 0,0-dineopentyl-phosphoro-dithioate, for example, this was found to result in the formation of visible gold films on the bearing surfaces only and a significant diminution in the wear rate. Deposition is localised presumably because decomposition is induced at the contacting surface, whenever contact results in localised high temperature, so that under operating conditions a balance between film wear and film deposition is set up.

4.10 Gold as a solid film lubricant

Under pressures below 10^{-8} torr surface films on metals which are lost as a result of friction or other effects are not replaced, with the result that metal surfaces in contact with one another can easily cold weld to one another. In space this difficulty is increased since in some external situations non-metallic lubricants may suffer from radiation damage, chemical breakdown or evaporation. Because of its low shear strength, cold welding is not the serious problem with gold that it is with other metals in such situations, and it has therefore been applied as a solid film lubricant in many space situations involving moving parts, both alone as a surface coating on the bearing surfaces, and in combination with low volatility synthetic lubricants.[153-156] Low alloy nickel hardened gold plate was found superior to pure gold plate on stainless steel balls and races in one situation, and an undercoat of silver was found to extend its life. Burnishing of the gold surfaces with MoS_2 has been applied.

REFERENCES

1. Sloboda, M. H., Industrial gold brazing alloys: their present and future usefulness, *Gold Bulletin*, **4** (1971), 2–8.
2. Chaston, J. C. and Sloboda, M. H. (Johnson Matthey and Company Limited), "An Improved Gold Alloy". British Patent 681,484 (1952).
3. Hansen, M., "Constitution of Binary Alloys". McGraw Hill Book Company Inc., 1958.
4. Anon., Gold in silver brazing alloys: low melting points with increased ductility, *Gold Bulletin*, **5** (1972), 13.
5. Schwartz, M. M., Applications for gold brazing alloys, *Gold Bulletin*, **8** (1975), 102–110 (see also **9** (1976), 119–122).
6. Kirby, R. S. and Hanks, G. S., Brazing of nickel–chromium alloy 718, *Welding J.* **47** (1968), 97s–105s.

7. Hofman, E. E. *et al.*, "An Evaluation of the Corrosion and Oxidation Resistance of High Temperature Brazing Alloys". O.R.N.L. Report UC-25, 1956.
8. Colbus, J. and Zimmermann, K. F., Properties of gold–nickel alloy brazed joints in high temperature materials, *Gold Bulletin*, **7** (1974), 42/49 (cf. *Schweissen Schneiden*, **26** (1974), 209–212).
9. Wise, E. M. (Editor), "Gold: Recovery, Properties and Applications". Van Nostrand Company Inc., 1964, pp. 117–118.
10. Sloboda, M. H. and Boughton, J. D. (Johnson Matthey and Company Limited), "Brazing Alloys". British Patent 1,280,460 (1972).
11. Mizuhara, H., "Copper–Gold–Indium–Nickel Brazing Alloy Compositions". U.S. Patent 3,519,416 (1970) (*C.A.* **73** (1970), 58820r).
12. Johnson, Matthey and Company Limited, "Brazing Alloys". French Demande 2,015,975 (1970) (*C.A.* **74** (1971), 56777d), cf. British Patent 1,280,460 (1972).
13. Donelly, R. G. and Slaughter, G. M., The brazing of graphite, *Weld. J.* (Miami), **5** (1962), 461–469.
14. Nesse, T., Microstructure of brazed joints produced by the active brazing of graphite with a Au-Mo-Ni filler metal, *Prakt. Metallogr.* **11** (1974), 77–82 (*C.A.* **81** (1974), 95278f).
15. Barb, R. P. and Sutar, W. (General Electric Co.), "Nickel–Gold–Base Brazing Alloy". U.S. Patent 3,764,307 (1973) (*C.A.* **80** (1974), 99298q), cf. German Offen. 2,339,090 (1974).
16. Fairbanks, N. P., Barb, R. P. and Sutar, W., "Nickel–Gold–Brazing Alloy". U.S. Patent 3,853,548 (1974) (*C.A.* **82** (1975), 89468x).
17. Grigor'ev, A. T. *et al.*, Alloys of the gold–nickel–palladium system, *Russ. J. Inorg. Chem.* **7** (1962), 570–573.
18. "Brazing Filler Metal, High Temperature, 50Au-25Pd-25Ni", S.A.E. Aerospace Materials Spec. A.M.S. 4,784, 1968. "Brazing filler metal, high temperature, 30Au-34Pd-36Ni", S.A.E. Aerospace Materials Spec. A.M.S. 4,785, 1968. "Brazing filler metal, high temperature, 30Au-34Pd-22Ni, S.A.E. Aerospace Materials Spec. A.M.S. 4,786, 1968.
19. Zwingmann, G., Das Dreistoffsystem Silber–Gold–Germanium, *Metall.* **18** (1964), 726–727.
20. Kuprina, V. V., Equilibrium diagram of the silver–gold–silicon system, *Russ. J. Inorg. Chem.* **7** (1962), 833–834.
21. Coad, B. C. (Texas Instruments Inc.), "Low Melting Point Composite Materials Useful for Brazing, Soldering or the Like". U.S. Patent 3,382,054 (1968).
22. Finch, R. G., Gold in thick film micro-electronics, *Gold Bulletin*, **5** (1972), 26–30.
23. Hagon, P. J. (North American Rockwell Corporation), "Face Bonding Technique". U.S. Patent 3,484,933 (1969).
24. International Standard Electric Corporation, "Electron Tube with Bonded External Semiconductor Electrode". U.K. Patent 1,319,008 (1973).
25. International Business Machines Corporation, "Method and Apparatus for Fabricating Semiconductor Devices". U.S. Patent 1,331,794 (1973).
26. International Business Machines and Motorola Inc., "Semiconductor Circuit Arrangement and Method of Formation Thereof". U.K. Patent 1,252,803 (1971).
27. English Electric Company Limited, "Method of Brazing". U.K. Patent 1,315,319 (1973).
28. See *Gold Bulletin*, **4** (1971), 14.

28a. Rivlin, V. G., Waghorne, R. M. and Williams, G. I., The structure of gold alloys in the liquid state: novel quench products from some eutectic systems, *Gold Bulletin*, **9** (1976), 84–87 (cf. *J. Phys. F.* **6** (1976), 147–155).
29. E. I. du Pont de Nemours and Company, "Printed Circuits from Non-Migrating Solders". U.S. Patent 3,579,312 (1971) (*C.A.* **75** (1971), 55332p) (cf. U.S. Patent 3,472,653 (1967)).
30. Anon., A new gold—tin solder, *Gold Bulletin*, **5** (1972), 37.
31. Zimmerman, D. D., A new gold–tin alloy composition for hermetic package sealing and attachment of hybrid parts, *Solid State Technol.* **15** (1972), 44–46.
32. Ainsworth, P. A., Soft soldering gold coated surfaces: techniques for making reliable joints, *Gold Bulletin*, **4**(1971), 47–50.
33. Thwaites, C. J., Some aspects of soldering gold surfaces, *Electroplat. Met. Finish.* **26** (1973) (8), 10–14 and (9), 21–26.
34. Harmsen, U. and Meyer, C. L., Uber Weichlötungen an Gold, *Z. Metallk.* **56** (1965), 234–239.
35. Heinzel, H., Verhalten von Weichloten beim Löten von Gold, *Elecktronik*, (1973), 439–440.
36. Flegontov, Yu. N., "Brazing Alloy for Joining Gold Plated Products". U.S.S.R. Patent 406,673 (1973) (*C.A.* **81** (1974), 81543b).
37. International Business Machines Corporation, "Indium–Bismuth–Gold Alloys". U.S. Patent 3,323,912 (1967).
38. Piffle, L. (A. G. Brown Boveri et Cie), "Lötverbindung für thermoelektrische Einrichtungen". German Offen. 1,508,311 (1970).
39. Pyatyshev, V. I. *et al.*, "Solder for Soldering the Elements of Semiconducting Devices". U.S.S.R. Patent 432,999 (1974) (*C.A.* **82** (1975), 20910t).
40. Johnson, D. R. and Sawers, J. R. (E. I. du Pont de Nemours and Company), "Verfahren zur Vereinigung von Diamant und Metall". German Offen. 2,407,412 (1974).
41. Rochel, E., Loebich, O. and Raub, C. J., Niobium–gold phase diagram, *Z. Metallk.* **64** (1973), 359–361.
42. Funk, W. and Reinacher, G., Gold alloy spinnerets for the production of viscose rayon, *Gold Bulletin*, **7** (1974), 2–9.
43. Reents, W. and Eilfield, F. German Patent 221,572 (1909).
44. Grafried, E., Protzmann, K., Ruthardt, K. and Speidel, H., Festschrift "Hundert Jahre Heraeus". Hanau, 1951, p. 43.
45. Pummerer, R. (Editor), "Chemische Textilfasern, Filme und Folien". Stuttgart, 1953, p. 397.
46. Holzmann, H., "Degussa Metal-Berichte". Frankfurt, 1941, p. 136.
47. Dreaper, W. P., "Improvements in Alloys and Their Uses". British Patent 260,672 (1925).
48. G. Siebert G.m.b.H., "Vergütbare Goldlegierungen". German Patent 584,549 (1929) (released 1933).
49. Nowack, L., "Process for the Manufacture of Alloys of Precious Metals which are Amenable to Heat Treatment". U.S. Patent 1,946,231 (1930).
50. Goedecke, W., "Festschrift zum Fünfzigjahrigen Bestehen der Platinschmelze". G. Siebert G.m.b.H., Hanau, 1931, p. 99.
51. Nowack, L., Vergütbare Edelmetall-Legierungen, *Z. Metallk.* **22** (1930), 94–103.
52. W. C. Heraeus G.m.b.H., "Verfahren zum Vergütbarmachen von Goldlegierungen mit einem Gehalt an Platinmetallen und zur Vergütung solcher Legierungen". German Patent 554,502 (1932).

53. W. C. Heraeus G.m.b.H., "Platinmetallegierungen und Verfahren zu ihrer Vergutung". German Patent 535,688 (1928).
54. Grigorjew, A. T., *Ann. Inst. Platine,* **6** (1928), 184.
55. Grigorjew, A. T., Die Legierungen von Gold mit Platin, *Z. Anorg. Allg. Chem.* **178** (1929), 97–107.
56. Johansson, C. H. and Linde, J. O., Kristallstrucktur, electrischen Widerstand, Thermokräfte, Wärmeleitfahigkeit, magnetische Suszeptibilität, härte, und Vergütungserscheinungen des Systems Au-Pt mit dem Zustandsdiagramm, *Ann. Phys.* **5** (1930), 762–792.
57. Wictorin, C. G., "Dissertation". Stockholm, 1947 (quoted by Wise, reference 9, p. 125).
59. Darling, A. S., Mintern, R. A. and Chaston, J. C., *J. Inst. Met.* **81** (1952–3), 125 ff (cf. Darling, A. S., *Platinum Met. Rev.* **6** (2) (1962), 60).
59. W. C. Heraeus G.m.b.H., "Verwendung von Goldlegierungen für Kunstseidespinndüsen". German Patent 691,061 (1935).
60. Degussa, "Verwendung von Gold–Platin–Legierungen zur Herstellung von Spinndüsen". German Patent 873,145 (1936).
61. Degussa, "Verwendung von Gold–Platin–Legierungen zur Herstellung von Spinndüsen". German Patent 1,075,838 (1939) (released 1960).
62. Johnson Matthey and Company, "An Improved Alloy". British Patent 1,112,766 (1966).
63. Comptoir Lyon-Alemand Louyot et Cie, "Gold Alloys for Spinnerets". french Patent 2,133,178 (1972) (*C.A.* **78** (1973), 163107q).
64. Schmid, H., Grain size and hardening of gold–platinum alloys, *Metall.* **12** (1958), 612–619.
65. Raub, E. and Falkenburg, G., The system gold–platinum–rhodium and the binary peripheral systems, *Z. Metallk.* **55** (1964), 392–397.
66. Selman, G. L., Spender, M. R. and Darling, A. S., The wetting of platinum and its alloys by glass. III. Microstructure and mechanical properties of gold–rhodium–platinum alloys, *Platinum Met. Rev.* **10** (2) (1966), 54–59.
67. Selman, G. L., Spender, M. R. and Darling, A. S., The wetting of platinum and its alloys by glass. II. Rhodium–platinum alloys and the influence of gold, *Platinum Met. Rev.* **9** (1965), 92–99 and 130–135.
68. Darling, A. S. and Selman, G. L., "Platinum Alloys for Spinning Nozzles in Glass Fibre Manufacture". German Offen. 2,053,059 (1971) (*C.A.* **76** (1972), 49124b).
69. Lacroix, R. and Niney, C., A high strength gold alloy for use at high temperatures, *Gold Bulletin,* **6** (1973), 94–98.
70. Loebich, O., The optical properties of gold: A review of their technical utilisation, *Gold Bulletin,* **5** (1972), 2–10.
70a. Dickson, P. F. and Jones, M. C., Infrared spectral reflectances of metals at low temperatures, *N.B.S. Tech.* Note 348, 1966.
70b. Saeger, K. E. and Rodies, J., The colour of gold and its alloys: The mechanism of variation in optical properties, *Gold Bulletin,* **10** (1977), 10–14.
71. Pfestorf, G., Die Bestummung der optischen Konstanten von Metallen im sichtbaren und ultravioletten Teil des Spektrums, *Ann. Physik,* 4 Folge, **81** (1926), 906–928.
72. Knosp, H., Determination of micro reflection and optical constants of metals and refractory compounds, *Z. Metallk.* **60** (1969), 526–531.
73. Robin, S., Mesure de pouvoirs réflecteurs de couches métalliques épaisses (Au,

Pt, Cr) dans la région de Schumann. Détermination de constants optiques, *Compt. Rend.* **236** (1953), 674–676.

74. Otter, M., Optical constants of massive metals and their temperature dependence, *Z. Phys.* **161** (1961), 163–178 and 539–549.
75. Schulz, L. G., The optical constants of silver, gold and aluminium. I. The absorption coefficient K, *J. Opt. Soc. Am.* **44** (1954), 357–362.
76. Philip, R., Constantes optiques de quelques couches minces d'or, *Compt. Rend.* **247** (1958), 1104–1108.
77. Stabe, H., Der Lichtpunkt-Linienschreiber—ein vielseitig verwendbares, tragbares, Registriergerät mit sofort sichtbarer Photoschrift, *Feinwerktechnik,* **97** (1953), 199–203.
78. Köster, W. and Stahl, R., Effects of alloying, deformation, recrystallisation and short and long range order on the optical constants of noble metals and their alloys, *Z. Metallk.* **58** (1967), 768–777.
79. Kretzmann, R., Über optische Konstanten dicken Metallschichten im Sichtbaren und nahen Ultrarot, *Ann. Physik*, 5 Folge, **37** (1940), 303–325.

79a. McKay, J. A. and Rayne, J. A., Temperature dependence of the infra-red absorptivity of the noble metals, *Phys. Rev. B*, **13** (1976), 673–685.

80. Worthing, A. G., Spectral emissivities of tantalum, platinum, nickel and gold as a function of temperature, and the melting point of tantalum, *Phys. Rev.* **28** (1926), 174–189.
81. Pepperhoff, W., "Temperaturstrahlung". Darmstadt, 1956.
82. Stahl, R., Spranger, H. J. and Aubauer, H. P., Optical properties of the ordered phase $AuCu_3$, *Z. Metallk.* **60** (1969), 933–938.
83. Rivory, J., Optical properties of ordered and disordered Au-Cu alloys, *J. Phys. Paris*, Colloque C4, supplément au No. 5 **35** (1974), C4–51.
84. Weiss, D. E. and Muldawer, L., Optical properties of gold and dilute gold–zinc alloys, *Phys. Rev. B.* **10** (1974), 2254–2261.
85. Brinkworth, B. J., Gold films in solar energy utilisation, *Gold Bulletin*, **7** (1974), 35–38 (cf. Reiner, H. W., Black gold to trap the sun, *Plating*, **63** (1976), 21, 24–25 and 29).
86. Brinkworth, B. J., "Solar Energy for Man". Compton Press, 1972, I.S.B.N. 9001 93131.
87. Fuschillo, N., Lalevic, B., Slusark, W. and Delahoy, A., Optical properties of thin Au-Cr films and their application to solar energy conversion, *J. Vac. Sci. Technol.* **12** (1975), 84–87.
88. Langley, R. C., Gold coatings for temperature control in space exploration, *Gold Bulletin*, **4** (1971), 62–66.
89. Tugwell, G. L., Industrial applications for the noble metals, *Metal Prog.* **88** (1965), 73–78.
90. Langley, R. C., Gold films for the control and utilisation of radiant energy, *Gold Bulletin*, **8** (1975), 34–40.
91. Langley, R. C. (Engelhard Industries Inc.), "Infrarotstrahler und Verfahren zur Herstellung desselben". German Offen., 1,540,740 (1971).
92. Lambert, J. L., Higher efficiency for industrial infra-red heating processes, *Electr. Rev. London*, **194** (1974), 681–684.
93. Lambert, J. L., Gold reflectors in infra-red heating, *Gold Bulletin*, **8** (1975), 111.
94. Anon., Gold reflectors for lasers, *Gold Bulletin*, **9** (1976), 44 (cf. *Appl. Opt.* **14** (1975), 1783–1787).
95. Groth, R. and Reichelt, W., Gold coated glass in the building industry: The design of interference systems for heat reflection, *Gold Bulletin*, **7** (1974), 62–68.

96. Schröder, H., Coating of large surface areas of glass for modification of radiation transmissivity, *Glastech. Ber.* **39** (1966), 156–163.
97. Kard, P. G., Analytical theory of optical properties of multilayer dielectric coatings, *Opt. Spektrosk.* **2** (1957), 236–244 (*C.A.* **51** (1957), 9323), cf. *Dokl. Akad. Nauk. S.S.S.R.* **108** (1956), 60–63.
98. "Infrastop-Sonnenschutzgläser". Flachglas A.G. Delog-Detag.
99. Liao, S. Y., Light transmittance and RF shielding effectiveness of a gold film on a glass substrate, *I.E.E.E. Trans.* **E.M.C.–17** (1975), 211–216.
100. Gillham, E. J., Preston, J. S. and Williams, B. E., A study of transparent, highly conducting gold films, *Philos. Mag.* **46** (1955), 1051–1057.
101. Ennos, A. E., Highly-conducting gold films prepared by vacuum evaporation, *Br. J. Appl. Phys.* **8** (1957), 113–117.
102. Kouchi, S. and Mukaiyama, T., "Transparent Gold on Glass Conductor with a Small Temperature Coefficient of Resistance". Jap. Patent 72 13,674 (1972) (*C.A.* **77** (1972), 155837s) and Jap. Patent 72 13,675 (1972) (*C.A.* **77** (1972), 155844s).
103. Missel, L., Gold plating in the space industry, *Electroplat. Met. Finish.* **26** (1973), 17–21.
104. Pfund, A. H., Bismuth black and its applications, *Rev. Sci. Instrum.* **1** (1930), 397–399.
105. Pfund, A. H., The optical properties of metallic and crystalline powders, *J. Opt. Soc. Am.* **23** (1933), 375–378.
106. Harris, L. and Beasley, J. K., The infra-red properties of gold-smoke deposits, *J. Opt. Soc. Am.* **42** (1952), 134–140.
107. Harris, L. and Loeb, A. L., Conductance and relaxation time of electrons in gold blacks from transmission and reflection measurements in the far infra-red, *J. Opt. Soc. Am.* **43** (1953), 1114–1118.
108. Harris, L. and Cuff, K. F., Reflectance of gold black deposits and some other materials of low reflectance from 254μm to 1100μm. The scattering-unit size in gold black deposits, *J. Opt. Soc. Am.* **46** (1956), 160–163.
109. Harris, L., Transmittance and reflectance of gold black deposits in the 15–100mμ region, *J. Opt. Soc. Am.* **51** (1961), 80–82.
110. Harris, L. and Fowler, P., Absorptance of gold in the far infra-red, *J. Opt. Soc. Am.* **51** (1961), 164–167.
111. Harris, L., "The Optical Properties of Metal Blacks and Carbon Blacks". The Eppley Foundation for Research, Newport, Rhode Island, Monograph Series No. 1, December 1967.
112. Zaeschmar, G. and Nedoluha, A., Theory of the optical properties of gold blacks, *J. Opt. Soc. Am.* **62** (1972), 348–352.
113. Ando, E., Infra-red transmission through gold black deposits, *Jap. J. Appl. Phys.* **11** (7) (1972), 986–991.
114. Ito, K., Formation of metallic blacks by evaporation, *Shinku*, **16** (1973), 163–167 (*C.A.* **79** (1973), 111226d).
115. Fritsche, L., Wolf, F. and Schaber, A., Structure and formation of black, evaporated bismuth films, *Z. Naturforsch.* **16a** (1961), 31–36.
116. Mizushima, Y., Mechanism of formation of black evaporated metal films, *Z. Naturforsch.* **16a** (1961), 1260–1261.
117. Kimoto, K. *et al.*, Election microscope study on fine metal particles prepared by evaporation in argon gas at low pressure, *Jap. J. Appl. Phys.* **2** (1963), 702–713.

118. Harris, L., McGinnies, R. T. and Siegel, B. M., The preparation and optical properties of gold blacks, *J. Opt. Soc. Am.* **38** (1948), 582–588.
119. Wilkinson, P. G., The properties of evaporated gold and tungsten oxides, *J. Appl. Phys.* **22** (1951), 226–232.
120. Chandrasekhar, "Radiative Transfer". Dover Publications, New York, 1960.
121. Sintsov, V. N., Properties of gold layers, *Zh. Prikl. Spektroscopii, Akad. Nauk Belorussk. S.S.R.* **4** (1966), 503–508 (*C.A.* **65** (1966), 14650).
122. Anon., Gold alloy spark plugs: easier starting and longer life, *Gold Bulletin*, **6** (1973), 68.
123. Erb, R. A., Thelen, E. and Varker, R. D., "Use of Silver Surfaces to Promote Dropwise Condensation". U.S. Patent 3,289,754 (1956).
124. Erb, R. A., "use of Gold Surfaces to Promote Dropwise Condensation". U.S. Patent 3,289,753 (1966).
125. Erb, R. A., "Method and Apparatus for Dropwise Condensation". U.S. Patent 3,298,427 (1967).
126. Erb, R. A., Dropwise condensation on gold: improving heat transfer in steam condensers, *Gold Bulletin*, **6** (1973), 2–6.
127. Bernett, M. K. and Zisman, W. A., Confirmation of spontaneous spreading of water on pure gold, *J. Phys. Chem.* **74** (1970), 2309–2317.
128. Wilkins, D. G., Bramley, L. A. and Read, S. M., Dropwise and filmwise condensation of water vapour on gold, *Am. Inst. Chem. Eng. J.* **19** (1973), 119–123.
129. Anon., Thermal fuses for the protection of electric furnaces: high temperature properties of gold–palladium alloys, *Gold Bulletin*, **8** (1975), 112–113.
130. A.S.M. Committee on Precious Metals, American Society of Metals. "Metals Handbook", 8th Edition, 1961, p. 1176.
131. Nippon Denso Company Limited, "High Temperature Fuse Device". U.S. Patent 3,906,423 (1975).
132. Philpott, J. E., Gold bursting discs for the protection of chemical plant, *Gold Bulletin*, **7** (1974), 97–99.
133. Walls, W. F., Gold capsules for high pressure experiments, *Gold Bulletin*, **6** (1973), 12–14.
134. Jones, G. R., Gold containers for single crystal growth: production of lead germanate for infra-red detectors, *Gold Bulletin*, **7** (1974), 69–71.
135. Bechtold, G., Michael, I. and Pruemmer, R., Zur Gold–Innenbeschichtung von Autoklaven durch Explosivplattieren, *Metall.* **29** (1975), 685–687.
136. Sieverts, A. and Krumbhaar, W., Über die Löslichkeit von Gasen in Metallen und Legierungen, *Ber. deutsch. chem. Gesell.* **43** (1910), 893–900.
137. Toole, F. J. and Johnson, F. M. G., The solubility of oxygen in gold and in certain silver–gold alloys, *J. Phys. Chem.* **47** (1933), 331–346.
138. Brandes, E. A., Gold coatings to protect chromium alloys against nitridation, *Gold Bulletin*, **8** (1975), 41–47.
139. Fulmer Research Institute, U.K. Patent Application 30,555/4, 4th July, 19745.
140. Tardif, H. P. and Marquis, H., Protection of steel from hydrogen by surface coatings, *Can. Metall. Q.* **1** (2) (1962), 153–171.
141. Rhines, F., "Gold as a diffusion stop". Proceedings of the First Annual Symposium, the Committee for Research on the Properties and Uses of Gold Inc., held at Sel-Rex Corporation, Nutley, N.J., March 1965.
142. Kitamura, M., "Hydrogen Permeable Fuel Cell Electrode". U.S. Patent 3,536,533 (1970) (*C.A* **74** (1971), 27516n).

143. Shirogami, T., Oxygen-hydrogen fuel cells. VIII. Hydrogen permeable electrode of Pd–Ag–Au alloy, *C.A.* **78** (1973), 51574h.
144. Grevstad, P. E., "Development of Advanced Fuel Cell System". Report 1972, N.A.S.A.–C.R.–121,136 (222 pp.) (*C.A.* **80** (1974), 55207e).
145. Anon., Gold metallisation of beryllia: heat dissipation from electronic components, *Gold Bulletin*, **4** (1971), 69.
146. Greenwood, J. C., "Semiconductor Device—Mounting Arrangement".British Patent 1,289,932 (1972) (*C.A.* **77** (1972), 145430q).
147. Hudson, P. R. W., The thermal resistivity of diamond heat-sink bond materials, *J. Phys. D.* **9** (1976), 225–232.
148. Kehrer, H. P., "Protection Against Whisker Formation of Tin Layers". German Offen. 1,903,907 (1970) (*C.A.* **73** (1970), 90704s).
149. Girling, D. S., Gold plating in submarine telephone cable repeaters: reliable capacitors for extended service, *Gold Bulletin*, **6** (1973), 69–71.
150. Lawton, S. L. and Kokotailo, G. T., The crystal and molecular structures of zinc and cadmium 0,0-Diisopropylphosphorodithioates, *Inorg. Chem.* **8** (1969), 2410–2421.
151. Dickert, J. J. and Rowe, C. N., Deposition of gold on rubbing surfaces, *Nature (London)*, Phys. Sci. **231** (1971), 87–88.
152. Dickert, J. J. and Rowe, C. N., "Gold Phosphorodithioate Salts as Anti-wear Agents in Lubricating Compositions". U.S. Patent 3 554 908 (1971).
153. Flom, D. G. and Haltner, A. J., "Lubricants for the Space Environment". General Electric Company, R 68 SD 6, March (1968) (Presented to the Materials Conference of the Am. Inst. Chem. Eng., March 1968).
154. Abbott, W. H. and Bartlett, E. S., Research in metallurgical characteristics and performance of materials used in sliding electrical contacts, *N. A. S. A. C. R.*–1447, October 1969.
155. Lewis, P., "Evaluation of Dry Film Lubricated Ball Bearings for use in a Spatial Environment". Rep. 61 GL 48, General Electric Company, 1961.
156. Evans, H. E. and Flatley, T. W., "Bearings for Vacuum Operation—Retainer Material and Design". N.A.S.A. Tech. Note D 1339, 1962.

7
The Bonding of Gold and Gold Alloys to One Another, to Other Metals, to Glass and to Ceramics

1. Introduction

Irrespective of the method by which gold and gold alloy coatings are applied in decorative, electronics and radiation-control applications, bonding of the coating to the substrate, whether of metal, glass or ceramic, is of fundamental importance. Moreover in electronic circuit construction, gold films and fine gold wires are bonded to one another, and also to other metal films and wires, especially films and wires of aluminium. Some consideration of the bonding of gold and gold alloys to one another, to other metals and to glass and to ceramics is therefore appropriate, if not essential, as an introduction to discussion of gold and gold alloy coatings.

2. Solid Phase Bonding of Gold to Gold and to Other Metals and Alloys

2.1 Background

The solid-phase welding of metals has been reviewed by Tylecote,[1] who refers not only to examples of the pressure or impact welding of gold in gold articles of very early manufacture,[2] but also discusses the many theoretical considerations which are involved in attempts to achieve an understanding of the mechanisms of bonding processes of this type. These include the attractive forces involved in the bonding process, the close contact and deformation required to make these forces operative, and the adverse effects of surface films and lubricants. They include also the diffusion across the interface which is called into play, and the effects of pressure, deformation, temperature, time and other factors on the role played by such diffusion in the bonding.

As will be seen, solid-phase bonding of gold is effected in practice by a variety of techniques, in which the conditions are selected to suit the circumstances. At one extreme, cold welding calls for high deformations. It involves little diffusion of matter across the interface, and makes use of the interatomic forces of the metals themselves. In accord with this the interface after bonding is often unbroken by intergranular growth and appears as one continuous boundary, or as a line of grain boundaries. High pressures are needed for bonding under these conditions in order to break up oxide films by lateral tension and to extrude clean metal through the resulting cracks. At the high pressures, metallic bonding forces between the two clean metal surfaces are called into play. At the other extreme, cases may arise where high pressures or deformations are undesirable, and only low pressures in combination with longer times and/or increased temperatures can be applied. In such cases, bonding (diffusion bonding) occurs primarily as a result of diffusion. In between these extremes, there are a number of processes, such as hot welding, ultrasonic welding, friction welding, high energy-rate or explosive welding, in which various methods are employed for breaking up surface films or for increasing atomic mobilities and so accelerating bonding, and in which the relative importance of the deformation and diffusion mechanisms depends on circumstances.

Gold undoubtedly owes much of the ease with which it can be bonded in the solid phase by processes such as these to the fact that obstructive oxide films do not develop on its surface.[3] Thus although Finch *et al.*[4] concluded that oxygen is absorbed by gold in such a manner that its desorption is difficult, such adsorption apparently does not create obstacles to solid phase bonding. Scratch brushing or annealing, although they may be necessary pretreatments in some circumstances, are by no means essential in order to prepare a gold surface for bonding.

The ductility and ease of deformation of gold must also contribute to its ease of bonding, since this implies that such surface films as may be present on a gold surface are much more easily disrupted and contact between clean gold surfaces more easily established.

2.2 Diffusion bonding of gold

Diffusion bonding, that is, bonding in which diffusion plays a major role, is effected without appreciable pressure and therefore without significant plastic deformation, and at temperatures below the melting points of the metals being joined.

The fact that gold surfaces do not develop obstructive oxide films and can be cleaned easily by annealing and other treatments means that gold is particularly susceptible to diffusion bonding. In fact early demonstrations of the occurrence of diffusion and bonding between metal surfaces in prolonged contact with one another at low temperatures and pressures were carried out[5–7] using gold or a gold alloy as one of the metal partners, and Barham and Jones[8] studied the behaviour of gold surfaces in contact with one another. In the latter studies it was found that if the test material after bonding at 100°C was heated to 800°C, complete recrystallisation of the gold across the interfacial zone occurred. Self-diffusion in gold has been studied and reviewed by Duhl *et al.*[9]

An important feature of diffusion bonding is that it may be assisted or "activated" by the presence of a second metal placed between the two surfaces to be joined. This results from the fact that the coefficient of diffusion of one metal in another is usually far greater than its coefficient of self-diffusion.[10]

As regards gold in diffusion bonding, therefore, two types of process are to be distinguished. In the first of these gold may be one or both of the metals being bonded, and in the other, gold may be applied in a thin film as an activator or promotor of the bonding of other metals or alloys.

The most important application of diffusion bonding of gold without use of a promoting metal is the microwelding of gold and other metal lead wires to gold or other metal surfaces in microelectronics. In this[11] the lead wire is positioned on the metal surface, and a split electrode pressed down upon it so as to provide electrical contacts. Direct current is then pulsed between the electrodes through the lead wire and the surface metal, so as to produce heating and to promote diffusion of the interface. Strict control of the current pulses (and of the spacing between the electrodes) ensures that no excessive thermal shock is imposed on the interface between the surface metal and its substrate, which might result in damage arising from differences in their thermal expansion coefficients. Such control also ensures that melting does not occur. Since the forces applied by the electrodes are limited to about 200–700 grams weight bonding occurs predominantly by diffusion.

The work of Ornellas and Catalano[12] illustrates the bonding of gold using a "promoting metal" at the interface. These authors explored the use of mercury, indium and gallium as interfacial diffusing species to promote bonding of gold to gold. They found that mercury was the most effective, gallium and indium giving poorer bonds, an effect which they considered as due either to oxidation or the formation of brittle intermetallic phases at the interface.

The use of gold as a promoter of bonding between other metals is illustrated by the known technique of applying gold or gold alloy coatings to the surfaces of other metals in what is called gold-diffusion joining. In such processes, the merit of gold lies partly in its resistance to oxidation, and partly in the ease with which it can be applied — either as electroplate or as foil — to the surfaces to be joined. It is not effective, however, in combination with metals such as lead and aluminium, with which it forms brittle intermetallic compounds. As an example of this type of process, gold diffusion seals between copper members can be effected at 450°C by placing a washer or wire ring of gold or Au-Cu alloy between the flat surfaces of the members and applying pressure during the heat treatment.[13]

It should be stressed that the depths to which the promoting metal diffuses into the partner metals in establishing diffusion bonds may be very small. For example, electron microprobe analyses across an Au-Au bond established using mercury as diffusing species[12] indicated that the mercury had penetrated no deeper than 2.5×10^{-5}m into the gold surfaces.

Attention must also be drawn to the fact that in some instances it becomes difficult to distinguish between this type of bonding and normal brazing, since if temperatures are high enough, diffusion may well lead to the formation of interfacial phases which are molten under the conditions of bonding. It is also necessary to remember that diffusion, although not the dominant bonding mechanism in other bonding processes, is nevertheless always a factor to be taken into account. Its role in soldered joints at temperatures between 23°C and 212°C, for example, has been studied by Creydt and Fichter[14] and by Heinzel,[15] and Wirsing[16] has attributed the increase in the as-bonded strength of many joints which results from ageing treatments to diffusion processes.

2.3 Cold bonding of gold

Because of the very positive effects of increased temperature upon the pressure welding of gold, most published information concerning the pressure welding of gold relates to hot bonding. In fact, very little[16a] has been reported on the cold pressure bonding of gold to gold outside the dental literature. This arises from extensive use in earlier years of gold foil for the filling of teeth. In practice the hard rolled foil was simply packed into the cavity, either with or without prior heating over a spirit flame. The flame treatment could be regarded as an annealing and surface cleaning operation, which undoubtedly promoted pressure welding of the foil in situ.[17] Sintered pellets of powdered or mat gold have been used in place of gold foil.[18]

The pressures and times required for cold bonding of gold to gold depend upon the pre-treatment of the gold. In butt-welding tests, for example, Hofmann and Schüller[19] found that scratch brushing enabled cold bonding to occur with lower deformation than did annealing, an effect which they attributed to the higher energy state of the scratch-brushed surface. On the other hand, Tylecote[20] found that hard-rolled sheet failed to weld at low deformations, but that if gold sheet was degreased and annealed at 900°C for thirty minutes, then bonding occurred at welding pressures of 28–57 tons/in^2 (432–880 MN/m^2) corresponding to deformations of 24–48%. Tylecote concluded that in the case of gold to gold bonding, absorbed films of moisture or gas have detrimental effects on welding, and that these can be broken up on degreased and scratch-brushed specimens only by deformations exceeding 20% or by heat treatment.

Little information has apparently been recorded concerning cold bonding of gold or gold alloys to other metals, though some work on Au-Pt alloys has been recorded.[17]

An interesting observation has been recorded by Wise[21] which illustrates the ease with which clean gold surfaces weld to one another. If the pressure is allowed to rise (1–3 mm of He or N_2) during vacuum evaporation of gold then as described earlier the gold condenses not as a bright film, but as "gold black", the absorptive properties of which in the infrared[22] have prompted its use on receivers of infrared bolometers. If such gold black is deposited on a bright gold film condensed immediately beforehand at low pressures, then mere running of the finger over its surface causes the gold black to bond to the bright substrate gold.

2.4 Hot bonding and ultrasonic bonding

Basic features[23] *and industrial techniques*[24]

Because the pressures and times required to break up surface films and promote bonding of gold at ambient temperatures are high, it is normal to assist bonding either by heating the joint area (thermocompressive welding)[25–27] or by supplying mechanical energy to it by ultrasonic means (ultrasonic welding).

Ultrasonic bonding is particularly successful where either one or both the metals concerned are coated with oxide layers. Such layers, and other surface films, are disrupted by the vibratory displacements occurring during the bonding operation, so that surface cleaning of the surfaces to be joined is not critical. Moreover it has the advantage that no preheating is required, so that possible damage to or contamination of parts being assembled from this source is eliminated. This does not mean that that temperature in the joint area is not raised, since investigations have shown that elastic hysteresis in the stressed portion of the weld zone during ultrasonic bonding causes considerable temperature rises there. These rises are transient and highly localised, however, for normal welding times, and where dissimilar metals are being joined, this implies that the likelihood of equilibrium conditions being established and intermetallic compounds formed is reduced. A significant feature of the ultrasonic process is that the high-frequency mechanical vibrations, apart from disrupting surface films, also exercise a plasticizing effect on the metal crystal structures. This is independent of any thermally-induced plasticization, and plays a significant role in promoting bonding. Another significant feature is that there is an optimum clamping force for each combination of metals, at which bonding is achieved with minimum vibrational energy. Control of ultrasonic bonding is therefore effected by varying three parameters, namely the clamping force, the power measured in terms of the displacement of the bonding tip, and the bonding time.

In thermocompressive bonding the parameters which are susceptible to variation are also essentially three in number, namely, force, temperature and time, and the most effective combination of these for any two bonding surfaces must be determined by experience for each geometry which is involved. For gold-gold thermocompression bonding Ahmed and Svitak[23] have reviewed and reported upon the fundamental mechanical parameters which influenced bond formation. Using slip line techniques from plasticity theory, the normal stresses, shear stresses, displacements and strains were determined at the interface for the case of the bonding of a gold ribbon lead to a gold metallised substrate. The predicted values of these parameters were then compared with experimental observations of them. These observations indicated that there is a readily predicted central dead zone where bonding is difficult, that the magnitude of the normal stress above that required for contact is unimportant, that mutual extension of the two bonding surfaces in contact is sufficient for bond formation after some time at any one temperature, that the required surface extension is reduced considerably by simultaneous application of an interfacial shear stress, and that neither sliding nor high interfacial shear stresses result in bonding unless both surfaces are deformed.

Extensive experience of both thermocompressive and ultrasonic bonding has been built up in the electronics industry, where bonding of fine gold and/or aluminium lead wires to conductive metals and alloys such as aluminium, gold, silver, nickel, copper, tantalum, Chromel and Nichrome may be required. These applications and the techniques used for them (e.g. ball or nail head, wedge, stitch or scissors and bird-beak thermocompressive bonding for wires, and flip-chip bonding in integrated circuit production) are well described by McCormick.[11] Ultrasonic methods are also used extensively for bonding gold-plated copper wires to gold-plated copper or circuit boards. For such operations, the quality of the gold plating both of the wires and the boards is important.[28]

Bonding of gold wires to aluminium surfaces, and of aluminium wires to gold surfaces by thermocompressive and ultrasonic bonding techniques is of particular importance, and metallurgical problems which arise in this connection are discussed later.

2.5 Friction bonding and explosive bonding

Although no reference has been noted to the deliberate use of friction for the bonding of gold to itself or other metals, such bonding is a well recognised phenomenon in certain situations where gold surfaces are required to slide over one another or over other metal surfaces. Since the sliding properties of gold and of gold alloy deposits are critical to the performance of much electrical equipment such as connectors, instrument slip rings, and switches, the wear and friction of gold in such devices has been extensively studied.

In brief, the position is that sliding damage to gold occurs primarily either by adhesive (bonding) or abrasive mechanisms, though fracture may become a significant source of damage to gold coatings in some situations. The adhesive wear process has been studied most closely in relation to wear of sliding gold contacts where it is the main wear process. It is characterised, for example, in the case of a flat gold surface pressed against the gold surface of a rotating turntable by the formation at the outset of a prow which comes from the larger partner and forms between the two surfaces. It is composed initially of particles formed by the shearing-off of asperities on the opposing surfaces, many of which may cold weld to one another before shear occurs. As a result the initial prow material is composed of work-hardened metal particles, cold-bonded to one another in considerable degree. It appears to build up further by cold bonding to projections on the softer opposing surface and by shearing of the bonded material preferentially on the weaker side of the bond.

The whole process is well documented[29–31] not only for electrodeposited gold, but also for electrodeposited gold alloys.[32,33]

Only one application of explosive bonding of gold has been noted. It has been used for gold coating of the interior surfaces of autoclaves.

3. Liquid Phase Bonding of Gold to Gold and to Other Metals and Alloys

Under this heading there will be discussed joints in which bonding is effected either by the local melting of one or both of the joint components, or by the application of specially added metals or alloys of lower melting points, that is by soldering or brazing.

3.1 Bonding with melting of one or both the joint components

This type of bonding is applied[11] extensively in the electronics field, though applications involving metals other than gold predominate. Techniques vary according as to whether the metal is fully melted (fusion welding) or merely "plasticised" (forged welding). The latter is in effect a variant on thermocompression bonding, and is the more commonly used technique in electronic bonding. Techniques are usually differentiated according to the methods employed for supplying energy to the bond area.

In resistance bonding the electrical resistance at the interface area is used to generate the necessary heat. It can employ electrodes on opposite sides of the joint (opposed electrode welding, pincer welding, series step welding) or electrodes on the same side of the joint (parallel gap welding, series welding).[34] In percussive arc welding, the two joint components are brought together either during or immediately after a period of electric arcing between them, which melts their surfaces. Laser[35] and electron beams have also been employed in certain instances, but their use is restricted by difficulties which often arise in focusing of the beams.

In so far as gold may be one of the metals involved in processes of these types, the problems which arise are more practical than fundamental. Thus in the bonding of gold-plated Kovar (an Fe-Co-Ni alloy of low thermal expansion) wire lead-outs to the copper circuit tracks of printed circuits, the problems[36] arise not from the presence of the gold plating, but rather from the differences in conductivity between the Kovar and the copper and the difficulty in arranging a welding schedule and technique which will ensure that the gold melts preferentially and without excessive alloying with the copper. Similarly in the bonding of silicon chips to gold-plated Kovar headers, if heating is excessive or prolonged then the gold is completely converted to the Au-Si eutectic and there is danger of cracks developing between this eutectic and the silicon chip. This serves to emphasise the general rule that the properties of bonds formed with melting depend upon the nature of the phases which are developed in the interface area.

3.2 Bonding of gold with soft solders

The individual components of electronic circuits may often be held in store for appreciable periods before they are incorporated into final assemblies. Where this last operation involves the making of soldered connections, it is essential that the solderability of the points which have to be bonded together remains unimpaired. For this reason they may be gold-plated. In practice, therefore, the soft soldering properties of electroplated golds are of great importance. The solders used are Sn-Pb solders of approximately eutectic composition (63Sn-37Pb) and basically two processes are employed for making joints, dip-soldering and re-flow soldering.[11] In dip-soldering, heat and solder are applied simultaneously to the pre-fluxed metal surfaces. In re-flow soldering, joining is effected in two stages. In the first the joint surfaces of the metals to be bonded are coated with solder, for example, by dipping or by passing the fluxed work through a standing wave of clean liquid solder created by pumping (wave soldering). Other processes such as solder-slinging or roll-coating or electroplating may also be used.[11] In the second stage, bonding is effected between the two surfaces by the application of heat, for example, by resistance soldering by the parallel-gap technique.

Where one of the surfaces to be bonded is easily solderable, as may be the case if it is gold plated, it may not be solder coated.

The solderability of gold plating was reviewed by Thwaites in 1973[37] and will be discussed only briefly here. Because of intermittent problems of low mechanical strength and reliability encountered with soldered joints to gold plated surfaces, such joints were studied in considerable detail in the 1960's.

Although pure gold coatings were found to provide excellent surfaces for soldering, the solderability of the hard and bright acid golds used in some applications was found to be affected by a number of factors. One of these was the presence in such plated golds of alloying metals and of carbonaceous materials. These affect the wettability of the plate by the solder and therefore the strength of the joint. Another is the fact that certain plated golds in contact with solder release gas which imparts porosity and weakness to the joint. The actual source of the gas has not been established, but it may be either hydrogen codeposited with the gold or a product of the action of heat on codeposited carbonaceous material.

Apart from the above, metallurgical problems were also identified, particularly where the volume of solder applied was small and the gold plating was relatively thick ($> 1.3\ \mu m$), so that more than about 4% of gold was dissolved in the solder in the joint. The phases formed when gold is added in increasing percentage to a 60%–40%Pb solder are:

%Au in solder	Solid phases at ambient temperatures
0–20%	$Pb + Sn + AuSn_4$
20–34%	$Pb + AuSn_2 + AuSn_4$
34–50%	$Pb + AuSn + AuSn_2$

All these phases may therefore be expected to occur in a soldered joint, depending upon the local concentrations of gold which develop within it.

In practice, the rates of gold dissolution in molten Sn-Pb solder are exceedingly high[14,15] and this leads to rapid formation of gold-tin intermetallics, firstly as a layer on the gold surface, and quickly thereafter as isolated crystals in the bulk of the solder. Although it has been suggested that the intermetallic compound layer formed initially may be AuSn, together with $AuSn_2$, by far the most striking feature in all joints which contain sufficient gold is the occurrence in the bulk of the solder of acicular crystals of $AuSn_4$ in which some of the tin may be replaced by lead. It would appear as if the rate at which the gold diffuses into the solder is so great that intermetallic phases richer in gold than $AuSn_4$ do not survive the soldering process.

The effects of these reactions on the properties of soldered joints appear to arise in two ways. On the one hand, separation of the intermetallics from the gold surface appears to take place on cooling, probably as a result of differential rates of contraction. On the other hand, the presence of these intermetallics in solder has been shown, at gold concentrations exceeding 4%, to have a deleterious effect on the impact strengths of the solder.

In general, therefore, it would appear beneficial to keep gold-plating thicknesses and therefore gold concentrations to a minimum in all soldered joints. There are limits to the extent to which the thickness of gold platings can be reduced however, since, firstly thinner coatings may develop tarnish films before they are soldered, and, secondly, deleterious effects may result from exposure of the substrate to the solder. In the latter case protection of the substrate by an undercoat of Sn-Ni has been suggested.

Alternatively, removal of gold locally from the areas at which soldered joints have to be made can be considered. Such removal can be effected by the use of two solder baths, one for dissolution of the gold and the other for the soldering operation itself. Where a soldering iron is employed, the initially applied solder can be "wicked" away and replaced by fresh solder.

Attempts have been made to overcome the difficulties in other ways. Thus Lee[38] has shown that the rate of dissolution of an electroplated 85%Au-15%Ni alloy in molten Sn-Pb solder is not fast enough to give rise to brittle Au-Sn intermetallics, and he has reported constant shear strengths for solder joints to such platings, even when their thicknesses were as great as 5 microns. Formation of Sn-Ni intermetallic at the surface appears to inhibit diffusion of gold into the solder. An additional advantage of such platings was the fact that their solderability was not significantly affected by ageing at 110°C over a four-week period. Apparently diffusion of substrate copper through the gold was reduced by the presence of nickel in the gold plating.[38] Electroplating (or cladding) with an 85Au-15Ni alloy can therefore offer advantages for situations where reliable soldered joints are required using Sn-Pb solders.

Modifications of solder composition can also provide a means for avoiding difficulties. Thus Braun[39] studied the reaction rates with gold wire of a number of Sn-Pb alloys to which additions of other metals such as indium, zinc, antimony, bismuth, gold and thallium had been made. He found that a 53Sn-29Pb-17In-0.5Zn alloy showed much reduced reaction rates, and that its melting range (122–156°C) compared very favourably with that (183°C) for normal Sn-Pb solder. Indium-containing solders are expensive, however, and apparently give lower joint strengths than can be obtained with Sn-Pb solders.[40,41]

This approach has been extended by Heinzel[42] and by Heinzel and Saeger,[43] who have studied the rates of dissolution not only of gold wire, but also of a variety of plated golds in tin, in 60Sn-40Pb, in 93Sn-7Au, in 49Sn-33Pb-18Cd and in the Braun alloy referred to above. They found that whereas rapid dissolution of gold wire by the 60Sn-40Pb alloy began at 240°C, it set in only above 260°C with the two latter alloys. The same trends were observed in the experiments with electroplated golds, but there were differences in behaviour between different types of electroplated gold. A 67Au-33Cu electroplate showed resistance to dissolution by all solder compositions. Although electroplates of 99.99Au, 99.6Au-0.4Co, 99.8Au-0.2Co, and 99.8Au-0.2Ni were significantly more resistant to dissolution by the latter two alloys than by normal solder, the gold-nickel electroplate was the most resistant. Heinzel and Saeger have also reported on the wetting of pure and plated gold wires. Pure gold was wetted best in all instances, and there was little difference between the wettability of wires coated with different types of electroplated gold. The Sn-Pb-In-Zn alloy was found, however, to have good wetting powers over the temperature range where it dissolved gold slowly, whereas the Sn-Pb-Cd alloy began to wet gold surfaces rapidly only at temperatures about 80°C above its melting point. Other alloys have also been examined, including both a 90Sn-10Cd alloy[44] and a 49Sn-33Pb-18Cd alloy.[43]

It may be noted that the high rate of solution of gold into Sn-Pb solder alloy makes it especially difficult to solder connections to thick gold conductor films printed on ceramic substrates. Soldering conditions must be controlled carefully if, for example, connections are to be soldered between Kovar lead frames and thick printed and fired gold pads, since residual undissolved gold is necessary for effective bonding. Other solders such as Au-Sn solders or Sn-Pb solders modified by addition of indium as described above give better results in such cases.

4. Bonding of Gold to Glass

4.1 Areas of application

The ability of gold to bond to glass has been long known and exploited in mosaic art, in which the gilding of the glass tesserae was achieved earlier by cold application of fine gold leaf to the glass, which was covered with powdered glass and fired.[45] In present-day practice the gold leaf (24 carat) is applied in squares (10 cm × 10 cm) to a circular damp glass surface (1 mm thick) over which molten glass (200 g) of the required colour is poured. The material is then fired at 1200°C, annealed, and cut into strips from which the artists cuts tesserae of the size required.

Similar bonding of gold to glass has been exploited in the assembly of vacuum tubes, the leads to which are normally of Ni-Cr-Fe or Fe-Co-Ni alloys, the thermal expansion characteristics of which match those of glass. These alloys have electrical resistivities which are great enough to cause local heating of glass-to-metal seals using these alloys. This can be overcome and the current carrying capacity of Kovar in glass seals increased substantially by gold coating of the Kovar. The coating must be thick (25–30 microns) because of the high rate of diffusion of gold into the Kovar at the sealing temperatures.[46–49]

Greatest importance attaches, however, to the bonding of gold to glass in the production of reflector and absorber coatings and in the frit bonding of gold which is involved in the use of thick film conductor pastes for the printing of conducting circuits.

4.2 Mechanism of bonding

Bonding between glass and a metal is generally considered a result of the presence of an oxide layer that is formed on the metal. In the bonding process[50] this oxide diffuses into and reacts with the molten glass and so establishes a link between the metal and the glass. The thickness and character of the oxide layer is therefore of significance and must be controlled carefully in bonding most base metals to glass. These processes are reflected[51] in such interfacial characteristics as contact angles, wettability, and interfacial and surface energies, the stronger bonds occurring with larger reductions in interfacial energy. This mechanism is readily understandable in the case of base metals, most of which carry oxide films of significant thickness on their surfaces. It is less obviously acceptable in the case of the noble metals. The thickness of oxide film which are necessary, however, are clearly very small. Gomer has pointed out that the removal of oxygen adsorbed on platinum requires a vacuum of 10^{-9} torr and a temperature approaching the melting point. In accord with this Volpe *et al.*[52] found that the contact angle between platinum and sodium disilicate glass remained constant to temperatures up to 1000°C at oxygen pressures of 10^{-3} torr. Moreover, Cherniak and Naidus[53] found that there was a direct correlation between resistance to wetting and resistance to oxidation in a series of platinum metal alloys.

The bonding of gold spheres to plane glass surfaces at different sintering temperatures has been studied by Polke[54] by following the increase in the contact area between the spheres and the glass as bonding developed. The work of de Bruin, Moodie and Warble,[55] is also of considerable interest in this connection. These investigators have described bonding of ceramic oxides, quartz and glass surfaces, separated by foils of metals such as platinum, palladium, gold and nickel, titanium, iron and copper, when they were pressed lightly together (100 kNm^{-2}) and heated to temperatures which were below the melting point of the lowest melting component.

Examination of the bonds revealed that where noble metal foils were used to promote bonding, interfacial reaction was localised at the interface, with the metals maintaining a sharp discontinuity at the interface. On the other hand, where base metal foils were used, the interfacial reaction was not localised and a "diffusional" type of interface was observed.

While these authors do not attempt to ascribe mechanisms to these bonding reactions, it does not seem inconsistent to assume on the basis of their data that the reactions are both reaction bonding in type, and that the sharp discontinuity at the interface when noble metal foils are used is a result of the limited extent to which oxide films are developed on these metals. It is perhaps significant in this respect that the gold foil which they used was not of the highest purity, in that it contained only 99.3% gold with copper, silver and arsenic impurities.

Klomp,[57] however, has reported on the bonding of platinum (and such metals as iron, copper, aluminium and lead) to alumina ceramics when heated in hydrogen atmospheres. Strong bonds were obtained with platinum at 1500°C in hydrogen, using pressures as low as 0.2 kg/cm.[2] Klomp postulated that the initial bond was formed by adsorption of evaporated metal on the ceramic surface, with decrease in interfacial energy. The adsorbed layer was visualised as being built up and bonding established through diffusion of metal atoms. In support of this mechanism he found that metals having a vapour pressure below 10^{-11} torr at their melting temperature showed no tendency to bond with alumina. It seems possible, therefore, that more than one mechanism may contribute to the formation of bonds between metals and glass or ceramics, depending on whether the metal is reactive or non-reactive, and that "non-reactive" bonding of the type postulated by Klomp may operate in the case of the noble metals.

Knowledge and experience of the bonding of gold to glass has also been built up in other areas, such as the vacuum deposition of gold on glass and the hermetic sealing of electronic devices. Only the former will be discussed here.

4.3 Gold film deposited on glass by vacuum processes

Whether it is evaporated or sputtered, gold exhibits only "marginal" adhesion to glass surfaces, and various procedures have been described for improving its adhesion, and modifying its properties. One of the best known is the method of Colbert *et al.*[58] in which the glass is first coated with aluminium in vacuum, and the aluminium film oxidised by an oxygen glow discharge. If gold is now deposited on this "activated" surface, it adheres far more tenaciously to the surface than it does to untreated glass. Moreover, the gold film is relatively resistant to abrasion. Pretreatment of glass surfaces by such metals as chromium,[59] titanium, molybdenum, antimony, bismuth, indium, lead or rhodium is also effective in promoting bonding.

The fact that a trace of rhodium is essential to the production of bright coatings by pyrolysis of liquid golds is significant. During such pyrolysis the thin gold films which are formed apparently contain[60] an extremely homogeneous dispersion of rhodium oxide. The effect of this oxide in preventing agglomeration of the deposited gold on ceramic or glass substrates is almost certainly in part a reflection of its activity as a bonding agent between the gold and those substrates.

Bismuth oxide has also found special application as a bonding agent not only in liquid gold formulations, but also in the production of conductive gold films on glass. If in the latter process the glass is precoated by evaporation with Bi_2O_3, then thin (0.3–0.6 μm) gold films which are subsequently applied, whether by sputtering[61] or by evaporation,[62] are so even and uniform that they can be heated electrically, and the coated glass therefore used in windscreens to control icing and misting conditions.[63]

Studies of evaporated bismuth oxide-gold films on glass have shown[61] that the oxide is condensed on the glass as a polycrystalline layer which is not modified by heat treatment. The gold condenses on the oxide also in polycrystalline and non-orientated form and the effect of heat treatment is to reduce the lattice misfits and to increase the gold crystal dimension in the plane of the film from about 2 μm to about 3 μm. On NaCl substrates, thin (0.3 μm) evaporated gold films have been observed to undergo a similar grain coarsening simply on ageing, with many grains containing bubbles of insoluble gas.[64] There is evidence that sputtered gold films also contain occluded gas and that this is lost in heat treatment. Where gold in alloyed form is applied to glass surfaces, as in the production of many infrared reflectant glasses, the base metals used in the alloy may play a significant role in the bonding of the coating to the glass.

5. Bonding of Gold to Ceramic Materials

5.1 Bonding mechanisms and techniques

As has been described in Chapter IV when discussing the bonding of dental porcelains to dental casting alloys, more than one factor may be involved in the bonding of metals to ceramic materials. Although mechanical interlocking may contribute to bond strength, the main bonding occurs as a result of interfacial diffusion. This leads to reaction bonding between the ceramic phase and oxides associated with the metal phase, so that the mechanism may be little different in character from that which has been discussed in relation to gold–glass bonding above. Alternatively (or additionally), conditions may be so designed, that glassy phases are formed at the interface which permeate the interfacial area and act as reactive cements.[65]

In the bonding of gold to ceramics in industry the processes which are used include the following:

(a) Liquid gold processes. In these the gold is generated by pyrolysis of organogold compounds applied to the ceramic surface. Organo-base metal compounds which generate oxides on pyrolysis are embodied in the liquid gold solution as applied, and as a result reaction bonding plays a major role in bonding the gold to the ceramic.

(b) Thick film processes using glass frit. In these, gold powder is mixed into a paste with glass powder in an organic medium and applied to the ceramic surface. On firing the gold is bonded to the ceramic surface by the cementing action of the glass.

(c) Thick film processes (glass-free). In these gold powder is mixed into a paste with an organic binder along with small amounts of base metals or base metal oxides which will promote reaction bonding of the gold to a ceramic surface when it is applied to it and fired.

(d) Thick film processes based on pre-metallisation of the ceramic surface. The classical example here is the well established molybdenum-manganese ("molymanganese") process; the ceramic substrate is normally alumina, and the gold is not bonded directly to it, but to a film of Mo-Mn previously established upon it. The merits of the process lie in the strength of the bonding between the Mo-Mn and the ceramic, the conductivity of the film and the fact that it can be used as a base material on to which leads may be brazed and lids attached in the production of integrated microcircuits.

The Mo-Mn metallisation of the alumina surface is effected by treating it with a blended mixture of molybdenum and manganese powders, inorganic additives and an organic bonder, and then firing in a hydrogen atmosphere. Gold may then be deposited on the metallised surface either by electroplating or by electroless methods.

(e) Thin film processes based on evaporation or sputtering of gold. The principles involved here are the same as those involved in the application of vacuum coatings to glass. An undercoat of aluminium, chromium, titanium or molybdenum is first applied in order to promote reaction bonding of the gold to the ceramic surface.

The use of insulating epoxy adhesives for bonding of gold plated Kovar to ceramics has recently been studied.[66]

REFERENCES

1. Tylecote, R. F., "The Solid Phase Welding of Metals". Edward Arnold Limited, London, 1968 (cf. Mohamed, H. A. and Washburn, J., Mechanism of solid state pressure welding, *Weld. Res.* (Miami), 1975, Sept., pp. 302S–310S.
2. Tylecote, R. F., "Metallurgy in Archaeology". London, 1962.
3. Anon., Oxygen coatings on gold, *Gold Bulletin*, **4** (1971), 24.
4. Finch, G. P., Quarrel, A. G. and Wilman, H., "Electron Diffraction and Surface Structure". *Trans. Faraday Soc.,* Vol. 31, 1935, pp. 1051 ff.
5. Roberts-Austen, W. C., On the diffusion of metals. Part I. Diffusion of molten metals, *Phil. Trans. Roy. Soc.*, London, **187** (1896), 383–415.
6. Roberts-Austen, W. C., On the diffusion of gold in solid lead at the ordinary temperature, *Proc. Roy. Soc.* **67** (1900), 101–105.
7. Spring, W., Über das Vorkommen gewisser für den Flüssigkeits- oder Gaszustand charakterischen Eigenschaften bei festen Metallen, *Z. Phys. Chem.* **15** (1894), 65–78.
8. Barham, B. S. and Jones, W. D., Crystal growth across interfaces, *Met. Ind., London*, **50** (1937), 181–182.
9. Duhl, D. N., Hirano, K. and Cohen, M., Diffusion of iron, cobalt and nickel in gold, *Acta. Metall.* **11** (1963), 1–6.
10. Hevesy, G. and Obrutsheva, A., Self-diffusion in solid metals, *Nature*, **115** (1925), 674–675.
11. McCormick, J. E., "Handbook of Materials and Processes for Electronics", Edited by C. A. Harper. McGraw Hill Book Company, 1970, Chapter 13.
12. Ornellas, D. L. and Catalano, E., Diffusion bonding of gold to gold, *Rev. Sci. Instrum.* **45** (1974), 955.

13. Kohl, W. H., "Handbook of Materials and Techniques for Vacuum Devices". Reinhold Publishing Corporation, New York, 1967, p. 373.
14. Creydt, M. and Fichter, R., Diffusion in galvanisch aufgebrachten Schichten und Weichloten bei Temperaturen Zwischen 23°C und 212°C, *Metall.* **25** (1971), 1124–1127.
15. Heinzel, H., Verhalten von Weichloten beim Löten von Gold, *Elektronik*, **22** (1973), 439–440.
16. Wirsing, C. E., An ultrasonic bond, monometallic, gold interconnection technique for integrated packages, *Solid State Technol.* **16** (10) (1973), 48–50.
16a. Mohamed, H. A. and Washburn, J., Mechanism of solid state pressure welding, *Weld. Res. (Miami)* (1975), 302S–310S.
17. Phillips, R. W., "Skinner's Science of Dental Materials", 7th Edition. W. B. Saunders Company, Philadelphia, 1973.
18. American Dental Association, "Guide to Dental Materials and Devices", 7th Edition. Chicago, Ill., 1974–1975.
19. Hofmann, W. and Schüller, H. J. Weitrentwicklung der Kaltpressschweissung, *Z. Metallk.* **49** (1958), 302–311.
20. Tylecote, R. F., Investigations on pressure welding, *Br. Weld. J.* **1** (1954), 117–135.
21. Wise, E. M., "Gold: Recovery, Properties and Applications". D. van Nostrand Company, Princeton, N.J., 1964, p. 211.
22. See Ref. 13, p. 235.
23. Ahmed, N. and Svitak, J. J., Characterisation of gold–gold thermocompression bonding, *Solid State Technol.* **18** (1975), 25–32.
24. Philips, L. S., Microbonding techniques, *Microelectron. Reliab.* **5** (1966), 197–201.
25. Baker, D. and Jones, R., New developments in thermocompression bonding, *Microelectron. Reliab.* **5** (1966), 229–234.
26. Ellington, T. S., Lead frame bonding, *Solid State Technol.* **16** (1973), 59–62.
27. Hayasaka, T. and Hattori, S., On the mechanism of thermocompression bonding of gold beam leads, *Rev. Electr. Commun. Lab. (Tokyo)*, **23** (1975), 344–352.
28. Missel, L., Gold plating of P.C. boards and wire: Plating techniques for ultrasonic bonding, *Met. Finish.* **73** (1975), 69–73.
29. Antler, M., in "Gold Plating Technology", Edited by F. H. Reid and W. Goldie. Electrochemical Publications, Ayr, Scotland, 1974, pp. 259–294.
30. Antler, M., The evaluation of gold contacts: Relationship between testing methods and performance in service, *Gold Bulletin*, **5** (1972), 85–89.
31. Solomon, A. J. and Antler, M., Mechanisms of wear of gold electrodeposits, *Plating*, **57** (1970), 812–816.
32. Anon., Wear resistance of electrodeposited gold: The effects of surface films and of polymer co-deposits, *Gold Bulletin*, **7** (1974), 10–14.
33. Anon., The wear of electrodeposited alloy golds, *Gold Bulletin*, **8** (1976), 6.
34. Knowlson, P. M., Fundamentals of parallel-gap joining, *Microelectron. Reliab.* **5** (1966), 203–206.
35. Smith, D. L., Burnett, A. P. and Gordon, T. E., Laser welding of gold alloys, *J. Dent. Res.* **51** (1972), 161–167.
36. Ref. 29, pp. 255–258.
37. Thwaites, C. J., Some aspects of soldering gold surfaces, *Electroplat. Met. Finish.* **26** (1973), 10–14 and 21–26.
38. Lee, W. K., Improvement of solder connections by gold alloy plating, *Plating*, **58**

(1971), 997–1001 (cf. Roggia, D., "Soldering in Thin Film Circuits", Mikroelektron., Manuscriptdruck Vortr. Mikroelektron. -Kongr., 5th, 1972) (*C.A.* **82** (1975), 6824m).
39. Braun, J. D., An improved soft solder for use with gold wire, *Trans. Am. Soc. Met. Q.* **56** (1963), 870 and **57** (1964), 568–571 (cf. U.S. Patent 3,226,226 (1965)).
40. Weil, R., Diehl, R. F. and Rinker, E. C., Solderability of some gold electrodeposits, *Plating*, **52** (1965), 1142–1148.
41. Brewer, D. H., *Weld. J. (Miami)*, **49** (1970) (10), 465S–470S.
42. Heinzel, H., "Ablegierverhalten und Benetzungsfähigkeit verschiedener Weichlote beim Löten von Gold", Festschrift "50 Jahre DODUCO", Pforzheim, 1972.
43. Heinzel, H. and Saeger, K. E., The wettability and dissolution behaviour of gold in soft soldering, *Gold Bulletin*, **9** (1976), 7–11.
44. Duckett, R. and Ackroyd, M. D., Private communication.
45. Bustacchini, G., Gold in mosaic art and technique, *Gold Bulletin*, **6** (1973), 52–56.
46. Turnbull, J. C., A high conductivity glass-to-metal seal, *R.C.A. Review*, **13** (1952), 291–299.
47. Ref. 13, p. 429.
48. Düsing, W., Einschmelzleiter mit veredelter Oberfläche für Mikrowellenröhren, *Telefunken–Ztg.* **30** (1957), 264–269.
49. Düsing, W., Herstellung und Verschmelzung von Einschmelzleitern mit veredelter Oberfläche, *Glastech. Ber.* **31** (1958), 137–142.
50. Ref. 13, p. 389.
51. Pask, J. A. and Fulrath, R. M., Fundamentals of glass-to-metal bonding. VIII. Nature of wetting and adherence, *J. Am. Ceram. Soc.* **45** (1962), 592–596.
52. Volpe, M. L., Fulrath, R. M. and Pask, J. A., Fundamentals of glass-to-metal bonding. IV. Wettability of gold and platinum by molten sodium disilicate, *J. Am. Ceram. Soc.* **42** (1959), 102–106.
53. Cherniak, M. G. and Naidus, G. G., *J. Text. Physics Acad. Sci. U.S.S.R.* **57**, 2268 ff.
54. Polke, R., Sinterhaftung von Gold und Glas, *Silic. Ind.* **37** (1972), 227–231.
55. de Bruin, H. J., Moodie, A. F. and Warble, C. E., Ceramic-metal reaction welding, *J. Mater. Sci.* **7** (1972), 909–918.
56. de Bruin, H. J., Moodie, A. F. and Warble, C. E., Reaction welding of ceramics using transition metal intermediates, *J. Austr. Ceram. Soc.* **7** (1971), 57–58.
57. Klomp, J. T., Solid state bonding of metals to ceramics, *Sci. Ceram.* **5** (1970), 501–521.
58. Colbert, W. J., Weinrich, A. R. and Morgan, W. L., "Light Transmissive Electricity Conducting Article". U.S. Patent 2,628,927 (1953).
59. Strong, J. and Wingot, J., quoted by Wise, E., Ref. 21, p. 207.
60. Langley, R. C., Gold films for the control and utilisation of radiant energy, *Gold Bulletin*, **8** (1975), 34–40.
61. Gillham, E. J., Preston, J. S. and Williams, B. E., A study of transparent, highly conducting gold films, *Philos. Mag.* **46** (1955), 1051–1065.
62. Ennos, A. E., Highly conducting gold films prepared by vacuum evaporation, *Br. J. App. Phys.* **8** (1957), 113–117.
63. Loebich, O., The optical properties of gold: A review of their technical utilisation, *Gold Bulletin*, **5** (1972), 2–10.
64. Andrew, R., Grain coarsening and gas bubbles in aged gold films, *Thin Solid Films*, **29** (1975), 53–57.

65. Clarke, J. F., Ritz, J. W. and Girard, E. H., "State of the Art Review of Ceramic to Metal Joining". Tech. Rept. AFML–TR–65–143, Air Force Materials Laboratory, Research and Development Division, Wright-Patterson Air Force Base, Ohio, U.S.A., May 1965.
66. Cadenhead, R. L., Substrate attach epoxies, *Solid State Technol.* **18** (1975), 53–63.

8 Electroplated Golds

1. Introduction

Hunt[1] has provided a delightful account of the polemics and acrimonious debate which accompanied the development of the electroplating of gold as an industrial process. This should be read in conjunction with a parallel historical review of gold plating by Weisberg.[2] In the early stages (ca. 1840–1860) developments tended to be centred on the plating of heavy deposits and upon electroforming, but with the passing of the Victorian influence the demand for such technology fell away, gold plating came to be practised more and more by jewellers for decorative purposes, and much of the early knowledge of heavy plating, electroforming and large scale operations was apparently lost or forgotten. Apart from the empirical development of alloy coatings for watch-cases, there were virtually no significant scientific or technical developments in gold plating over the next eighty years.

It was the growth of the electronics industry in the early 1940s which finally stimulated renewed interest in the science and technology of gold plating. Methods were required for the plating of bright gold deposits which did not require polishing or burnishing, and which conformed to precise specifications in respect of thickness and other properties. As Weisberg has summarised the position, "the 1940s marked the development of chemical and electrical controls to ensure these results. The 1950s brought the rediscovery of solutions and techniques for thick, uniform and fine-grained gold deposits, and the introduction of bright acid and neutral gold solutions. Following this in the 1960s, acid gold and gold alloy systems were developed to produce deposits with specific physical properties, such as ductility, corrosion and wear resistance, purity and so forth. During this period, the non-cyanide gold electrolytes were also reintroduced in bright and hard plating modifications."

In the 1970s, the course of events has been dictated partly by the growing sophistication of the electronics industry and partly by economic pressures resulting from much increased gold prices. This has led to the introduction of equipment for continuous and selective plating of stripes and spots on metal substrate strips, and a move away from rack and barrel plating of components in some situations. It has also called for increasing attention to the feasibility and merits of the plating of gold alloys rather than gold, and has stimulated developments aimed at increasing speeds of deposition from baths with high levelling and throwing powers.

The 1960s and 1970s have also seen a tremendous growth in the quantity of literature on the electrodeposition of gold and its alloys. Gold plating technology has been reviewed in a text edited by Reid and Goldie,[3] while the technology of gold plating solutions has been reviewed by Page.[4,5] Moreover, a bibliographic review of the electrodeposition of gold alloys has become available[6] and the plating of gold and gold alloys is discussed in various texts[7–10] and handbooks, and in a recent review by Raub.[12] The uses of gold platings, together with their properties and methods for measuring them, have been the subject of a critical bibliography (1940–1968) published by the American Electroplaters' Society,[11] and have been discussed also by Reid and Goldie.[3]

In the treatment which follows an attempt is made to complement available publications as far as possible and no endeavour is made to cover all aspects of gold and gold alloy electroplating processes.

2. Chemical considerations: Gold Complexes and their Stabilities

Since gold is always electrodeposited from aqueous solutions of gold complexes, a knowledge of the characteristics of these complexes and in particular their stabilities, as well as the characteristics and stabilities of the complexes of any metals which it may be desired to co-deposit with gold in alloyed form, is fundamental to any consideration of the electrodeposition of gold and its alloys.

Although complexes of gold(V) are known, and evidence for the occurrence of gold(II) in certain complexes has been reported, the stable oxidation states of gold are gold(I) and gold(III). Neither of the simple ions Au^+ or Au^{3+}, however, occurs in the free state to any significant extent, and gold compounds whether in solution or in other forms are covalently constituted, usually as complexes formed by the bonding to a central Au^+ or Au^{3+} cation of a number of ligands, which may either be ions such as Cl^-, Br^-, OH^-, CN^-, SCN^-, SO_3^{2-}, $S_2O_3^{2-}$ or uncharged molecules such as NH_3, H_2O, Ph_3P, $(NH_2)_2CS$, etc.

With few exceptions gold(III) complexes are 4-coordinate, containing four ligands covalently linked to the central Au^{3+} ion in a square-planar configuration. This stereochemistry is readily understood in terms of the spin-paired d^8 electronic configuration of the Au^{3+} ion, which leads to dsp^2 hybridization. Gold(I) complexes, on the other hand are, with some exceptions which are not encountered in electroplating media, 2-coordinate with the two ligands linked to the central Au^+ ion in a linear configuration. A straightforward explanation of this cannot be given in terms of its d^{10} configuration which might be expected to result in sp^3 hybridization and a tetrahedral configuration.

The formation of such complexes between a free metal cation M^{z+} and n ligands L^{y-} (z and y denote the electronic charges on the ion and the ligand, respectively) occurs by an equilibrium reaction:

$$M^{z+} + nL^{y-} \rightleftarrows ML_n^{z-ny}$$

for which the equilibrium constant[13]

$$\beta_n = \frac{\{ML_n^{z-ny}\}}{\{M^{z+}\}\{L^{y-}\}^n}$$

provides a measure of the stability of the complex, and which is therefore known as its stability constant. ({ }) denote the activity of the species enclosed, in moles per litre). The stability constants of gold(I) and gold(III) complexes are, as will be seen, important properties of such complexes when the electrodeposition of gold from them is considered, since they determine the concentration levels at which simple Au^+ and Au^{3+} ions become available for discharge at the cathode. Variations in the nature of the complex from which gold is electrodeposited can therefore affect materially the conditions under which deposition occurs. They can, as will be seen, so affect the deposition of gold that it will behave under certain conditions like less noble metals, and deposit with them in alloyed form.

What is known concerning the relative stabilities of gold(I) and gold(III) complexes can be presented usefully in both qualitative and quantitative terms.

Qualitatively, a first useful generalization is the division of metal ions into A and B types, or as they are also known, hard and soft ions.[14] The most stable complexes of A-type or hard metal ions are formed with the more electronegative elements in any group in the periodic table, while those of B-type or soft metal ions are formed with the less electronegative members. For their complexes with halides, therefore, the order of stability for A-type metal ions is $F^- >> Cl^- > Br^- > I^-$, whereas B-type metal ions it is $I^- > Br^- >> F^-$. Metals developing B-type ion behaviour lie in a triangle in the periodic table and centred approximately on gold:

Cr	Mn	Fe	Co	Ni	Cu(III) $\underline{\text{Cu(I)}}$	Zn	Ga	Ge
Mo	Tc	Ru	Rh	Pd	$\underline{\text{Ag(I)}}$	Cd	In	Sn
W	Re	$\underline{\text{Os}}$	$\underline{\text{Ir}}$	$\underline{\text{Pt}}$	$\underline{\text{Au}}$	$\underline{\text{Hg}}$	$\underline{\text{Te}}$	$\underline{\text{Pb}}$

(The symbols for B-type metals are underlined).

Both gold (I) and gold(III), the two stable oxidation states of gold, are strongly B-type and extremely soft in behaviour. Not only is the order of the stabilities of their halide complexes that mentioned above as being characteristic of B-type metals, but also stability orders occur such as:

$$\underline{Se}C(NH_2)_2 > \underline{S}C(NH_2)_2 > \underline{O}C(NH_2)_2$$

$$\text{and } \underline{C}N^- > \underline{N}H_3 > H_2\underline{O}$$

for its complexes with selenourea, thiourea and urea and cyanide, ammonia and water, respectively (in each ligand the donor atom is underlined). This is in line with the general experience that the orders of stability of complexes formed by B-type metal ions vary according to the donor atom involved approximately as follows:

	C	N	O	F	↑ electronegativity increasing
		P	S	Cl	
		As	Se	Bi	
stability ↓	Sb	Te	I		
← increasing					

In the absence of ligands stronger than H_2O, auric Au^{3+} and aurous Au^+ ions are, therefore, to be seen as occurring in aqueous solution not as the bare ions, but as their weak aquo complexes $Au(H_2O)_2^+$ and $Au(H_2O)_4^{3+}$, respectively. These complexes, being of low stability, readily undergo ligand replacement reactions in the presence of ligands stronger than H_2O.

Quantitatively, the stability constants of gold(I) complexes can be calculated from their standard reduction potentials determined potentiometrically via Nernst plots,[15] and the value of the $Au^+(aq)/Au(s)$ standard reduction potential. The latter value (see later), however, is not easy to determine experimentally and the values assigned to it are normally calculated from thermochemical data.

The overall stability constants for a number of complexes of gold(I) and gold(III) are listed in Table XXXIX. The values for most of the complexes included in this Table have been obtained from a critical review of all available data,[16] and the values for the stability constants have been referred to infinite dilution. In regard to the data[17] for other complexes, however, there is doubt regarding the ionic strengths of the media to which the data apply. The values for these complexes should be considered as being only approximate.

Where such stability constants are not directly available, as is the case for a number of gold complexes, the method developed by Hancock and Finkelstein[14,20] for estimating the standard reduction potentials ($E°$ values) of gold(I) and gold(III) complexes can be applied. These investigators found that there is a linear relationship between the $E°$ values of gold complexes and the $E°$ values of corresponding complexes of the same electronic configuration. A diagram such as Figure 43, for example, can be used to estimate the $E°$ values for gold(I) complexes in cases where the $E°$ values of the corresponding silver complexes are known. For example, the value for the Au^+ aquo ion, which cannot be determined by experiment because of the instability of this ion, can be estimated from the determined $E°$ value of + 0.799 V for the Ag^+ aquo ion to be + 1.67 V.

In the same way the linear relationships between the $E°$ values for Pd(II) and Pt(II) complexes and corresponding complexes of the electronically similar gold(III) complexes, namely,

Table XXXIX. Overall Stability Constants for a Selection of Complexes of Au(I) and Au(III)

Au(I)			Au(III)		
Complex	β	Ref	Complex	β	Ref
$Au(CN)_2^-$	2×10^{38}	16	$Au(CN)_4^-$	$\sim10^{56}$	16
AuS^-	$\sim10^{40}$	17	$Au(OH)_4^-$	$\sim10^{55}$	17
$Au(SO_3)_2^{3-}$	$\sim10^{30}$	17	AuI_4^-	5×10^{47}	16
$Au(S_2O_3)_2^{3-}$	5×10^{28}	16	$Au(SCN)_4^-$	10^{42}	16
$Au(CS(NH_2)_2)_2^+$	$\sim10^{25}$	17*	$AuBr_4^-$	10^{32}	16
$Au(OH)_2^-$	$\sim10^{21}$	17	$AuCl_4^-$	10^{26}	16
AuI_2^-	4×10^{19}	16	$Au(SO_4)_2^-$	$\sim10^{6}$	17
$Au(SGN)_2^-$	1.3×10^{17}	16			
$AuBr_2^-$	10^{12}	16			
$AuCl_2^-$	10^{9}	16			

*This value has been based upon a value of $E^\circ = 1\ 0.380 \pm 0.020$ V[(18)] for the reaction $Au[CS(NH_2)_2]^+ + e \rightleftarrows 1\ Au + 2CS(NH_2)$. Subsequent work indicates that this value may be somewhat high,[(19)] as the gold surface is rapidly contaminated in acidic solutions containing thiourea.

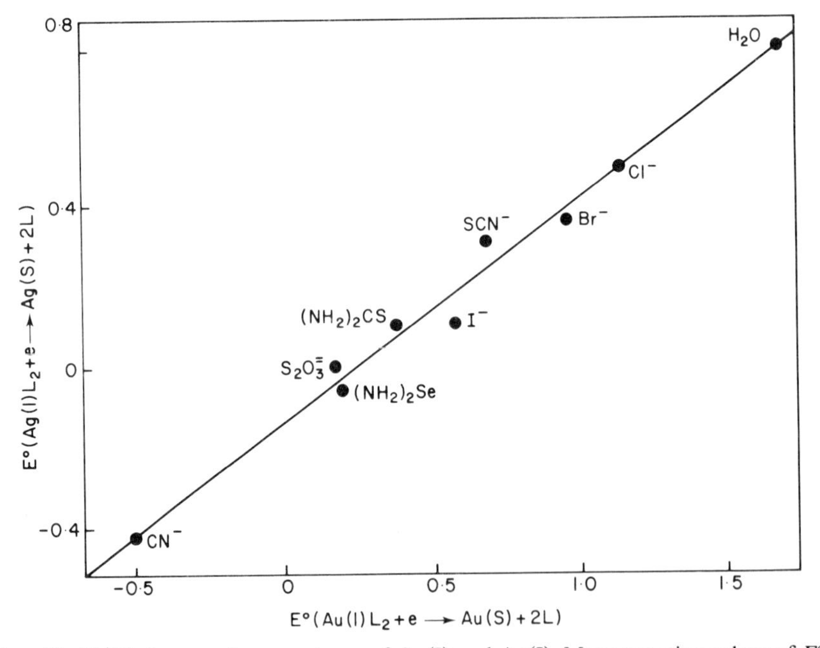

FIG. 43: E°/E° diagram for complexes of Au(I) and Ag(I). More negative values of E° indicate higher stabilities in the complex. It can be seen that the stabilities of corresponding complexes of the two ions are linearly related. (After Finkelstein and Hancock.[14])

$$E^{\circ}_{Au(III)} = 1.05\,E^{\circ}_{Pd(II)} + 0.35\ \mathrm{V}$$
$$\text{and } E^{\circ}_{Au(III)} = 0.93\,E^{\circ}_{Pt(II)} + 0.32\ \mathrm{V}$$

can be used to estimate a mean value of + 1.39 V for the standard reduction potential of the Au^{3+} aquo ion.

As a further example of the usefulness of these diagrams, Figure 44 can be used to compare the relative stabilities of various complexes of gold(I) and gold(III). It reveals that, for the greater part, gold (III) complexes with softer ligands have higher E° values than gold(I) complexes with such ligands, and that they will therefore be less stable and more easily reduced. Gold(III) complexes with harder ligands, on the other hand, have lower E° values than the corresponding gold(I) complexes and will therefore be more stable. $AuCl_4^-$ is therefore more stable than $AuCl_2^-$, which tends to disproportionate as follows:

$$3AuCl_2^- = Au(S) + AuCl_4^- + AuCl_4^-$$

3. Thermodynamic Aspects of the Deposition of Gold from Gold Complexes

The driving force for the precipitation of a metal from a solution of its ions can be expressed quantitatively in terms of its reduction potential.[20] For a metal M in contact with a solution of its ions M^{n+}, deposition of the metal occurs by the reaction

$$M^{n+} + ne = M$$

and the reduction potential E is given by the Nernst equation

$$E = E^{\circ} - \frac{RT}{nF} \ln \frac{\{M\}}{\{M^{n+}\}}$$

where R = the gas constant = 8.134 $JK^{-1}\ mol^{-1}$
T = the absolute temperature (°C + 273)
E° = the standard reduction potential for the reaction on the hydrogen scale
F = the Faraday constant = 96490 c/mol

This equation reduces to

$$E = E^{\circ} + \frac{RT}{nF} \ln [M^{n+}]$$

if the convention is adopted that the activity of the solid phase (M) is unity and if it is assumed that under the conditions of deposition of gold, the concentration $[M^n]$ is equivalent to the activity (M^n). The greater the value of E the greater is the tendency for the ion M^{n+} to be reduced to, and precipitated as, the metal. Conversely, the smaller the value of E, the greater the tendency for the reverse reaction, and, if the product of reaction is soluble, for the metal to dissolve.

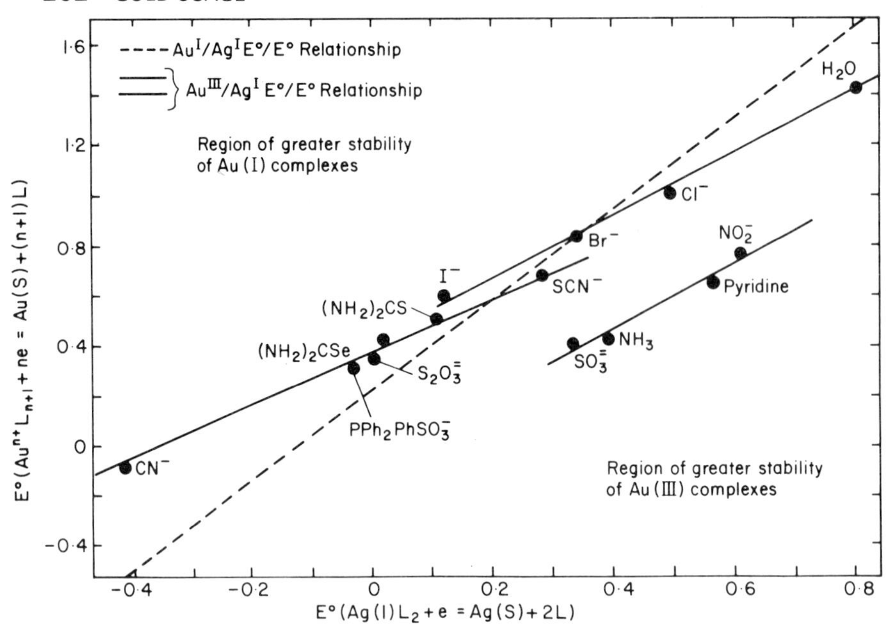

FIG. 44. *E/E* diagram comparing the stabilities of complexes of Au(III), Au(I) and Ag(I). This shows clearly which ligands prefer to form auric complexes and which favour the aurous state. (After Finkelstein and Hancock.[14])

If these relationships are applied to the deposition of gold from simple Au^{3} and Au^{3+} ions, then at 25°C for

$$Au^{+} + e = Au$$

$$E = 1.730 + 0.0591 \log [Au^{+}] \quad (1)$$

and for $Au^{3+} + 3e = Au$

$$E = 1.498 + 0.0197 \log [Au^{3+}] \quad (2)$$

Because of the instability of the Au^{+} and Au^{3+} ions, or aquo-complexes, the values for E° used in these equations cannot be measured directly, and such values as are available for E° for these reactions are calculated from the free energy changes, determined by procedures which are discussed by Page[4] and by Gedansky and Hepler,[18] using the equation

$$E^{\circ} = \frac{-\Delta G}{nF}$$

in which ΔG is the Gibbs free-energy change.

The values 1.730 and 1.498 used in equations (1) and (2) are those quoted by Finkelstein,[16] and their magnitudes in comparison with the standard reduction potentials of other metals accord with the fact that both Au^{+} and Au^{3+} are very susceptible to reduction to the metal. The extent to which gold is noble in comparison with other metals is illustrated by the E° values in Table XL, in which are listed also E° values for various complexes of gold(I) and gold(III). It will be noted that standard reduction potentials for deposition of gold(I) from its complexes are less than that from Au^+. In other words, gold(I) behaves less nobly in the complexed than the uncomplexed form. A point of interest which will be referred to later, however, is that the E° value for the formation of silver from $Ag(CN)_2^-$ is greater than that for the formation of gold from $Au(CN)_2^-$; thus, for example, silver is more noble than gold in cyanide solutions.

Table XL*. Standard Reduction Potentials for the Formation of Gold and some other Metals from their Ions or Complex Ions

Redox system	E°	T (°C)	Method
$Au^+ + e \rightleftarrows Au$	+1.73		ΔG
$AuCl_2^- + e \rightleftarrows Au + 2Cl^-$	+1.15	25	emf
$AuBr_2^- + e \rightleftarrows Au + 2Br^-$	+0.93	25	emf
$Au(CN)_2^- + e \rightleftarrows Au + 2CN^-$	−0.6	ord.	emf
$Au(SCN)_2^- + e \rightleftarrows Au + 2SCN^-$	+0.7	17	emf
$Au(CS(NH_2)_2)_2^+ + e \rightleftarrows Au + 2CS(NH_2)_2$	+0.35	30	emf
$Au^{3+} + 3e \rightleftarrows Au$	+1.50	—	ΔG
$AuCl_4^{3-} + 3e \rightleftarrows Au + 4Cl^-$	+1.00	25	emf
$AuBr_4^{3-} + 3e$ Au + $4Br^-$	+0.85	25	emf
$Au(SCN)_4^{3-} + 3e \rightleftarrows Au + 4SCN^-$	+0.66	17	emf
$Ag^+ + e \rightleftarrows Ag$	+0.80	25	emf
$Ag(CN)_2^- + e \rightleftarrows Ag$	−0.31	—	emf
$Cu^+ + e \rightleftarrows Cu$	+0.52	25	ΔG
$Cu^{2+} + 2e \rightleftarrows Cu$	+0.34	25	emf
$Fe^{2+} + 2e \rightleftarrows Fe$	−0.44	25	emf
$Co^{2+} + 2e \rightleftarrows Co$	−0.29	25	emf
$Ni^{2+} + 2e \rightleftarrows Ni$	−0.25	25	emf
$Zn^{2+} + 2e \rightleftarrows Zn$	−0.76	25	emf
$Cd^{2+} + 2e \rightleftarrows Cd$	−0.40	25	emf
$Pd^{2+} + 2e \rightleftarrows Pd$	+0.92	25	emf
$PtCl_4^{2-} + 2e \rightleftarrows Pt + 4Cl^-$	+0.73	25	emf

*The data are from Charlot, Collumeau and Marchon[22], with the exception of those for Au^+, Au^{3+}, and $Au[CS(NH_2)_2]_2^+$; the information in regard to the latter is from Groenewald[19].

As indicated in equations (1) and (2), the E values for Au^+ and Au^{3+} solutions depend not only on the E° values for these ions but also upon the values of $[Au^+]$ and $[Au^{3+}]$ and, through their effects on these, complexing agents can modify the conditions under which gold is deposited from plating baths. In the presence of cyanide ions, for example, Au+ forms $Au(CN)_2^+$ by the reaction

$$Au^+ + 2CN^- \rightleftharpoons Au(CN)_2^-$$

for which the stability constant $\beta_2 = 2 \times 10^{38}$

that is, $$\frac{[Au(CN)_2^-]}{[Au^+]} = [CN^-]^2 \times 2 \times 10^{38}$$

The addition of cyanide to a solution containing Au^+ ions therefore decreases the proportion of gold present as Au^+ ions in direct proportion to the value of the stability constant of the complex formed, and to the square of its concentration. If, for example, the concentration of CN^- ions added is 10^{-2}M, then the concentration of gold present as Au^+ ions is reduced by the factor

$$\frac{[Au(CN)_2^-]}{[Au^+]} = 2 \times 10^{34}$$

In terms of equation (1) above, this will reduce the value of E by 0.0591 log $2 \times 10^{34} = 2.03$ V. This reduces the equilibrium potential of the system to the levels which commonly occur in the deposition of gold.

Depending upon their concentrations and the stability constants of the complexes which they form, complexing agents such as CN^-, which are necessary in order to achieve practical concentrations of gold in gold plating baths can therefore have a profound effect upon the conditions under which gold can be electrodeposited.

4. Electrochemical aspects of the Plating of Gold and of Gold Alloys

4.1 Current efficiencies and the codeposition of hydrogen

In the plating process the possibility that release of hydrogen at the cathode may compete with the release of the metal or metals being plated must always be borne in mind, since it will affect the current efficiency of the desired metal plating process; codeposition of hydrogen may also affect the properties of both the plated and the substrate metal. Bright acid golds plated from acid cyanide solution containing cobalt or nickel, for example, are characterised by the fact that hydrogen codeposits with the gold during the plating process. Current efficiencies are therefore low (36–50%) and the deposits tend to contain significant quantities of "occluded" hydrogen.

The question arises as to when hydrogen codeposition may be expected. The equilibrium potential (E_h) at which the onset of hydrogen codeposition is possible thermodynamically can be obtained from the Nernst equation applied to either of the following equations

$$2H_3O^+ + 2e = H_2(gas) + 2H_2O \text{ (acid media)}$$

or $$2H_2O + 2e = H_2(gas) + 2OH^- \text{ (alkali media)}$$

In either case, by taking into account the ionic product of water $Kw = [H_3O^+] . [OH^-] = 10^{-14}$, it can be shown that $E_h = 0.059$ pH. The evolution of hydrogen is thus possible at far lower negative potentials in acid media than it is in alkali media.

Whether and to what extent hydrogen codeposition will in fact occur depends not only upon the equilibrium potentials of the hydrogen-evolution and metal-deposition reactions but also upon kinetic factors[23] which depend upon a variety of other factors such as the pH of the solution, the current density, the material of the cathode and the nature of its surface. The excess energy required to maintain the codeposition of hydrogen at a measurable rate is known as the hydrogen overvoltage, the value of which differs markedly from metal to metal (see Table XLI);[24] there are, however, few data available regarding gold alloy systems which are applicable to the deposition of gold alloys.

Table XLI. Values for the Overvoltage at Various Electrodes During the Cathode Reaction $2H_3O^+ + 2e \rightarrow H_2 + 2H_2O$ in N HC1, with a Current Density of 10^{-3} A/cm[2,24]

Ag	−0.59 V	In	− 0.80 V
Au	−0.12 V	Ni	− 0.32 V
Be	−0.63 V	Pb	−0.85 V
Cd	−0.99 V	Pd	−0.09 V
Cr	−0.42 V	Pt	−0.29 V
Cu	−0.50 V	Ta	−0.20 V
Fe	−0.40 V	T1	−0.80 V
Ga	−0.68 V	W	−0.11 V
Hg	−1.03 V	Zn	−0.80 V

In a sense, however, codeposition of hydrogen with gold is to be regarded as essentially analogous to the codeposition of gold with other metals. As will be seen later, current density-potential considerations are of little use in predicting the behaviour of alloy baths, and they must be regarded with at least the same reserve in considering the question of codeposition of hydrogen.

The widespread and rather complicated effects of hydrogen codeposition on the crystallography of metal deposition have been summarised effectively by Bockris and Reddy.[23] Adsorption of hydrogen on some crystal planes may result in a preferred orientation of the deposit crystals. Hydrogen may also alter the mechanical properties of the deposit. In gold plating codeposition of hydrogen often leads to the deposition of the metal in a powdery or "burned" form. Finally, hydrogen deposition results in the removal of H_3O^+ ions from the diffusion layer surrounding the electrode, and if this layer becomes sufficiently alkaline the precipitation of metallic hydroxides on the surface of the electrode becomes possible.

4.2 Current density-potential relationships and limiting current densities

A typical current-density or polarisation curve for a cathode in a plating bath is shown in Figure 45. As increasingly negative potentials are applied, the current increases initially, but then levels off to form a plateau in the curve, which is called the limiting or diffusion current. At this point the concentration polarisation has reached a maximum since all the depositing ions are plated out immediately they reach the surface of the cathode; the concentration of these ions on the surface of the cathode becomes effectively zero. If the negative potential is increased further, hydrogen evolution also takes place on the cathode, and the current increases once more; however the current efficiency of metal plating decreases.

An expression for the limiting current (I_{lim}) can be obtained[23] from Fick's law, according to which the rate of transport of an electrolysed substance by diffusion is proportional to the concentration gradient (dC/dx), that is, the flux

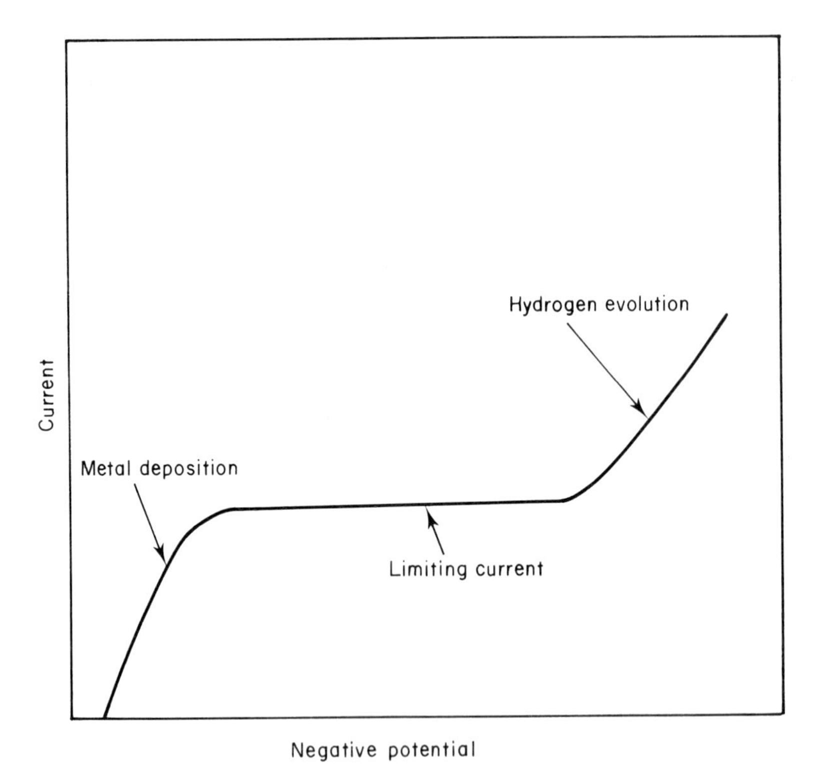

FIG. 45. A typical current-potential curve for cathodic deposition of metals.

$J = D\ dC/dx$, where D is the diffusion coefficient of the metal ions. But $J = (1/A)\ (dM/dt)$, where A = area, M is the number of moles and t = time. Since $I_{lim}/nF = dM/dt$,

$$I_{lim} = nFAD(dC/dx)$$

If it is assumed that in the stationary state the diffusion is limited to a layer of constant thickness δ around the electrode (Nernst's hypothesis), then

$$I_{lim} = nFAD\left(\frac{C - C_s}{\delta}\right)$$

where C is the concentration of the metal ions in the bulk of the solution, and C_s is the concentration on the surface of the electrode. Since, as mentioned previously, $C_s \approx 0$, this reduces to

$$I_{lim} = nFADC/\delta \qquad \text{(A)}$$

According to equation (A) the limiting current may be increased in a number of ways. The most obvious is to increase the concentration of the metal ions in the solution, an effect which can be realised either by increasing the concentration of the metal complex in the bath, or by increasing the dissociation of this complex by decreasing the concentration of complexing agent or by an alteration of pH. Secondly, increased agitation of the bath will result in a decrease of the thickness of the diffusion layer (δ) around the electrode, leading to an increase in I_{lim}. The term δ is a rather complex function of the coefficient of diffusion, the viscosity of the fluid and the fluid velocity; this dependence changes according to whether laminar or turbulent conditions prevail.[25]

The effect of greater agitation was demonstrated in a patented process[21] for electroplating and electroforming of gold, according to which exceptionally high current densities can be used when a gold plating bath is exposed to an ultrasonic field.

The limiting current is also affected by a change in temperature; this is not immediately obvious from equation (A) which can, however, be rewritten as $I_{lim} = kC$, where the rate constant for the reaction $k = nFAD/\delta$. The rate constants of reactions vary exponentially with temperatures[25] according to a typical Arrhenius relationship where $k = Ze^{-\varepsilon/RT}$, in which Z is a constant and is the activation energy. Thus I_{lim} should increase with temperature, although the extent of increase for a diffusion-controlled process is usually not very large.

Raub[27] and others have illustrated the effect of such measures in studies of the electrodeposition of both gold and its alloys. The cathodic polarisation curves in Figure 46 illustrate the scale of the effects of temperature and agitation for two $KAu(CN)_2$ electrolytes each containing 2 g/litre gold. It also illustrates the pronounced effect on such cyanide baths of decreasing the concentration of the CN^- complexant by decreasing the pH of the solution. With the greater extent of dissociation of the $Au(CN)_2^-$ ion which the lower CN^- ion concentration in the acid cyanide bath permits, the gold behaves much more

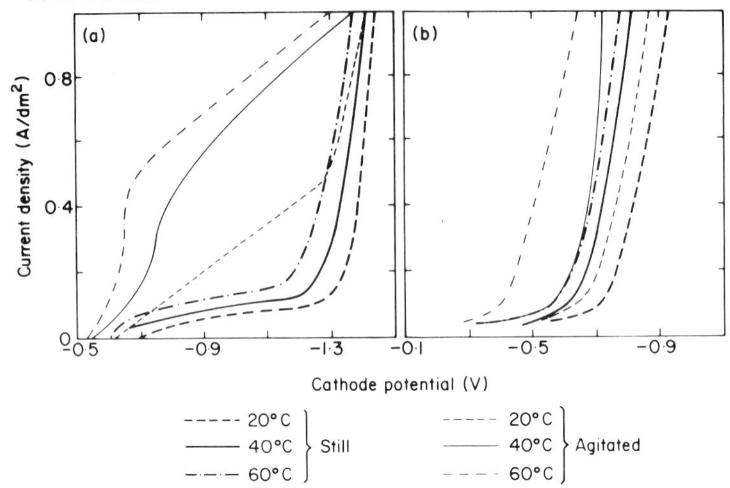

FIG. 46. Cathode potential-current density curves for $KAu(Cn)_2$ electrolytes.
(a) Au 2 g/litre, KCN 8 g/litre, pH11.0
(b) Au 2 g/litre, citric acid 40 g/litre, sodium citrate 60 g/litre, pH 4.2.
(Data of Raub.[22])

nobly, and the limiting current is reached at significantly less negative cathode potentials under all conditions.

In addition, it can also be seen that when gold is plated from alkaline cyanide media, the deposition of metal alone occurs below the level of the limiting current, although hydrogen codeposition at the cathode sets in as the limiting current is reached, with loss in current efficiency. With acid cyanide baths, however, hydrogen is codeposited with the gold even at the lowest levels of cathodic current, despite the fact that gold is deposited at lower negative potentials in acid solution than is possible in alkaline solutions.

Because of the deterioration of the deposit of gold that often occurs at higher current density levels, due to hydrogen codeposition, it is often convenient in gold plating technology to refer to the "limiting current density" as that current density above which the bath does not give good deposits with acceptable current efficiency, rather than to the formal definition of limiting current density in terms of current density-potential curves.

4.3 Current density-potential relationships in gold alloy deposition

In a lecture to the American Electroplaters' Society in 1965 Brenner[28] drew attention to the fallacies underlying early attempts to determine the conditions for codepositing two metals, and perhaps even the composition of the alloys, from the current density-potential (cd-ptl) relations of the metals when deposited separately. This idea was apparently originated by Kremann in 1914 and has appeared in many subsequent discussions of alloy deposition. The basis of Kremann's approach was a postulated additivity relationship between the cd-ptl curves of two metals during codeposition.

Although there is, as pointed out by Brenner, no evidence to support the validity of this postulated relationship as a basis for predicting the behaviour of two metals during codeposition, considerable information can nevertheless be obtained about the true cd-ptl relations of two metals in codeposition from a knowledge of the composition of the alloys deposited at various potentials and the derivation of partial cd-ptl (polarisation) curves for the codepositing metals.

The application of this latter technique can be illustrated from the study by Andreeva *et al.*[29] of the codeposition of copper (and hydrogen) with gold from alkaline cyanide baths from the following conditions:

Au as $KAu(Cn)_2$	5.5–0.2 g/litre	(0.028–0.001 M)
Cu as CuCn	5.5–0.2 g/litre	(0.086–0.003 M)
KCN (free)	5.5 g/litre	(0.084 M)
Temperature	60°C	
Current density	0.15 A/dm^2	

Their findings are illustrated in Figs 47–50.

The gold content in the deposit increases with increasing ratio of Au:Cu in the electrolyte (Fig. 47); it will be seen (Fig. 48) that the cd-ptl curve for co-discharge of Au^+, Cu^+ and H^+ ions lies between the curves for the two components, but nearer the curve for co-discharge of Au^+ and H^+, and almost merging with the latter up to the limiting current of 0.25 $A/dm.^2$ Copper was deposited with gold only at current densities above this level, codeposition of gold and copper setting in, with the formation of alloy, once the potential of the cathode reached that for copper deposition. Under the conditions used, simultaneous deposition of gold and copper resulted in the formation of two phases, one Au-Cu solid solution, the other being pure copper.

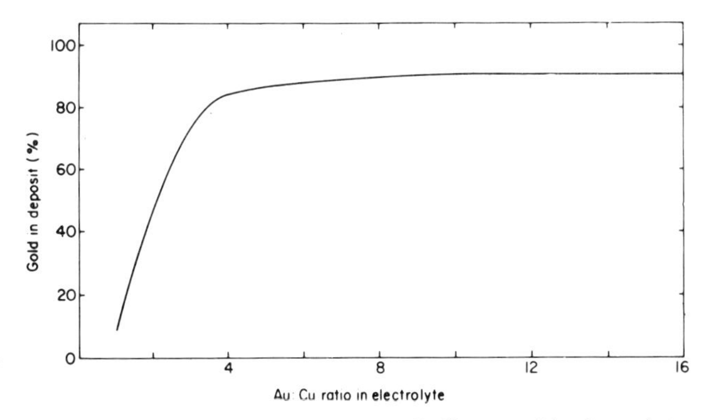

FIG. 47: The effect of the Au:Cu ratio in an alkaline cyanide electrolyte upon the gold content of the deposited alloy. (After Andreeva *et al.*[29])

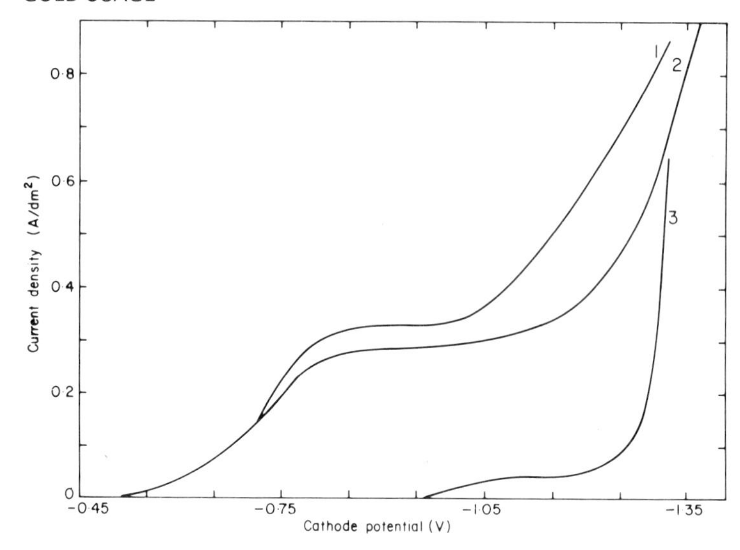

FIG. 48: Current-potential curves for deposition of
(1) gold,
(2) Au-Cu alloy, and
(3) copper with hydrogen
from an electrolyte containing 5.5 g/litre free KCN. Electrolyte composition in g/litre:

Curve 1	Au 5.5
Curve 2	Au 5.5, Cu 0.85
Curve 3	Cu 0.85

(After Andreeva *et al.*[29])

The explanation by Andreeva *et al.* that the partial cd-ptl curves for gold (curves 1 and 2 of Fig. 49) and for copper (curves 3 and 4 of Fig. 49) can be related, in the case of gold, to an increased overpotential at which the solid solution separates out on the free and highly dispersed copper with which it is deposited continuously, and in the case of copper, to a decrease in overpotential as a consequence of the separation of so much of this metal as free copper and not Au-Cu, is perhaps an over-simplification. According to Brenner[8] the codeposition of metals should be related in a complex manner to changes in both the equilibrium potentials and overpotentials which occur as a result of the presence of alloyed metal surfaces. They noted that the steady potentials of the Au-Cu deposits (see Fig. 50) were closely related to their compositions.

Analogous information relating to the codeposition of gold and copper from alkaline cyanide baths was reported much earlier by Raub and Sautter,[30] who also made detailed studies of the effects of changes in free cyanide content and of pH upon Au-Cu alloy deposition. They found that as the pH was reduced from 10.5 to 7, the polarisation curves became straighter, the plateau in Fig. 48 virtually disappearing so that there was no longer any region of limiting

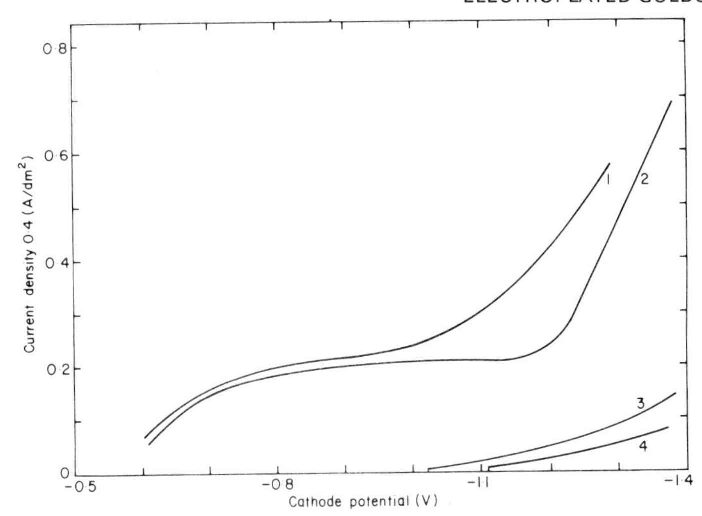

FIG. 49. Partial current-potential curves for deposition of gold (curves 1 and 2) and copper (curves 3 and 4) from an alkaline cyanide electrolyte containing 5.5 g/litre free KCN. Electrolyte composition in g/litre:

Curve 1	Au 5.5
Curves 2 and 3	Au 5.5, Cu 0.85
Curve 4	Cu 0.85

(After Andreeva *et al.*[29])

current. Alloy deposition (and hydrogen evolution) were found accordingly to begin under these conditions at low current densities, and the copper content of the deposit to increase as the current density increased. At the same time, however, the percentage of copper in the deposit which was free as opposed to being in solid solution, increased to values of up to 15%. Even at low current

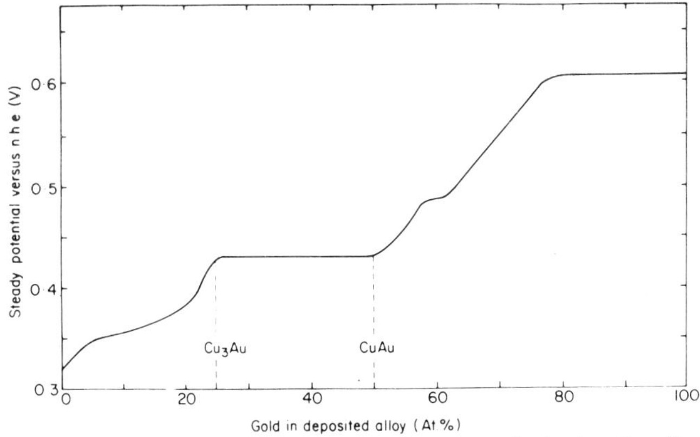

FIG. 50. Steady potential-composition diagram for electrolytic Au-Cu alloy. (After Andreeva *et al.*[29])

densities the percentage of the deposited copper which was in separately crystallised form was high; in deposits containing between 14 and 40% copper, the percentage of free copper was as high as 36%.

The effects of lowering the free cyanide content were essentially similar to those of lowering the pH. In both instances the copper behaved more nobly and no limiting current density effect was observed. The addition of sulphur compounds which depolarise the deposition of copper from cyanide solutions did not modify this effect.

The above description of the mechanism of the codeposition of gold, copper and hydrogen from cyanide baths has been given in some detail in order to exemplify the manner in which cd-ptl curves can be used for the more *scientific* study of gold alloy-plating processes. They have been applied, especially by Russian investigators, in studies of the plating of gold with a range of other metals.

Except in so far as the results of these studies provide a background to the processes involved, they do not appear to have found application in the design of alloy plating baths, most of which are still developed mostly on the basis of technical experience and trained intuition.[31]

4.4 The effects of current (or potential) modulation

Current modulation is well known as a means for improving the quality of electrodeposited coatings. Most of what has been published in regard to its application to the plating of gold and gold alloys, however, is in patent specifications.

Various types of modulation have been claimed, including periodic reversal, and the use of various modified forms of direct current, achieved by superposition of ac on dc, pulsing and the application of cyclically varying potentials.

During the flow of reversed current, polarisation effects are at least partially reduced, bath impurities which may have attached themselves to the plated surface are dislodged, and electropolishing conditions may be established. Periodic reversal, however, has the disadvantage that current efficiencies are inevitably low if forward to reverse current rates are not kept high. Yampolsky,[32] in a review of Russian precious metal plating, mentioned periodical reversal of current as being applicable in the plating of gold from alkaline cyanide baths, but gave few details. The technique has been cited as yielding gold deposits with low internal stress, and it is applied by at least one company in the high velocity plating of integrated circuits in which the use of a square wave periodically reversed current with a forward to reverse current ratio of 30:1 has been reported.[33] Hischmann,[34] Heilmann[35,36] and Degussa[37] have also claimed benefits from its use in the plating of gold alloys, particularly thick (20 μm) Au-Cu-Cd alloys, and Hänsel *et al.*[38] have claimed the use of a periodically reversed current as a means of producing tarnish-resistant and "polished" coatings of Au-Cu-Ag alloys. In the reversed current or polishing cycle, free copper in the electrodeposit is apparently dissolved preferentially. Similar claims in regard to other gold alloys have been made by Aliprandini *et al.*[39]

As regards current modulation without reversal various techniques have been used. Thus the superposition of ac on dc has been investigated by Berterolle,[40] who described satisfactory deposits up to 46 μm in thickness from alkaline cyanide gold baths.

Other approaches claimed include those of Winkler[41–42] who claimed advantages in the plating of metal alloys from the application of periodically changing voltages to the cathode and anode(s) in the bath and exemplified its use in the plating of Au-Cu-Ni alloys; and Dettke *et al.*[43] who have made much more specific claims in respect of the use of cyclically varying potentials for the plating of a wide range of gold alloys. Equipment for the supply of asymmetrical alternating current for gold plating has been described.[44]

Both Avila and Brown[45] and Cheh[46] have reported on pulsed plating of gold. Avila and Brown mention as advantages of pulse plating the formation of dense or highly conductive deposits, increased plating rates, elimination of hydrogen embrittlement problems, reduced needs for additives and improved resist performance resulting from the absence of hydrogen on the cathode. Such hydrogen tends to lift off the masking material. Pulse-plated gold coatings deposited from a phosphate-buffered cyanide bath have also been reported[47] as being more ductile than coatings formed by the use of direct current.

Brown has reported[48] on the use by one company of pulse plating in hybrid circuit production. Pulsed plating has also been reported as a means for reducing the grain size and porosity of soft gold deposits.[49] The effects of varying wave forms in pulsed plating of soft golds on the properties of the deposits has also been reported on by Rehrig,[51] who found[52] that with asymmetric ac, and especially with pulsed currents, the deposits were less porous than they were with dc at thicknesses less than 3 μm. Whereas with dc grain growth was confined to nucleation sites established early in the growth of the film, the coatings obtained with asymmetric ac and with pulsed currents showed a greater degree of nucleation and smaller grain size, as well as enhanced intergranular growth.

5. Kinetics and Mechanisms of Gold Plating

Kinetic studies of gold deposition have been focused mainly on deposition of gold from gold(I) cyanide solutions.[52] Maja[53] concluded that the mechanism involved direct reduction of $Au(CN)_2^-$, and Cheh and Sard[54] accepted this same mechanism as being operative from the results of galvanostatic studies of the process at a gold rotating disc electrode using cyanide (pH 10.8), citrate (pH 5.0) and phosphate (pH 7.5) buffered solutions. Using a potentiostatic linear sweep technique at a rotating gold electrode, however, Harrison and Thompson[55] found from a cathodic shift in potential with increasing free cyanide concentration, that the reaction was first-order in respect of cyanide. The reducible species was seen by them as being the neutral complex AuCN, with gold being released from the $Au(CN)_2^-$ complex in two stages, as follows:

$$Au(CN)_2^- \rightleftarrows AuCN + CN^- \quad (1)$$

$$AuCn + e \rightarrow Au + CN^- \text{ (slow)} \quad (2)$$

The same conclusion was reached by MacArthur[56] from experiments at low overpotentials in cyclic voltammetric studies using citrate and cyanide buffered gold(I) cyanide solutions and a stationary gold electrode. He obtained evidence for the adsorption of the AuCN intermediate followed by its reduction *in situ* to the metal. At higher overpotentials, however, and on the basis of diffusion-controlled peak currents, direct reduction of $Au(CN)_2^-$ was also postulated, namely,

$$Au(CN)_2^- + e \rightarrow Au + 2CN^- \quad (3)$$

Harrison and Thompson[52] pointed out, however, that tne two reactions (1) and (2) under these conditions would produce similar experimental observations, and they have used these latter reactions in the interpretation of data obtained in potentiostatic pulse and ac impedance studies of gold deposition,[57] and in the discussion of the incorporation of CN in gold deposits.

Bek and Zelinskii[58] on the other hand have used chrono-potentiometry and polarography to study gold deposition from gold(I) thiocyanate solutions and have claimed the rate determining step in this case to be diffusion of the discharging gold thiocyanate.

Gold deposition from gold(III) chloride[59] and cyanide[60] solutions has also been studied but has little relevance to practical gold plating systems.

6. Electrocrystallisation Processes

6.1 The basic model

A comparison between crystal growth of metals from the vapour phase and from electrolytes is useful in gaining a perspective on electrocrystallisation processes. Frank and his collaborators[61] were the first to point out that the steps on the surface of a crystal presented favourable sites for crystal growth and concluded that surface diffusion of metal atoms to such sites played a primary role in the growth of crystals from the vapour phase. Further, Frank[62] was the first to suggest that crystal growth must be associated with imperfections in the crystal surface, and in particular with screw dislocations. These were seen as providing self-perpetuating steps on the crystal surface as sites for continuing growth, thus eliminating the need for assuming two- or three-dimensional nucleation as an essential step in the crystal growth process.

The basic model[53] for electrolytic crystal growth is similar to that for growth from the vapour phase, namely, deposition of metal on flat surfaces followed by surface diffusion to crystal growth sites. The exact form in which metal first becomes attached to the surface is not known, but it is thought that in the so-called "adion" state the metal may not be entirely free of either ligands or of ionic charge. In the initial stages of plating, the crystal growth sites must be provided by the substrate which can thereby affect the manner in which crystal growth occurs. Thus if the inter-atomic distances in a particular lattice plane of the substrate metal are close to those in a lattice plane of the metal being deposited, then the latter may grow with the same structure as the former, a process which is known as epitaxial growth. Once started, epitaxial growth of the deposit may continue over a significant thickness until the "normal" structure of the depositing metal is established. If epitaxial growth does not occur, the interfacial zone is characterised by the presence of what are called interfacial dislocations. The driving force for electrolytic growth can also affect the manner in which deposition is initiated, and if the applied overpotential is high enough, for example, three-dimensional nucleation may occur and any relationship between the structures of substrate and coating metals is rapidly lost.

The rate of crystal growth is seen as controlled by mass transport in the liquid medium. Because of this and because of the various processes such as complex dissociation which must occur in the double layer before metal from it is released to the cathode surface, crystal growth from electrolytes tends to be more complex than it is from the vapour phase.

This is particularly so under practical plating conditions, where the structure, appearance and other properties of the deposit such as uniformity, smoothness, hardness, etc. may be affected by a variety of factors. These include geometrical parameters which affect the current density distribution over the cathode, the deposition and incorporation of impurities or additives in the deposit, as well as the rate of transport of metal to the cathode surface. These factors are of special importance in gold plating, because gold concentrations in electrolytes are usually low and current density distribution therefore critical, and because specifications in respect of such properties as minimum thickness, porosity, internal stress, wear and corrosion resistance, solderability, etc. have to be met, while at the same time maximum economy in the use of gold is exercised.

6.2 Parameters in gold plating which increase surface irregularities

Since uniform, smooth and bright coatings are the objective in most gold plating, it is important to appreciate that deposition of the metal in such form is not always the norm, and that amplification of surface irregularities, growth of dendrites and whiskers, and deposition of metal in powdered form all occur when metal deposition is controlled by diffusion.[64] Since the first two phenomena usually cause difficulty only in electroforming or in the plating of thick deposits, they are not of as great significance in gold plating technology as is the deposition of metal in powdered form. Nevertheless, some reference to them seems desirable for the sake of completeness.

Krichmar[65] first pointed out that in the deposition of metals, the inverse effect to that achieved by metal dissolution in electropolishing can occur under certain conditions, namely, an amplification of surface irregularities. He defined these conditions by analogy to those under which electropolishing effects are observed as prolonged cathodic reduction, under conditions close to those at which the process is completely diffusion-controlled. While this effect has been studied with other metals, no reference to it as a factor of any significance in gold deposition has been noted.

Two features apparently distinguish the formation of dendrites on a cathode surface from simple amplification of surface coarseness. The first of these is the highly-ordered structures of the dendrites which grow and branch in well-defined directions, while the second is that there seems to be, at least in the cases which have so far been studied, a critical overpotential value below which dendrites do not grow. Neither dendritic deposition nor the formation of whiskers are normally observed in gold plating processes, though dendritic growths may occur in electrolytic refining of gold from chloride baths.

At current densities which exceed a critical value, however, gold resembles other metals in that it is deposited in powder form. This phenomenon, which leads to the formation of "burnt" deposits, is associated with slow transport of the depositing ions and accordingly is promoted not only by high current densities but also by decreasing the concentration of depositing ions, the bath temperature and the extent of agitation of the bath, and increasing the concentration of indifferent electrolyte, and the viscosity of the solution. These changes in bath conditions which favour powder deposition also tend to decrease the particle size of the powder which is deposited.

The geometry of the plating bath may also be responsible for the formation of powdery deposits on areas of the cathode which are exposed to higher current densities because of their position relative to the anode. Two explanations have been given[63] as to why crystal growth under these conditions does not occur by the same processes as those involved in amplification of surface roughness and the growth of dendrites. One explanation associates such growth with the high nucleation rates which are a feature of decreasing concentration, while the other visualises that under these conditions electrons from the cathode start meeting the metal ions at points separated from the cathode surface with the result that there is opportunity for three-dimensional nucleation of the metal.

6.3 Parameters in gold plating which decrease surface irregularities: Macrothrowing power, microthrowing power, levelling and brightening

Because of the importance which so often attaches to uniformity, smoothness and brightness of gold deposits, the factors which enhance these properties in plating processes are of the greatest importance.

Macrothrowing power

The capacity of a bath to produce coatings of uniform thickness over articles of irregular shape is referred to as its macrothrowing power, and is of particular significance in regard to uniform plating over rough or macro-profiles. It has been the subject of a number of reviews,[66,67] and is determined primarily by two factors. The first of these is the *primary current distribution.* A non-uniform current distribution arises from the appreciable differences in the resistances of the electrolyte between the anode and different positions on the substrate or cathode surface, and since the current distribution will favour the path of least resistance, the surface will become progressively roughened as the metal deposits preferentially on the peaks on the substrate surface nearest the anode. Much can be done[68] by way of design to ensure that the primary current distribution, and, therefore, the gold deposition, are as uniform and even as possible.

Additionally, the tendency towards roughening of the surface will, in general, be counteracted partially by the secondary current distribution, which is set up as soon as current begins to flow in a plating bath. The current distribution over the cathode is modified because polarisation occurs at each point on its surface to an extent which is determined by the current density at that point. The effect is that any attempt to maintain a larger current density over the peaks than in the recesses results immediately in an increase in the overpotential in the recesses, the result being an increase in the uniformity of the distribution of the current density.

Another factor which may modify further the uniformity of distribution of metal on the cathode is the relationship between the current efficiencies for the process and the current densities.[66] If, for example, current efficiency decreases with increasing current density, the effects of polarisation are reinforced and metal distribution is more uniform than secondary current distribution and *vice versa.*

Comparative studies of the macrothrowing power of various types of gold plating solution has been carried out by Foulke and Johnson,[69] using the Haring-Blum cell. In this cell two cathodes are arranged at different distances from a central anode in a rectangular cell and the throwing power TP of an electrolyte is expressed empirically by the Field formula:[70]

$$TP(\%) = \frac{100(K-M)}{K+M-2}$$

where M is the ratio of the quantities of metal deposited on the two cathodes (the secondary current distribution ratio) and K is the primary current distribution ratio, which is usually taken as the ratio of the distances of the two cathodes from the central anode;[3] values of TP for various typical gold plating electrolytes have been listed by Foulke.[3]

Levelling and microthrowing power

During electrodeposition progressive reduction of surface irregularities can occur by two levelling mechanisms.

Geometric levelling occurs under conditions of uniform current distribution when material is deposited relatively more rapidly on concave areas of recesses than at the convex areas of the peaks at the surface; this effect is greater the smaller the roughness of the profile which is being coated, and the greater the thickness of the coating.[66]

Far more rapid action arises, however, from "true" or "electrochemical" levelling, which is due to a non-uniform current distribution characterised by recess-to-peak coating thickness ratios exceeding unity. Such ratios are observed more particularly at roughness profiles having amplitudes of less than about 500 μm, so that this effect is often termed the microthrowing power of the plating medium; it is usually measured in terms of the relative rates of build-up of deposit thickness at the high and low points of a standard surface microprofile such as a micro-groove record "mother".[71]

True levelling effects were first described by Meyer,[72] Schmellenmeier[73] and Gardam[74] and result from the presence, often in minute concentrations, of a variety of organic substances in the plating bath. There has been considerable debate as to the manner in which these levelling agents produce their effects. A critical point is that any theory of their mode of action must explain why they are specifically effective in the levelling of microprofiles.

Present evidence[9,66] favours the view that true levelling agents act mainly by diffusion controlled inhibition. Over microprofiles, the diffusion boundary does not follow the profile contour as it does with macroprofiles, and is closer to the metal surface over peaks of the profile than over the recesses. Access of both the levelling agent and of metal complex ions is therefore slower to the surfaces of the recesses than to those of the peaks. There is a considerable body of evidence which supports the view that it is these differential rates of access to peaked and recessed areas which result in inhibition of metal deposition on the peaks during levelling.

A wide variety of apparently unrelated substances act as levelling agents in gold baths and most tend to be incorporated to some extent into the deposit. Most work on levelling action has been carried out with metals other than gold.

Brightening

A matter of some interest[63] is the fact that levelling agents do not necessarily produce bright deposits, that is, deposits for which there is a high ratio of specularly to diffusely reflected light. Where plating baths do not yield sufficiently bright deposits it has been found that the addition to them of what are known as brightening agents is helpful. The mechanism by which these brightening agents modify the crystallisation of the metal so as to impart to it a bright surface is not known. The growth of crystal faces at different rates, however, leads to faceting of the growing metal surface, and therefore tends to roughen it and increase diffuse reflection from its surface. There is therefore reason to conclude that uniformity of growth at crystallographic level is an important factor in brightening. Other factors such as the decreased grain size, which is observed in many bright deposits may, however, also be important. There is general agreement that organic brighteners resemble organic levellers in producing their effects as a result of adsorption processes at the metal surface. There is much that is still little known, however, about the interrelationships between these two effects.

The importance of grain refinement in the brightening of gold electrodeposits is underlined by the extensive use that has been made of small additions of other metals and semi-metals, and particularly arsenic and antimony, to gold plating baths as one means for obtaining brighter (and harder) deposits. Although the amounts of such metals used may be so small that they are not incorporated in the deposits to a significant extent, they do have pronounced grain refining and brightening effects on the deposits. Arsenic, for example, in concentrations of 15 p.p.m. in a gold sulphite bath has been found to convert a very dull, smooth brownish deposit into a bright, shiny and very smooth one.[75] Metals such as nickel, cobalt and indium, which are also widely used in gold alloy plating baths, tend to produce their grain refining effects at higher concentrations.

The hardness of electroplated gold has also been reported as being improved by exposure of the plating bath to an ultrasonic field during the deposition process.[26,76]

6.4 Modes of growth of pure gold deposits

Of the studies which have been conducted on the growth and structure of pure gold electrodeposits those of Craig *et al.*,[75] of Cheh and Sard[54] and Huettner and Sanwald[77] will be described briefly.

The deposits studied by Craig *et al.*[75] were made on copper cathodes from fourteen types of plating solutions, some of which were newly prepared, and some replenished baths from factory production lines. Thick (127 μm) deposits were examined by X-ray, SEM, gravimetric an electrical techniques. It was found that the purity, microstructure and apparent density of the deposits depended on the type of plating solution used and its age; the ageing effects were associated with build up of impurities or exhaustion of additives. The surface microstructures included course grain structures with deeply penetrating grain boundaries from hot alkaline cyanide solutions, while the deposits from other cyanide baths were of non-uniform (5–50 μm) grains with tight grain boundaries. From a pure sulphite bath, however, the deposit had a small and uniform (0.2–5 μm) grain size and grain boundaries were not tightly closed, and when as little as 15 p.p.m. of arsenic was added to this bath, the surface of the gold deposit was featureless even when viewed at 30,000X.

The preferred orientation of each deposit was studied by examining the deposits both from the substrate and bath sides. In this way it was found that from most new baths without hardeners or brighteners, deposits tended to nucleate and grow (111) initially, with this orientation being replaced by others and particularly (311) as deposition proceeded. Special hardeners or brighteners in new baths, apart from decreasing the grain size, affected the orientation of the gold crystals and the appearance of the deposits. With use and regeneration of baths, however, differences between deposits tended to diminish.

Cheh and Sard's studies,[54] on the other hand, were carried out on rotating disk electrodes using only pure dilute [1–2 g $KAu(CN)_2$/litre] cyanide-, citrate- and phosphate-buffered solutions of pH's 10.8, 5.0 and 7.5 respectively, free from additives. The cyanide system showed a tendency to form an outward growth type structure consisting of fine (0.1–1.0 μm) projections or spikes which gave a brown appearance to deposits. This tendency was enhanced by increases in current density or by decreases in speed of rotation, and was less pronounced in the citrate and in phosphate systems. Smooth layer types of growth were observed at low over-potentials. It was concluded that the morphological structures of their deposits were governed by the electrode potential below the limiting current density, and by mass transport near or above this limiting current. The initial stages of deposition were complex, consisting of a substrate-dominated stage which resulted in epitaxial layers several hundred angstroms thick, followed by another stage where three-dimensional growth set in, initiated at structural defects in the epitaxial layer.

The fact that cyanide electrolysis products, formed during the plating of gold from cyanide baths can induce gross changes in the morphology of gold plated from such baths,[77] emphasises the considerable effects which bath constituents can have upon the forms in which gold is electrodeposited. The properties of a range of gold electrodeposits have been reported by Duva and Foulke.[78]

Under practical conditions, the use of addition agents is of the greatest importance, since it tends to reduce or eliminate substrate effects, and to promote the formation of fine-grained and often layered coatings.

A reduced grain size and a high rate of sidewise crystal growth can be very effective in increasing the rate of pore closure during gold plating, and one company[49] has developed a gold plating bath to produce deposits of reduced porosity along these lines.

6.5 The structure of gold alloy deposits

Electrodeposited binary alloys may or may not be of the same structure as those formed metallurgically.[79] Electrodeposited Au-Cu alloys provide what is perhaps a classical example of this phenomenon. According to the conditions under which they are deposited they can, as described earlier, consist either of Au-Cu solid solutions, which show little Au-Cu ordering when freshly deposited, or of mixtures of such solid solutions with copper, despite the fact that the latter mixtures are thermodynamically unstable under the conditions under which they are formed.[29,80] Au-Cu electrodeposits, because of their content of free copper, may therefore be susceptible to tarnish, unless they are heat treated in order to bring them to the fully solid solution equilibrium state. An appreciation of these facts was of great importance in the early development of Au-Cu alloy coatings for watch cases.[81]

The reverse effect, namely, formation of super-saturated solid solutions has also been observed. Examples are the electrodeposited Au-Sb alloys, in which hardness and electrical resistance data indicate the formation of solid solutions containing up to 3–4 atomic % Sb.[82] Anomalies have also been observed in regard to the phases present in electrodeposited Au-Ni,[79,83] Au-Co[83] and Au-Fe[83-85] alloys, though Au-Ag appears to deposit always, at least from cyanide baths, in the expected fully solid solution form.[79]

With small concentrations (for example, less than 1% Co) the gold lattice is apparently preserved in the deposited alloys. With larger concentrations of alloying metal in the electrodeposits (for example, about 15% Ni, or about 35% Cu in Au-Cu deposits from certain sulphite baths) solid solution phases appear whose lattice constants differ from those of gold.[86]

Fedot'ev and Vyacheslavov[79] suggest that the possibility of the formation of supersaturated solid solutions increases with decreasing difference between the electronegativities of the components of the alloys. The conditions under which electrodeposition takes place are clearly a most important factor, for in a large number of studies it has been observed[22] that the codepositing metals separate in "mixed crystal" or solid solution form only if the less noble metal begins to separate at the lowest current densities, and not only after the limiting current density for deposition of the nobler metal has been exceeded. Where the deposition potentials of the two metals concerned are far apart therefore as is the case, for example, with gold and copper in *alkaline* cyanide media, the tendency is for the two metals to deposit individually. Where on the other hand the deposition potentials of the two metals are relatively close together, as is the case for gold and copper in cyanide media of lower pH in which copper behaves more nobly, then copper begins to deposit with gold at current densities well below the limiting current density for gold. Under these conditions, and provided copper concentrations are not too high, the gold and copper separate largely in solid solution form.

In ternary gold alloy deposits, similar but more complex considerations arise, as is exemplified by the discussion of the properties and structures of Au-Cu-Cd alloy deposits by Dettke and Ludwig.[87]

7. Electroplating of Pure Gold from Cyanide Baths

7.1 Introduction and historical development

Cyanide baths today are virtually exclusively gold(I) cyanide baths, and may be replenished where necessary by additions of $KAu(CN)_2$, which is commercially available, rather than by the dissolution of gold anodes. Because of its lower solubility, the sodium salt is not used, and for this reason use of the sodium salts is to be avoided in general in the making up of cyanide baths. In the circumstances little purpose would be served by discussion of "ferrocyanide" baths or of baths containing the gold(III) cyanide complex $Au(CN)_4^-$, which are documented elsewhere[4] and are now mainly of historical interest.

Prior to 1944 it was difficult to obtain bright and thick coatings of gold and its alloys directly with the plating baths available at the time. Buffing and polishing of the deposits were essential parts of any plating process.

In 1944, however, Hischmann[34,88,89] claimed processes by which it was possible to obtain bright thick coatings of gold alloyed with copper, nickel, zinc and cobalt. The baths contained phosphates and sulphites as conducting salts and buffers, and strict procedures ensured that they remained constant in composition. One important step was the addition at intervals of barium cyanide in order to precipitate any carbonates formed by cyanide decomposition. Moreover a modulated current was used to minimise polarisation effects ("Aurodur" process).

In 1953 Ostrow[90] found that smooth and thick semi-bright to bright deposits of gold with such metals as copper, nickel and silver could be obtained from alkaline cyanide baths in which the base metals were present as their complexes with aliphatic hydroxy amines and diamines. The technique of using more than one complexant in gold alloy plating baths dates from this time.

A number of other developments were taking place at this time. Thus Raub[91] and Raub and Sautter[30] were revealing many of the facts relating to deposition of gold and its alloys from cyanide baths which have been described above, and Raub had also made his pioneering observations on the constitution of Au-Cu electrodeposits and their importance in determining the properties of such deposits. Important reviews of gold and gold alloy plating by Raub and Sautter,[30] Wullhorst,[92] Loebich[93] and Parker[94] undoubtedly assisted developments. It was in this period that the findings of Volk became better known. Volk had recognised that free cyanide concentrations should be kept low if base metals were to be deposited with gold, and had established[95] that alkaline cyanide gold baths remained stable when buffered to neutral pH levels by such reagents as primary and secondary phosphates, and that they could be used for the plating of Au-Ag, Au-Cu, etc. alloys. In this period also came the patents of Spreter and Mermillod.[96,97] In the first of these was disclosed the addition of oxidising agents such as nitroguanidine and nitrourea to gold plating baths as a means for preventing evolution of hydrogen at the cathode, and the resulting surface irregularities and brittleness which were caused in the deposits. In the second there was disclosed the use of gold alloy plating baths in which base metals such as copper, zinc and nickel were present not as their cyanide or amine complexes, but as their chelate complexes with such amino carboxylic acids as EDTA. These baths found considerable use by the Swiss watch industry, largely as a result of studies[81,98] of the constitution of the deposits obtained, and the effects of heat treatment in increasing their resistance to wear and to corrosion. The Spreter-Mermillod gold bath of Dérobert et Cie de Genève has been described.[81,99]

In the 1960s came a further series of reviews by Raub[100] and by Korovin,[101] and the disclosure by Rinker and Duva[102] and by Erhardt[103] that the $Au(CN)_2^-$ ion is stable in acid media down to a pH as low as 3 before decomposition with precipitation of AuCN occurs. This opened the way to the development of the modern acid bright gold and alloy gold baths. At the same time an appreciation developed of the difficulties arising from the accumulation of cyanide in baths of the Spreter-Mermillod type, which were accordingly adapted to operation at lower pH levels.[104,105]

Throughout the course of the above events, the main interest in all developments was the plating of bright thin deposits, an objective which was realised most easily by codepositing other metals with the gold. With increasing knowledge of grain refining, levelling and brightening, and with the growth of a demand from the electronics industry for bright high purity gold electrodeposits, however, baths for the production of such deposits were soon developed, which operated not only under alkaline conditions in the presence of excess cyanide, but also under netural and acid conditions.

The same degree of attention has not been focused in recent years upon the plating of thick gold and gold alloy deposits and upon the electroforming of gold. These topics are discussed in a later section on electroforming.

7.2 Plating of pure gold from $Au(CN)_2^-$ in the presence of excess cyanide: Alkaline baths

As already described, the main effect of excess cyanide in an $Au(CN)_2^-$ bath is to reduce the concentration of free (i.e. aquo-complexed) Au^+ ions, and thereby to make the reduction potential for the gold deposition more negative. Where other metals are present in the bath, as in gold alloy plating, it may also, as in the case of silver and copper, for example, act as a complexing agent for these, and in this way exercise a significant influence in the reduction potentials for their deposition also. It is therefore a critical factor in the plating of decorative gold alloy coatings from alkaline cyanide baths.

In addition to this primary role, however, excess cyanide can play a number of other roles. Where gold anodes are used, for example, it promotes dissolution and prevents passivation of the anode. It also increases the throwing power and the conductivity of the bath,[9] and in the plating of gold itself, has a marked effect on the structure of the deposited metal. Thus Bauer[96] found that with small amounts of excess cyanide, gold deposits were coarsely crystalline, treed and relatively dull in appearance. As the amount of excess cyanide increased the deposits first became smoother and lost their crystalline appearance, but later became very coarsely crystalline again. The best results were obtained with molar ratios of cyanide to gold of 20:1, regardless of gold concentration. Similar results have been recorded by Gore and Seegmiller.[107] Even under optimum conditions, however, pure gold coatings from such baths tend to be coarse-grained and matt in appearance and must be brightened by buffing and polishing.

These direct effects of excess cyanide in the deposition of gold from $Au(CN)_2^-$ solutions tend to be obscured in operating baths by secondary effects brought about by the generation and accumulation of cyanide decomposition products such as formates, carbonates, cyanates, urea, ammonia and "polymeric materials". These are formed by hydrolytic and anodic oxidation processes, of which the former in particular are accelerated in hot plating baths. These processes have been reviewed by Page.[4]

The generation of the "polymeric material" as a result of reactions of the cyanide ion at the anode can affect the quality of the electrodeposit if this material is later codeposited with, or occluded in, gold plated at the cathode. Periodic or continuous clarification of plating baths is therefore advantageous.

Of the salts (usually potassium) which may be incorporated as conductive salts or buffers in baths containing excess cyanide, the more important include phosphates, polyphosphates, organic salts such as sodium potassium tartrate, hydroxides, borates and carbonates. Some of these also act as complexing agents for base metal impurities in the baths, and where very pure gold deposits are required chelating agents such as EDTA may also be added to reduce contamination of the coating with other metals. Although salts such as sulphites, thiosulphates, thiocyanates and sulphides have been added, they seem where used today to function more as brightening agents than as conducting agents. Grain refining agents and organic brighteners may also be employed.

The compositions of some typical cyanide gold plating solutions containing excess cyanide are listed in Table XLII. It appears, however, that such baths for plating of pure gold are, to an increasing extent, being replaced by baths containing no excess cyanide, operated under neutral pH conditions.

7.3 Plating of pure gold from $Au(CN)_2^-$ with no deliberate excess of cyanide: Neutral and acid baths

The matt, or at best, semi-bright gold deposits from $Au(CN)_2^-$ in the presence of excess cyanide are unsuited for many applications. The plating of bright and pure golds from $Au(CN)_2^-$ in the absence of excess cyanide was therefore a most important development.

The first baths of this type were the so-called neutral baths, operated in the pH range 5.5–8.0, and were developed in the light of experience in the plating of gold alloys under these conditions, by the methods of Volk[55] and of Spreter and Mermillod.[96,97] They are used extensively in barrel plating[109] of electric components where their good throwing power and metal distribution, and the brightness, smoothness and relative tarnish resistance of the coatings which they yield offer distinct advantages.[110] Typical bath formulations are listed in Table XLIII, which is reproduced from Reid and Goldie.[3]

Apart from the absence of excess cyanide, and the use of grain refining agents such as sodium arsenite, of a variety of levelling and brightening agents, and of complexing agents to prevent codeposition of base metals such as copper, they do not differ greatly in their formulation from the alkaline cyanide baths listed in Table XLII.

As with neutral gold baths, the focal point of initial interest in plating from acid cyanide baths was the production of hard and bright deposits. Since such deposits were first achieved by the use of alloying metals, it was not until high-purity gold deposits were called for in semi-conductor applications that attention was turned to the plating of pure golds from such media. In the first baths organic acids such as citric acid and their salts were employed as buffering agents. The ability of such acids to form complexes with a range of base metals was an additional and significant advantage. The use of additives for grain refinement and for hardening of the deposits is a feature of most bath formulations and a wide variety of materials for this purpose have been claimed in the patent literature. For example, thallium is claimed[115] as a grain-refining and brightening agent in addition to arsenic, and claims have been made that the addition of various alums[116] and of hydrazine sulphate[117] hardens the deposits.

Table XLII. Typical Alkaline Cyanide Electrolytes for Plating Pure Gold*

	Ref. 26	Ref. 108	Ref. 9**	Ref. 9***
Au (as $KAu(CN)_2$)	56	40	41	82
KCN	60	7.5	20	30
Conducting	—	—	20	30
Salt†	Optional	—	(K_2HPO_4)	(K_2HPO_4)
K_2CO_3	—	—	20	30
KOH	—	40	—	—
pH	10.5–11.5	>13	11-11.5	11-11.5
Temperature (°C)	60	75	50–60	50–65
Current density (A/dm²)	1.0††	4.5	0.1–0.5	0.1–0.5
Wetting agent†††	0.1	—	—	—

* Concentrations in g/litre
** For barrel plating
*** For rack plating
† K_2HPO_4, $(NH_4)_2HPO_4$, sodium potassium tartrate, etc.
†† With ultrasonic agitation
††† Pentafluoro-octanoic acid as alkali salt

Table XLIII. Typical Neutral Cyanide Electrolytes for Plating Pure Gold

Constituents	Bath composition (g/litre) and operating conditions				
	Ref. 111	Ref. 112	Ref. 102	Ref. 113	Ref. 114
Au (as $KAu(CN)_2$)	10–31	10–20	7–18	6	8.2
Potassium dihydrogen phosphate	60	—	82	—	—
Potassium pyrophosphate	—	—	—	—	150
Citrate	60*	60–125**	50–75†	90††	
As (as $NaAsO_2$)	0.02–0.04	—	—	—	—
Sodium thiosulphate pentahydrate	5–10	—	—	—	—
Ethyl dihydrogen phosphate	—	30–60	—	—	—
Amino trimethyl phosphonic acid	—	—	—	80	—
Benzyl alcohol (optional)	—	—	—	—	0.05fl-wt
pH	5.5–8.0	6–8	5–6.5	6.0	7–8
Temperature (°C)	60	70	45–100	65	60
Current density (A/dm²)	0.1–1.5	0.1–0.3	0.1–0.4	0.1–0.5	0.1

* potassium salt
** potassium or ammonium salt
† ammonium salt
†† water soluble citrate

(From Reid and Goldie[3])

Table XLIV. Typical Acid Cyanide Electrolytes for Plating Pure Gold

Constituents	Bath composition (g/litre) and operating conditions					
	Ref. 101	Ref. 102	Ref. 118	Ref. 115	Ref. 119	Ref. 117
Au (as $KAu(CN_2)$)	8	4–30	4	8	2.7	16
Potassium phosphate	—	—	45	100	—	—
Citric acid	40	40	—	—	—	55
Sodium citrate	40	—	—	—	—	135
Ammonium citrate	—	40	7.5	—	—	—
Potassium pyrophosphate	—	—	—	—	10	—
Superphosphoric acid (42% pyrophosphoric)	—	—	—	—	10	—
Arsenious oxide	—	—	—	—	0.2	—
Thallium (as sulphate)	—	—	—	0.0025	—	—
Hydrazine sulphate	—	—	—	—	—	6
pH	3–6	2.3–6.6	2.0	3–6	5.5–6.0	4.2
Temperature (°C)	Room	30–60	—	20–90	50–65	70
Current density (A/dm²)	0.5–1.0	0.1–0.8	1.0–2.0	0.1–1.0	0.5–1.0	1.0

(From Reid and Goldie[3])

Typical baths are listed in Table XLIV, which is also reproduced from Reid and Goldie.[3]

The gold plating of beam lead integrated circuits merits mention as exemplifying the problems which may be involved in the precision plating of gold, even after the normal plating variables such as temperature, agitation, conductivity and pH of the bath, and current density, anode-to-cathode area ratio, and use of additives have been optimised. As described by Figuli[33] the deposition of gold on to a wafer containing several hundred beam lead integrated circuits is a two-step plating process. In the first, or light gold, operation the entire circuit pattern, consisting of the inter-connecting paths and beam leads as defined by photoresist is gold plated. In the second, or heavy gold plating step, photoresist is used to define the beam leads, which are then plated further to the required thickness. At this stage the hardness of the plated gold must be kept below about 90 Knoop in order to ensure that the beam leads will be soft enough for easy bonding.

The success of this whole operation of depositing gold on the semiconductor is determined largely by the effectiveness of the photoresist. The types of defect which occur include underplating, gold balls, edge build up on beams and missing gold.

Underplating is the most common of these, and involves lateral deposition of gold under the photoresist. The usual cause is breakdown of the photoresist. Underplating can lead to short-circuiting between conductor paths on the chips. Gold balls on the other hand are heavy deposits of gold, often spherical in shape, which are deposited above the plane on the conducting paths. They result usually from pinholes in the resist used at the heavy plating stage, though they may result also from spikes of gold which develop during the light gold plating and which are not covered adequately by the heavy gold plating resist. They then act as areas of high current density and heavy gold deposition during the heavy gold plating stage. Edge build-up on the beam leads also usually occurs during the heavy plating stage and if not controlled can cause difficulties in later fabrication operations. Missing gold usually results from adherence of undeveloped resist to beam lead or conducting paths, or non-adherent resist depositing in such areas.

In so far as these defects result from degradation of the photoresist in the electrolytic bath to which it is exposed, their incidence increases with length of such exposure, and it is against this background that the development of high-velocity gold plating techniques has been undertaken.

The compositions of the electrolytes which have been evaluated and are used for such plating have not been fully revealed, though they apparently contain high concentration of gold, and in at least one context they are fed at high flow rates across the cathode in order to increase limiting current densities to the maximum. Advantage has also been found[33] from the use of a square-wave periodically reversed current, though with a very high forward-to-reverse current radio.

In the selective plating of spots, stripes and strips[120,121] similar principles apply, the baths being modified to operate at higher current densities, using higher temperatures, rapid flow rates, and highly conducting electrolytes.

8. Electroplating of Gold Alloys for Decorative Purposes from Cyanide Baths

8.1 Introduction

Electrodeposited coatings of gold alloys of various colours are used extensively by the jewellery and cosmetics industries, and such coatings were being applied essentially as an art long before the principles of gold alloy plating were understood. A wide variety of alloying metals, including copper, silver, nickel, cadmium, cobalt, tin, lead and arsenic have been employed. The thicknesses of the coatings vary considerably, from less than 0.5 μm in purely ornamental applications where high tarnish and wear resistance is not aimed at, to as much as 20 μm on objects from which long life and service are expected.

Specifications and legal requirements in respect of the thickness, caratage and description of the coatings exist in some countries and have been reviewed by Mills.[3] They are particularly significant with regard to coatings designed for longer service such as those on watchcases, for which legal requirements were laid down early in Switzerland. The position in other countries varies but there is a tendency in specifications and legislation for the term "gold electroplated" or its equivalent to be restricted to coatings of thicknesses and caratages which are such that the plated products can be expected to be reasonably tarnish-and-wear-resistant, and for the term "gilt" to be specified for the thinner and less resistant coatings.

Colour specifications are normally a matter for negotiation between supplier and purchaser, and are increasingly in terms of the standard colours for watch-case plating already described under carat golds.

8.2 Plating processes

Because gold platings are applied to a range of substrates, such as brass, copper, silver, gold, mild steel, nickel plate, stainless steel and zinc-bases alloys, pretreatment of the basis metal constitutes an integral part of the plating process. Suggested pretreatment and plating sequences have been formulated by Dickinson.[3]

For the plating itself, the compositions and operating conditions of some thirty baths for depositing coatings of different colours were tabulated and referenced by Brenner in 1963.[8] Others have been described by Thews [122] and by Korbelak.[123] As might be expected, all of these except a few which had been reported only recently at that date are alkaline cyanide baths containing excess cyanide, whereas typical baths described by Dickinson in 1974[3] are all of the neutral type. Invariably the gold concentrations in these baths are low (1–2 g/litre) in order to reduce drag-out losses which could otherwise be considerable with the high throughputs and short plating times (5–30 seconds) which are used. Colour variations and adjustments in the deposits from the one bath are normally achieved by varying the applied voltage, modulating the current, or changing the degree of agitation of the bath.

The metals codeposited with the gold in order to achieve the different colours are much the same as those which are alloyed with gold in order to obtain the different coloured carat golds, namely, copper, silver, cadmium, tin, zinc and cobalt. The colours of electrodeposited gold alloys cannot be predicted, however, from a knowledge of the colours of thermally prepared alloys of identical composition.

Even with the greater understanding of gold alloy plating which exists today, the operators of coloured gold plating baths depend heavily upon accumulated experience for control of the baths. It is therefore not unusual to apply the final 0.5 μm of a decorative coating separately and under more controlled conditions than the rest of the coating.

Gardam and Tidswell[124] have described the use of pulsed currents in plating decorative gold coatings.

8.3 Pink, rose and red gold coatings

When alkaline cyanide baths, which contain only gold and copper are used, the change in the colour of the deposit through reddish hues and finally to pink when the deposit contains 20% Cu is relatively rapid. The change is less sharp,[82,84] however, when nickel, silver, cadmium or palladium are present, and such metals, and particularly nickel and cadmium, are therefore frequently added to baths for plating pink and red golds. Other factors which can be used in the control of such baths are the cyanide concentration, an increase in which descreases the current efficiency of copper deposition and tends to make the deposit more yellow, the copper concentration, an increase in which increases copper deposition and makes the deposit redder, and temperature which when increased causes an increase in the efficiency of copper deposition more than that of gold deposition and also makes the deposit redder. The deposits from alkaline baths, in which large excesses of copper are usually used, are homogeneous, bright and ductile.

When neutral cyanide baths were developed they were found more difficult to control than alkaline cyanide baths, since the composition and colour of the deposits from them are very sensitive to changes in operating conditions. Also they and the other fully cyanided baths used at the time all showed ageing effects, with the current efficiency dropping rapidly in use, so that thick gilding presented problems. Although baths of the Spreter-Mermillod type,[96,97] in which the copper and nickel were present not as their complex cyanides but as their chelates with EDTA, overcame these difficulties to some extent, they were not ideal, since at pH 8 at which they were operated, the cyanide liberated at the cathode tended to replace EDTA in its copper complex, so that the baths gradually became more fully cyanided in type. Wiesner and Frey[105] have observed the same phenomenon in the plating of thick Au-Cu alloy coatings from baths of this type.

Table XLV. Compositions and Operating Conditions of Some Pink Gold Plating Baths*

(a) With cyanide as the only complexant	Ref. 127	Ref. 127	Refs. 128,130**
Au (as metal)	0.8 –2.0	1.5	5.1
Cu (as metal)	5–25	3.0	42
Ni (as metal)	0.1–0.3	0.2	—
Cd (as metal)	—	—	0.2
Free cyanide	9–60	1.2	15
Disodium phosphate	15–40	20	—
Sodium sulphite or bisulphite	0.5–2.0	0.5	—
pH	8–10	7.2	—
Temperature (°C)	70–80	75–80	60
Current density (A/dm^2)	0.3	0.3–0.5	0.75
Speed of deposition (μm/h)	12–16	15–20	—
Agitation m/min	2–3	>2	—
Coating caratage	18	18	18
(b) With complexants additional to cyanide			
	Ref. 104†	Ref. 3	Ref. 3††
Au (added as KAu(CN)^2)	1--8	2.0	2.0
Cu (added as EDTA complex)	9–10	2.0	2.5–5.0
Ag (added as KAg(CN)$_2$)	—	0.05	—
Sb (added as tartrate)	0.5–10	—	
Diptassium phosphate	20–120	10	10
Dipotassium EDTA	5–60	5	5
pH	4.5–6.0	6–8	6–8
Temperature (°C)	40–60	35–60	35–60
Current density (A/dm^2)	2	—	—
Coating caratage	>18	18	—

*Concentration of bath components in g/litre. Surface active agents and brighteners not included.

**The rose gold deposit from this bath is reported to be 75Au-18Cu-7Cd.

† Deposits claimed to contain 78-85Au/14-24Cu/0.5-1.5Sb. The alloys from this bath are improved by addition to it of 0.015 g/litre Pb.

†† Reported as yielding deep pink-red deposits.

Another type of neutral bath for the plating of pink golds and patented by Yamamura,[124] contains water-soluble pyridines as complexants additional to cyanide, as well as thiourea.

In acid baths of pH 4–6, such as those of Cathrein and Danemark[104,126] (see Table XLV) such difficulties do not arise, but the coatings deposited from them, although hard and bright, tend to be stressed and to be not as ductile as might be desired. Moreover these baths lack penetration and tend to deposit alloy which is richer in gold and therefore yellowish in cavities.

In reviews of the plating of pink golds, Bacquias[127,128] emphasises the advantages of working with alkaline baths with Cu:Au ratios which exceed 15. He claims that under these conditions the cyanide content of the bath is less critical and that bright and ductile deposits are formed. The fact that the cathode potential curves of a copper cyanide bath depend on cyanide concentration, and that with high enough CN^-:Cu concentration ratios copper deposition is inhibited, is probably relevant in this connection.

A proprietary plating bath which produces bright and ductile pink coatings of the composition 75Au-19Cu-6Cd has been developed.[129]

8.4 Green gold coatings

Green gold electrodeposits have traditionally been obtained from gold cyanide baths to which either silver or cadmium or silver and cadmium are added until a deposit of the desired colour is produced. Silver appears to be the more commonly used.

Shades of green are realised when the gold in the alloys is in the range 75–84%, over which both electrodeposited Au-Ag and Au-Cd alloys are fully solid solution in type. In green gold plating the objective therefore is to obtain deposits in this composition range.

Table XLVI. Compositions and Operating Conditions of Some Green Gold Plating Baths*

	Ref. 94	Ref. 3	Ref. 97
Au (as $KAu(CN)_2$)	2.0	1.0–2.0	1
Ag (as $KAg(CN)_2$)	0.2–0.4	0.125–0.4	—
Cd (as Cd-NTA)**	—	—	2
Free cyanide	15	—	—
K salt of NTA**	—	—	5
K_2EDTA	—	20	—
$Na_2HPO_4 \cdot 12H_2O$	15	—	—
K_2HPO_4	—	10	—
pH	Alkaline	6–8	8
Temperature (°C)	66–71	35–50	60
Current density (A/dm^2)	1.1–3.2	—	—

* Concentration of bath components in g/litre
** NTA=nitrilo-tricetate

Earlier baths for the purpose[8,94] were of the alkaline cyanide type, but neutral baths may also be used.[3]

Since in alkaline cyanide baths, silver behaves more nobly than gold, while cadmium behaves less nobly, the effects of increasing the current density on the composition of the alloy when plating Au-Ag alloys are the opposite to what they are in the plating of Au-Cd alloys. With the former a decrease in the Ag content of the deposit results, while with the latter there is an increase in the Cd content of the deposit.

Table XLVI gives details of one alkaline and one neutral Au-Ag green gold plating bath, and of one neutral Au-Cd green gold plating bath.

8.5 Yellow gold coatings

Yellow gold alloys provide coatings in a range of shades and are preferred to gold itself because of their greater hardness and resistance to wear. Thus even the early process of Bek and Thoma[131] involved heating of the dull gold deposits to alloy them to the substrate, followed by buffing and polishing.

A variety of metals have been used to produce a range of yellow and Hamilton shades. Nickel and cobalt, for example, have been used extensively in flash alkaline gold plating solutions.[132] The main effects of nickel are to lighten the colour of gold itself, to reduce the grain size and the porosity of the deposit, and to retard diffusion of substrate metals such as silver and copper through the plating.

The first fully bright and hard deposits from alkaline cyanide baths, however, were those obtained with the use of metallic brighteners such as silver,[133] tin[134,135] and antimony,[136] assisted in some cases by a variety of organic brighteners.[134] These found a measure of application, and the introduction of complexing agents additional to cyanide in neutral baths, referred to earlier, increased further the availability of thicker bright gold alloy deposits. Despite advantages resulting from thc use of low gold concentrations, good throwing power, high cathode efficiencies and satisfactory performance with either soluble or insoluble anodes, however, the alkaline baths aged and lost efficiency rapidly, were dificult to control and ceased to yield bright deposits when coating thicknesses exceeded about 5μm.

The acid baths introduced first by Rinker and Duva,[102] in which nickel (or cobalt) are incorporated as citrates and in which citric acid and its salts are used as both complexing and buffering agents have therefore proved attractive as a source of yellow decorative coatings. Wilson[136] has discussed in some detail the operating conditions developed for the large-scale plating of thick (>10 μm) fully bright deposits of about 23 carat Au-Ni alloys from a proprietary acid citrate-cyanide bath for which the working specification was the following, with compositions given in g/litre.

Au (as $KAu(CN)_2$)	7.5
Ni (as proprietary complex)	5.0
Citrate complexes as proprietary make-up to stabilise the pH at	4.5
Organic brighteners	
Temperature (°C)	50 ±5
Currency density (A/dm^2)	1.3–1.6
Plating time	about 25 mins for 5 μm deposit

Baths such as these are used extensively in plating for technical rather than decorative purposes and are discussed in greater detail later.

8.6 White coatings

The development of electroplated white golds has been retarded by the fact that such coatings have to compete with electroplated rhodium which can be applied with relative ease. Nevertheless Brenner[8] lists eight white gold cyanide plating baths, all of which are of the alkaline type, and which contain either tin or nickel as the alloying metal. Raub and Bihlmaier[137] have reported on such baths, and Menzel[138] patented baths containing nickel, cadmium and palladium. An acid cyanide bath has been patented[139] for the plating of lustrous yellowish white coatings of Au-Ni alloys.

More recently Greenspan[140] has described alkaline gold cyanide baths containing nickel and zinc from which bright white gold deposits with excellent tarnish and corrosion resistance can be deposited. These have compositions in the range 82–90Au/8–12Ni/2–6Zn and are claimed as being suitable for the plating of tableware. A typical bath of this type has the following composition in g/litre:

Au (as $KAu(CN)_2$)	1–2
Ni (as $K_2Ni(CN)_4$	8–10
Zn (as $K_2Zn(CN)_4$)	0.05–0.2
KCN	3–6
K_2HPO_4	20–50
KOH to adjust pH to	10–12

The nickel content of the bath must be at least four times the gold content. Operating conditions are temperature 40°–80°C and current density 1–11 A/dm^2, and substrates are preferably flash coated with nickel before plating. If the gold content of the deposit exceeds 90% the coating loses its brightness and takes on a yellowish colour.

Sulphite baths have also been patented[141] which are suitable for plating of bright white golds, the example quoted having the following composition in g/litre:

Au (as sodium gold sulphite)	5
Cd (as $CdSO_4$)	15
Ni (as $NiSO_4$)	2.5
Cu (as $CuSO_4$)	0.01
Sodium EDTA	130
Sodium sulphite	100
1–hydroxyethylidene–1, 1–diphosphonic acid	2.5
Proprietary product "Forlanon R" (ml)	0.5

This bath is operated with vigorous stirring at 58°C and a current density of 0.5–0.7 A/dm^2. The composition of the deposit is not disclosed. It is claimed that other metals such as indium, cobalt and iron can also be incorporated in baths of this type.

White gold coatings have been described[142,143] as formed from Au-Pd plating baths.

9. Electroplating of Gold Alloys from Cyanide Baths for Industrial Purposes

9.1 Introduction

The greatest use of plated golds in industry is in electronic circuitry.[144,145] As far as semiconductors are concerned pure gold deposits are required for transistor headers and lead frames, though traces of arsenic or antimony in the deposits may be allowed. Ease of die-bonding is important and the deposits must be stable on heating. Moreover, barrel plating or spot plating of components is called for. For these reasons it seems unlikely that gold alloys will easily find use in this sphere. In contact applications, resistance to tarnish and corrosion is all-important, so that gold alloys for use in this sphere must show outstanding performance in this respect. In printed circuits on the other hand, a critical specification is that for hardness which must not exceed 150Hv if good bondability is to be retained. This requirement is met by neutral or acid pure golds, but the hardnesses of most gold alloy deposits lie in the range 270–420 Hv.

It will be appreciated from the above that the replacement of pure gold by gold alloy coatings in electronic circuitry is not easy, particularly if it is borne in mind that the additional control costs involved in alloy plating must be met by overall savings resulting from the replacement. Unless specifications can be changed, these savings could be small since in some situations the most that could be envisaged would be the application of gold alloys as undercoats, which would be overplated with gold. Moreover, in instances where specifications require that coatings should be capable of withstanding exposure to elevated temperatures, only very high (>95%) gold content alloys of gold with base metals are likely to be suitable.

In the light of the above Mason[144] has discussed certain alloys which might be considered as replacements for gold. These include Au-Cu-Cd, Au-Ag, Au-Cu-Sb, Au-Cd and Au-Cu alloys.

In the meantime, however, the gold alloy platings which are probably used most widely in the electronic industry are the bright hard golds deposited from acid cyanide electrolytes containing cobalt and nickel. Although the amounts of cobalt and nickel which are codeposited with gold from most proprietary electrolytes of this type are so small (0.1–0.2%) that the base metals could be regarded essentially as brightening and hardening agents, rather than alloying metals, they will be treated as the latter in what follows.

For contact applications the 18 carat Au-Cu-Cd alloy coatings have been claimed to offer substantial savings over Au-Co alloy coatings, because of the uniformity with which they are deposited both in rack and barrel plating, and because of their lower density and caratage when deposits of equal thickness are compared.[146]

9.2 Electrodeposition of gold alloys containing cobalt or nickel from alkaline and neutral cyanide baths

Early studies of electrodeposition from alkaline $Au(CN)_2^-$ electrolytes containing cobalt appear to have yielded varied results,[137,147] but more recent studies[148] using radioactive cobalt have show that cobalt cannot be codeposited with gold under these conditions. The $Co(CN)_6^{3-}$ complex is a very stable one. Mermillod[149] in 1961, however, claimed that electrodeposition of Au-Co alloys with a wide range of compositions was possible from mixed cyanide-pyrophosphate electrolytes, a finding which was followed up by Fedot'ev *et al.* in 1968.[150] These investigators used neutral cyanide baths to which the cobalt was added as its pyrophosphate complex, together with excess $K_4P_2O_7$ and Rochelle salt and, chiefly by varying the Au:Co ratio in the electrolyte, succeeded in depositing Au-Co alloys over the whole composition range. The deposits were fine-grained mixtures of gold and cobalt, though the possibility was mooted that a solid solution of cobalt in gold with a maximum cobalt content of 2–2% was present. These electrodeposited Au-Co alloys have not been found to be of technical interest.

Electrodeposition from alkaline $Au(CN)_2^-$ baths containing nickel on the other hand was of early practical interest because of the use of such baths in the production of decorative golds *(q.v.)* and because of the hardening effect of small percentages of nickel on gold. It has been studied by a number of investigators.[137,149,151,154] According to Raub, who has recorded [152] polarisation curves for both acid and alkaline electrolytes, nickel begins to be deposited from alkaline electrolyte at lower temperatures only after the limiting current density for gold has been passed. Solid solution formation between gold and nickel is limited in the deposit which tends to be a heterogeneous mixture of gold and nickel. Only at lower pH's does nickel deposition begin at low current densities, and the composition of the deposited alloys is very dependent upon bath composition and the plating conditions. In general, the deposition of alloys of high nickel content is accompanied by the evolution of considerable amounts of hydrogen and very low current efficiencies.

Like cobalt, however, nickel can be plated out with gold in a wide range of alloy compositions from neutral cyanide-pyrophosphate-tartrate electrolytes.[155]

Of interest is the claim[156] that a bright, hard and stable black gold-base alloy can be plated from a cyanide bath containing gold and nickel, together with an amino acid such as glycine. It is reported as having a caratage of 17.

9.3 Electrodeposition of bright alloy golds containing cobalt or nickel from acid cyanide baths

The extensive use of nickel and/or cobalt in acid cyanide baths for the deposition of bright hard golds[3,9] gives a special interest to the processes of Au-Ni and Au-Co codeposition under these conditions. Baths of this type are typically buffered to a pH of 3.5 with citric acid/citrate mixtures, the citric acid acting also to form complexes with the nickel and cobalt from which base metal ions are released in a controlled manner.[157] There is a significant difference, however, between the relative stabilities of the citrate and cyanide complexes of nickel and cobalt.[158] The $Co(CN)_6^{3-}$ complex is very stable, as is evidenced by the difficulty of codepositing cobalt with gold from alkaline cyanide baths, and even in acid cyanide baths the Co-citrate complex tends to be converted gradually to the $Co(CN)_6^{3-}$ complex by cyanide liberated at the cathode. In this form, cobalt exerts no brightening or hardening effect and is not codeposited with the gold. The $Ni(CN)_4^{2-}$ complex is less stable than the Ni-citrate complex, however, so that the state and function of nickel in a citrate-cyanide bath remain unchanged during its operations.

Other complexing agents may be used either alone or in addition to citrates; examples are polyamines[159] and amino-carboxylic acids such as EDTA.[160] The EDTA complexes of cobalt and nickel are very stable and for best results when used in citrate baths the molar ratios of Co:EDTA and Ni:EDTA, and the current densities, should be kept within defined limits.[11] Such other complexing agents become essential if phosphoric acid/phosphate mixtures[161,162] are used for buffering the electrolyte. In their absence, for example, acid phosphate $Au(CN)_2^-$ baths containing nickel salts give white deposits of Au-Ni alloys, since the nickel is not strongly complexed and the discharge of base metal ions is not controlled. In their presence, however, bright hard gold deposits result.

Where pyrophosphates[163] are used as buffering agents it is necessary to keep the pH from falling or the temperature from rising since these factors promote hydrolysis of the pyrophosphate.

Typical bright acid gold electrolytes are listed in Table XLVII. The specifications for a commercial bath have already been quoted.[136] For bright crack-free deposits, current densities tend to be critical.

Table XLVII. Typical Acid $Au(CN)_2^-$ Electrolytes Containing Nickel and Cobalt for Plating of Bright Alloy Golds*†

	1[158]	2[160]	3[160]	4[164]
Au (as $KAu(CN)_2$)	4	8	12	12
Citric acid	120	25	—	105
Tripotassium citrate	—	35	—	—
Acetic acid/potassium acetate	—	—	60	—
Tetraethylene pentamine	20	—	—	—
KH_2PO_4	—	30	30	—
Ni as N,N′-ethylene diamine diacetate	—	3	—	—
Ni as citrate	2.5	—	—	—
Ni as KNi nitrilo-triacetate	—	—	3	—
Co as CoK_2EDTA	—	—	—	1
H_3PO_4	—	—	—	24
KOH	—	—	—	56
pH	4.0	3.2–4.0	3–5	3–3.5
Temperature (°C)	40	21–49	—	35
Current density (A/dm²)	2.1	—	6.5	0.54

*Concentrations are in g/litre.
† The base metal content of the deposits is low, the amount of Ni in deposits from electrolytes Nos. 2 and 3 being reported as 0.18% Ni.

Although they are hard and bright, alloy golds from acid electrolytes containing nickel and/or cobalt are not very ductile and cannot therefore be applied to surfaces which have subsequently to be shaped mechanically. They also tend to crack as a result of internal stress when thick, and to blister on heat treatment, and their soldering and bonding properties leave something to be desired.[356] Their deficiencies received much publicity as a result of a NASA report[165] which attributed the failure of count-down tests in a Saturn V launching to the failure of soldered connections to deposits of this type. This report recommended against the use of acid cyanide gold plating in circuitry of the type that had failed, and thus provided a strong incentive for greater understanding of this process and for the development of ways and means of overcoming its deficiencies.

As a result, a series of publications[158,166–187] has appeared in recent years, which has served to focus attention on the fact that in acid gold plating, metals such as cobalt, nickel, iron and perhaps others such as chromium and manganese influence the deposit not only as a result of their codeposition with the gold in small quantities, but also as a result of the fact that they promote incorporation of other components of the electrolyte, such as carbon, hydrogen, oxygen, nitrogen, potassium and sodium, in the deposit. Information relating to the manner in which they do this, the forms in which these elements are incorporated, the conditions favourable for their incorporation, and the nature of their effects on the performance of hard acid golds in situations in which they are applied, is now emerging.

Since knowledge in this field is developing so rapidly no attempt will be made to summarise the findings to date. A review[356] of the temperature-sensitive properties of gold and gold alloy electrodeposits is of particular interest, however, since it summarises, *inter alia*, certain important differences between the behaviour on heat treatment of Au-Ni and Au-Co coatings. These differences arise from the fact that cobalt has a much lower solubility in gold than nickel. In the as-deposited condition, therefore, the cobalt in Au-Co alloy coatings containing as little as 0.37% Co or less is present in supersaturated solution. On being heated such coatings undergo precipitation hardening, in the same way as Au-Co coatings[80] undergo hardening by ordering. This is evidenced not only by the changes in hardness and tensile strength which Au-Co coatings display on heat treatment, but also by changes in their magnetic susceptibility and electrical conductivity. Whereas Au-Co coatings are paramagnetic, for example, in the as-deposited solid solution form, they become ferromagnetic on heating. Au-Ni coatings do not behave in this way.

A number of patents,[134,149,187–228] have been granted covering cyanide baths, which contain not only nickel and/or cobalt but also other metals, most of which are essentially modifications of baths of the types which have been referred to above.

9.4 Electrodeposition of gold alloys containing iron

Like cobalt and nickel, iron can be codeposited in varying proportions with gold from suitable alkaline and neutral cyanide electrolytes. Thus Solev'eva[229] and Solev'eva and Lapshina[230] have reported on the use of ammoniacal cyanide-citrate baths for this purpose, Lokshtanova *et al.*[231] have reported on the use of cyanide-tartrate-borate baths, and Solev'eva and Shishakov[232] and Vyacheslavov *et al.*[233] and Fedot'ev *et al.*[234] have described the structures and properties of Au-Fe electrodeposits. Such deposits, however, do not appear to have found applications.

Iron also resembles cobalt and nickel in its effects upon the electrodeposition of gold from acid cyanide electrolytes. Thus Rinker and Duva[157] in their early patent on such electrolytes claimed iron as one of a number of metals which promoted the deposition of gold in hard and bright form. Although iron has likewise been claimed as a brightening and/or alloying metal in a number of patents[117,189,199,206,213,219,225] granted in subsequent years and covering the deposition of acid bright golds by different processes, it has been little used for this purpose, for the brightening effects are accompanied by the generation of high internal stresses in the deposits and a tendency for them to crack.[177]

More recent studies[177,179] have shown that iron, like cobalt and nickel, has the property of promoting the incorporation of carbonaceous material in gold deposits from acid cyanide baths, an effect which appears to be an integral part of the brightening action of iron, since if conditions are adjusted so that the carbon content of the plate falls below 0.1%, the electroplates become dull. Holt, Stanyer and Ellis[179] reported conditions, however, under which considerable codeposition of iron apparently occurred to yield exceptionally bright deposits of a pale yellow colour.

9.5 Electrodeposition of gold alloys containing copper

The electrodeposition of copper with gold from cyanide baths was reviewed by Brenner[8] in 1963, and has been discussed from various points of view above in earlier sections.

References to the plating of gold alloys from electrolytes containing copper are listed in Table XLVIII.

9.6 Electrodeposition of gold alloys containing silver

Scientific and technical experience of the plating of Au-Ag alloys up to 1960 has been reviewed by Brenner.[8] It includes the more basic studies of Field,[235] Grube,[236] Raub[91,237] and Fedot'ev *et al.*[238] on Au-Ag alloy deposition from alkaline cyanide baths. Because of the lower stability of its cyanide complex, silver behaves in such baths as the more easily deposited and therefore more noble metal. In accord with this a gold leaf introduced into a solution of $KAg(CN)_2$ is immediately covered by a deposit of silver formed by a displacement reaction.

If this is borne in mind, Au-Ag alloy deposition from such baths displays all the attributes of a "regular" alloy plating system. Thus dilution of the bath increases the content of the less noble metal gold in the deposit, and with increasing current density the gold content of the deposit increases until the Au:Ag ratio in the deposit approaches that in the bath.

Likewise increase in bath temperature or agitation of the bath increase the percentage of silver in the deposit. The effect of agitation is, in fact, so pronounced that Harr and Cafferty[239,240] found it a major obstacle to the consistent deposition of a given composition of alloy. An effect of bath temperature which is of practical significance is that when the temperature is below 65°C alloy deposits tend to be bright and brittle, while when it exceeds 80°C they tend to be ductile.

Current density-potential curves have been determined for Au, Ag and Au-Ag alloy deposition by both Grube[236] and Raub[91] and the curve for the alloy has been resolved into partial curves for silver and gold deposition, respectively. Interpretations of these are made difficult, however, by the fact that the deposition of both metals and alloys is accompanied by the evolution of hydrogen. At high current densities such as those used by Harr and Cafferty, the composition of the deposit is affected little by changes in the cyanide content of the bath from 8–75 g/litre.

Deposition of the less noble metal, gold, begins at current densities less than the limiting current density for silver, and the gold and silver are deposited in solid-solution form. In accord with the Au-Ag phase diagram, this solid-solution formation occurs over the full range of Au-Ag alloy composition.

Apart from the above findings relating to deposition from alkaline cyanide baths, most additional technical information concerning the electrodeposition of gold alloys from baths containing silver is embodied in patent specifications. References to such information are summarised in Table XLVIII.

9.7 Electrodeposition of gold alloys containing zinc and cadmium

Both zinc and cadmium form weaker cyanide complexes than, for example, either copper or nickel. They therefore codeposit readily with gold even from $Au(CN)_2^-$ baths containing excess cyanide, despite their less noble character. Codeposition of these metals with gold has been studied by Raub and Disam[241] and by Raub[153] using both alkaline cyanide and acid cyanide electrolytes. The acid electrolytes were citrate/cyanide baths to which the zinc and cadmium were added as their sulphates.

Codeposition of zinc with gold took place "normally" in that the cd/ptl curves were at more noble levels in the acid than the alkaline electrolytes and the gold content of the deposits fell with increase in current density. The effects of agitation under alkaline conditions on the polarisation curves were particularly marked. Under these conditions also, zinc was discharged either not at all or in small amounts at low current densities, whereas from acid electrolytes zinc discharge began even at the lowest current densities. In general, the zinc contents of the deposits from acid baths tended to be higher than those from alkaline baths.

In deposits from alkaline baths gold and the intermediate β' phase of the Au-Zn system were observed even when the zinc content was very small. Zinc was apparently not taken up in the gold lattice. The β' phase was replaced by γ_2 and

γ_1 phases at high zinc contents, with zinc accompanying the γ_1 phase at zinc contents above 75%. On the other hand, the β' phase was not observed in deposits from acid baths unless the zinc contents exceeded 20%, and it was inferred from this that Au-Zn solid solutions may well deposit from acid electrolytes.

Similar effects were observed in the case of cadmium, except that whereas under alkaline conditions cadmium deposited preferentially with increase in current density, the reverse held under acid conditions. This behaviour, and the fact that in alkaline baths deposition of cadmium begins at current densities below the limiting current density for gold deposition, distinguishes the behaviour of cadmium from that of zinc. In both alkaline and acid baths, Au-Cd alloys could be deposited in almost any desired composition.

When the deposits contained 1.5–28.8% cadmium, they appeared to be of the solid-solution type, with increasing disturbance of the gold lattice. Though evidence for a saturation limit of 9.5% Cd was obtained, the β-phase was identifiable along with the α-phase only when the cadmium content was 28.8–51.3%. When the cadmium content was greater than 51.3% only the ε and ε + Cd phases were observed. Once the cadmium content of the electrodeposit exceeded the saturation limit, the resistance of the deposits to tarnishing and corrosion fell off rapidly, as might be expected.

It seems that no uses have been found for electrodeposited Au-Zn alloys, but both zinc and cadmium are frequently mentioned as brightening and alloying metals in ternary and multiple metal gold alloy coatings. Electrodeposited gold alloys containing cadmium, for example, are of interest as decorative (pink and green) coatings, while the pale yellow ternary alloy 75Au-10Cd-15Cu has relatively good corrosion resistance and has attracted technical interest. Both zinc and cadmium have also been described as components of baths for the plating of white golds.[140,141]

Table XLVIII summarises references to the codeposition of gold with zinc and cadmium.

9.8 Electrodeposition of gold alloys containing gallium, indium or thallium

Experience of the behaviour of gallium and thallium salts when added to gold cyanide plating baths appears to be confined to instances where these salts have been incorporated essentially as additives and not as alloying metals. Schumpelt,[116] for example, found that the hardness of gold electrodeposits from acid cyanide baths was increased by the addition to the baths of gallium (or aluminium or chromium) alums. Moreover, the addition also of thallium even in low (1–25 mg/litre) concentrations increased the throwing power of the electrolyte greatly,[115] so that it became suitable for barrel plating of intricately-shaped objects, and the deposits were significantly brightened. Small amounts of thallium have similar effects in phosphate-containing alkaline cyanide baths.[242] The lemon-yellow deposits which are formed under these conditions are apparently free from thallium or gallium, though the formation of Au-Tl alloys under such conditions has been patented[243] and gallium and indium are mentioned as being codeposited with gold and tin when added to a bath for the deposition of Au-Sn alloys. Thallium, and lead and arsenic, have also been used as brightening agents in other types of baths.[245] Regarding the effects of thallium in gold plating baths, it is of interest to note that Tl^+ increases significantly the rates of chemical and electrochemical dissolution of gold in alkaline cyanide solutions.[246,247]

In contrast, the corrosion resistance of thermally prepared Au-In alloy coatings demonstrated by Ellsworth,[248] as well as the use[132] of an Au-In electrodeposit containing 0.1–0.5% In for producing a p-type germanium resistor, stimulated early interest in electrodeposited Au-In alloys. As a result (see Table XLVIII) indium is mentioned as a component of a variety of gold cyanide plating baths, sometimes as a brightening agent, and sometimes as an alloying metal. Although indium was reported by Foulke[249] to act like nickel and cobalt as a brightener in acid gold cyanide baths, it is not certain that this is so. Thus Holt, Ellis and Stanyer[179] did not observe the formation of bright Au-In deposits from an acide citrate-cyanide bath containing 10 g In/litre and 10 g Au/litre over a considerable range of temperatures and current densities. They also did not find that carbon was codeposited from such baths in significant amounts, and concluded that earlier observation[168] of carbon deposition from gold baths containing indium was a result of the presence of small amounts of iron, cobalt or nickel in the baths used. Indium apparently does not form a cyano-complex which is stable at low pH.

9.9 Electrodeposition of gold alloys containing tin or lead

Apart from an early report by Raub and Bihlmaier,[137] the available information concerning the electrodeposition of tin with gold is virtually confined to the patent literature. A difficulty with all Au-Sn alloy plating baths[145] is that of preventing the oxidation of tin(II) to tin(IV). While such oxidation can be limited by using soluble tin anodes, gold is deposited on these anodes unless they are isolated by semi-permeable diaphragms. These aspects of Au-Sn alloy deposition have restricted commercial development of these alloys, despite their attractive properties.

Raub and Bihlmaier found that Au-Sn alloys containing up to 30% Sn could be deposited from baths containing no free cyanide, and containing the tin as its stannate complex formed with KOH. Their interest and that of Kersten,[250] who also used alkaline electrolytes, was in the deposition of white electrodeposits which could be used in the jewellery industry. Later claims[117,134,192,195,199,219,244] in respect of Au-Sn alloy plating, however, have been based upon the use of both alkaline and acid cyanide electrolytes, in which the tin has at times been incorporated more with the object of obtaining brightening effects than of producing deposits containing significant amounts of tin.

Raub and Disam[251] and Raub[153] have reported on the deposition of lead with gold from both acid and alkaline cyanide baths. They found that the lead could conveniently be complexed in both types of bath by hydroxy acids, and that in these hydroxy acid-cyanide baths the lead behaved as the more noble metal. Factors which tended to increase the cathodic polarisation favoured separation of gold and factors which tended to decrease such polarisation favoured the separation of lead. In their studies the lead was added as basic carbonate and tartaric acid was used as complexing agent. They found no evidence for deposition of lead in solid solution with gold. When the lead content ranged from 2.8% to 35.7%, deposits consisted of Au + Au_2Pb, and with lead contents of 35.7 to 50.8% they contained Au_2Pb + $AuPb_2$; when the concentrations of lead were higher the only components were $AuPb_2$ + Pb. Other reports on the codeposition of lead with gold are embodied in patents.[126,199,245,252,253] The last two of these patents describe the deposition of gold from cyanide baths which contain lead essentially as an additive.

References to codeposition of tin and lead with gold from cyanide baths are summarised in Table XLVIII.

9.10 Electrodeposition of gold alloys containing antimony or bismuth

The incorporation of antimony in gold alloy electro-deposits has been employed in the production of antique gold finishes,[8] and in the production of germanium and silicon transistors.[132,254] It is now applied mostly in the production of bright and hardened golds for electrical and decorative uses.

These latter applications stem from the observations by Raub[255] and Parker relating to the hardness induced in gold deposits by quantities of antimony as much as 0.3–0.6%.

In their more scientific studies Raub *et al.*[256] and Fedot'ev[257] both added the antimony as its tartrate complex to alkaline gold cyanide baths and they have described the behaviour of such cyanide-tartrate baths in detail. The microhardness of an electrodeposit containing 1.86 atomic % Sb was found to be double that of pure gold. Both the hardness and the electrical resistance of the deposits increased to maxima when the alloy compositions were in the region of that of $AuSb_2$. Fedot'ev *et al.* recommended alloys with about 1.5–2.0% Sb for coating electrical contacts, and specified the following bath for obtaining such coatings:

Au (as $KAu(CN)_2$)	6–8 g/litre
Sb (as K(Sb0) tartrate)	0.3–1 g/litre
NaK tartrate	20–40 g/litre
KCN	8–10 g/litre
Temperature	50°C
Cathodic current density	0.25 A/dm^2
Anodes of Pt or Au	

Patents and other publications[101,104,126,196,199,201,195,227,244,258,259,260–266] indicate a continuing interest in the hardening and brightening properties of antimony in gold alloys of various types.

Electrodeposition of bismuth with gold was found by Raub[27] and by Raub and Disam[251] to take place from alkaline cyanide baths, which like those for deposition of Au-Pb alloys, contained hydroxy acids such as tartaric acid and citric acid as complexants for the bismuth. Like lead, bismuth behaves as the nobler metal under these conditions. A significant feature is the sensitivity of these baths to carbonates. The structures of the deposited alloys corresponded with what was expected from a knowledge of the phase diagram. Apart from these studies, there have been few claims[192,258,242,255,267] recorded in regard to the electrodeposition of Au–Bi alloy coatings which do not appear to have found any practical use (cf. Table XLVIII).

9.11 Electrodeposition of gold alloys containing platinum metals

Although the platinum metals form cyanide complexes, the addition of these metals to gold cyanide plating baths has not proved to be an effective means for electrodepositing gold alloys with such metals. Foulke[268] has commented that the palladium complex is so stable that Au-Pd alloys cannot be deposited on a production basis from baths replenished with $K_2Pd(CN)_4$. Moreover, even when added in combination with some other complexant, platinum metals may form their extremely stable cyanide complexes at the expense of cyanide present as $Au(CN)_2^-$[269]. Greenspan[163] attributes the relative ineffectiveness of palladium as a brightening agent in gold cyanide-pyrophosphate baths, for example, to the formation in the bath of the stable $Pd(CN)_4^{2-}$ complex from which palladium is not deposited in sufficient amount to cause brightening.

Amines or polyamines appear to be the most effective complexants for use with palladium in gold cyanide baths, as shown by the patents of Moore and Butler,[216,217] and others,[270,271] and as reported upon by Vinogradov *et al.*[272] Dialkyldithiocarbamate complexes have also been employed for this purpose.[258]

A wide variety of other conditions have been claimed[273,107,138,199,208,201,222,274–277] for producing alloys of gold with palladium and other platinum metals (cf. Table XLVIII). In many cases the effects achieved or aimed at have been essentially those of brightening or grain refining in the gold or gold alloy deposit.

9.12 Electrodeposition of gold alloys containing manganese, chromium, molybdenum, tungsten or uranium

Alloy coatings of these types have not proved to be of great interest to date, and references to the few patents relating to their deposition which have been noted in the literature have been incorporated in Table XLVIII.

Table XLVIII. Classified References to the Electrodeposition of Various Metals from Cyanide Baths

Metal components of electrolyte (Metals other than Au in alphabetical order)	References
Au, Ag	8, 90, 133, 159, 163, 199, 207, 215, 222, 236-240, 258, 278, 279, 281-284
Au, Ag, Cd	207, 286
Au, Ag, Co	43, 191, 198, 207
Au, Ag, Cu	38, 207, 211, 287-290
Au, Ag, Cd, Cu	36-38, 288, 291
Au, Ag, Cu, Ni	92, 211, 292
Au, Ag, In	207, 280, 293
Au, Ag, Mo	198
Au, Ag, Ni	43, 191, 198, 207, 214
Au, Ag, Ni, Pd	43, 277
Au, Ag, Sb	259, 260
Au, Bi	27, 192, 251, 256, 258, 267
Au, Bi, Sn	244
Au, Cd	43, 97, 153, 164, 192, 199, 213, 216, 217, 222, 241, 258, 294, 295
Au, Cd, Cu	35-37, 43, 87, 130, 146, 277, 284, 295-300
Au, Cd, Cu (Pd, Ni)	209
Au, Cd, Ni	138
Au, Cd, Sb	43
Au, Cd, Sn	244
Au, Cd, Pd, Zn	137, 145, 146, 149, 150, 157, 158, 163, 166-190, 192, 194-197
Au, Co	201-206, 213, 221, 225-227, 285, 294, 300, 302-304, 356
Au, Co, In	164, 192, 194, 196, 199, 225
Au, Co, In, Ni	199
Au, Co, Sb	194, 196, 225
Au, Co, Ni	345
Au, Cr	117, 92, 219, 301, 305
Au, Cr, Zn	219
Au, Cu	8, 27, 31, 43, 80, 81, 91, 96-99, 105, 125, 130, 140, 163, 189, 207, 216, 217, 222, 235, 236, 277, 284, 285, 303, 306-312

Table XLVIII (contd)

Au, Cu, Ni	41-43, 127, 128, 208, 209, 220, 222, 311, 313, 357
Au, Cu, Ni, Zn	97
Au, Cu, Pb	43
Au, Cu, Pb, Sb	126
Au, Cu, Pd	209
Au, Cu, Sb	104, 126, 213, 314, 315
Au, Cu, Sn	43
Au, Fe	117, 157, 177, 179, 189, 192, 197, 200, 206, 213, 219, 221, 225, 229-232
Au, In	157, 179, 192, 196, 207, 213, 216,, 217, 219, 248, 258
Au, In, Pt (Pd, Ru, Rh)	201
Au, Ir	276
Au, Mn	117, 192
Au, Mo	208
Au, Mo, Ni, Zn	316
Au, Ni	136, 137, 142, 149, 151-156, 189, 190, 192, 193, 195-197, 201-206, 212-226, 284, 285, 294, 301-304, 311, 356
Au, Ni, Pd	138
Au, Ni, Zn	192, 210, 98
Au, Os	276
Au, Pb	153, 192, 241, 245, 251-255
Au, Pb, Sn	244
Au, Pd	149, 163, 201, 216, 217, 222, 269-272, 276, 284
Au, Pd, Rh	138
Au, Pd, Zn	201
Au, Pt	201, 222, 269, 276
Au, Rh	201, 272, 273, 276, 317
Au, Ru	274
Au, Sb	101, 117, 132, 192, 195, 196, 199, 255-258, 261-266, 318, 319
Au, Sb, Sn	244
Au, Sn	117, 134, 137, 145, 192, 195, 219, 244, 250, 277, 284, 320, 321
Au, Sn, Zn	244
Au, U	208
Au, Zn	153, 164, 189, 192, 196, 199, 216, 217, 241, 258, 322, 323

10. Electroplating of Gold and Gold Alloys from other Types of Baths

10.1 The sulphite, amine-sulphite and other "mixed" sulphite complexes of gold

Although the plating of gold from sulphite baths was first patented by Woolrich[324] in 1842, information relating to the processes involved is, with very few exceptions, confined to patents granted over the period 1962 to date.

The complexes formed by gold(I) and gold(III) in sulphite media might be expected to be of the types $Au^{I}(SO_3)_2^{3-}$ and $Au^{III}(SO_3)_4^{5-}$, respectively. In practice attempts over the years to isolate and characterise gold sulphite complexes have proved surprisingly unrewarding[325] and as yet there is no unequivocal evidence for the preparation of an $Au^{I}(SO_3)_2^{3-}$ complex in the solid state. Oddo and Mingoia,[326] however, have isolated pure $Na_5[Au^{III}(SO_3)_4]\cdot 5H_2O$. Gold(III) complexes of this type are deep yellow in colour with a red-green fluorescence.

The existence in "pure" sulphite baths of the $Au^{I}(SO_3)_2^{3-}$ complex is nevertheless normally inferred. In one method[327] of preparing such baths, for example, the precipitate formed by adding ammonia to a gold chloride solution is simply dissolved in an alkaline metal sulphite solution, the gold(III) being reduced to gold(I) in the process. This reaction has been studied potentiometrically by Socha *et al.* over a range of pH conditions[328] and evidence for the formation of the $Au^{I}(SO_3)_2^{3-}$ complex was obtained. Recently, however, gold(I) amine-sulphite complexes have been found to have advantages over gold(I) sulphite complexes in plating baths. Unlike the gold(I) sulphite baths, which become unstable if the pH is lowered to the neutral region, baths containing gold(I) amine-sulphite complexes are stable at pH levels as low as 4.5, and are therefore particularly suited to the deposition of gold alloys. In the preparation[329] of such baths the gold may, if desired, be added in the form of solid gold(III) amine-sulphite complexes which are reduced *in situ* by warming of the bath before use. Thus the complex

$$Na^{+}\left[\begin{array}{l}CH_2\text{–}NH_2\\ \mid \\ CH_2\text{–}NH_2\end{array}\cdot Au^{III}\cdot(SO_3)_2\right]$$

can be precipitated by mixing solutions of gold(III) chloride and excess ethylene diamine, cooling, and then adding a saturated solution of sodium sulphite. A square-planar configuration for such complexes has been demonstrated by Dunard and Gerdil[330]. The gold(I) amine-sulphite complexes to which they presumably give rise on reduction do not appear to have been isolated.

Reliable information about the standard reduction potentials or stability constants of the gold(I) sulphite and amine-sulphite complexes is lacking. Thus three discordant values of 0.06 V,[331] 0.105 V[332] and 1.02 V[14] have been reported for the standard reduction potential for the $Au/Au(SO_3)_2^{3-}$ complex. Peschevitskii *et al.* claim that the direct determination of the E° of gold in sulphite solution has not given satisfactory results. The apparent ease with which the amine-sulphite complexes are formed indicates, however, that sulphite groups are much less firmly bonded than are cyanide groups when complexed to gold.

In accord with this Bradford and Brentford[333] have claimed the formation of a variety of mixed complexes of gold by heating the sulphite complex in solution with other complexing agents, among which nitrites, cyanates, citrates, tartrates, phosphites and arsenites were included. Only the mixed complex formed by reaction with sodium nitrite was isolated in the solid state, but no conclusive evidence for its composition or structure was obtained. The deposition of gold coatings from the solutions of the mixed complexes was described.

10.2 Plating of gold and gold alloys from sulphite and amine-sulphite baths

The 1962 patent of Smith[327] was concerned with the plating of pure gold, and the baths claimed contained per litre gold (3–4 g) as its sodium sulphite complex, disodium EDTA (40 ml of 40% solution) and, optionally, hexamethylene tetramine (40 g), though chelating agents other than EDTA were mentioned. Such baths were distinguished by their capacity to yield relatively thick bright and fine-grained gold coatings. The susceptibility of the sulphite baths to deterioration simply on exposure was a deterrent to their more extended use despite the further improvements to the deposits which it was shown by Shoushanian[334] could be effected by the incorporation in sulphite baths of two types of additives. The first of these included a number of semi-metal brighteners (arsenic, antimony, selenium, tellurium) used in amounts of 1–400 p.p.m., and the second included a number of alloying metals (cadmium, titanium, molybdenum, tungsten, lead, zinc, iron, nickel, cobalt, tin, indium, copper, manganese and vanadium) in amounts of 0.5–10 g/litre, which acted both as brightening and hardening agents. Page[4] has commented that presently-marketed proprietary baths of this type appear to contain arsenic and cadmium additives and that they can be operated at considerably higher current densities than usually recommended if the agitation rate is increased, though this would result in some increase in stress in the deposits produced. Even at lower current densities delayed "peeling" or "lifting" can occur if there is an inadequate supply of $Au(SO_3)_2^{3-}$ at the cathode surface during deposition.

These baths enable bright deposits covering a wide range (96–320 Knoop) of hardness to be obtained. Both they and the analogous baths claimed by Baker[335] for the deposition of softer (<130 Knoop) deposits are operated under alkaline ($pH > 9$) conditions.

The development of neutral baths based upon the use of amine-sulphite complexes began with the work of Smagunova *et al.*[336,337] who added amines, and particularly ethylene diamine, to gold sulphite baths in an attempt to make them more stable. The effects of such addition were explored in greater detail by Losi, Zuntini and Meyer[338, 339] who noted that when copper is added with a chelating agent such as EDTA to alkaline gold sulphite solutions it was difficult to obtain deposits which contained less than 75% Au and which were not very susceptible to tarnish. Moreover, the baths were very unstable and agitation was found to affect the composition of the deposits very markedly, so that control of the quality of the deposits was difficult. This same sensitivity to agitation of the bath had also been noted by Wiesner and Frey.[105] Losi *et al.*

therefore explored the effects of the addition of amines to such baths, and observed a considerable displacement of the cathode polarisation curve (see Fig. 51) of an electrolyte containing 0.05 mole/litre of gold(I) sulphite complex when an equivalent amount (0.05 mole/litre) of ethylene diamine was added to the bath. This addition also increased the pH range over which the electrolyte was stable, plating being possible over a pH range of 5–8. A very complete study was made of the deposition of Au-Cu alloys from a bath with the following composition in moles per litre:

Au (as sulphite complex)	0.051
Cu (as Na_2Cu EDTA)	0.039
Ethylene diamine	0.051
EDTA·Na_2H_2·$2H_2O$	0.055
Alkali sulphite	0.25

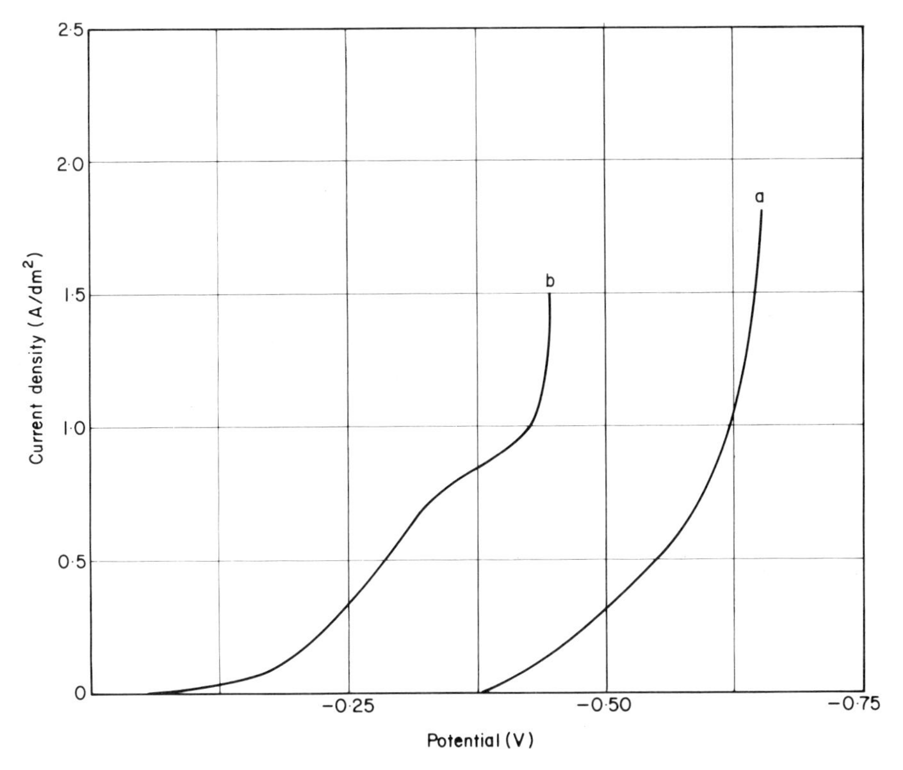

FIG. 51: Current density/voltage curves for a gold(I) sulphite electrolyte.
(a) Solution containing 0.05 mole/litre of gold sulphite.
(b) The same, but with 0.05 mole/litre of aliphatic polyamine added.
(From Losi *et al.*[338])

Further information about amine-sulphite baths of this type for the plating of gold and gold alloys is confined to patents[329,340,342] by Zuntini, Aliprandini, Gioria, Meyer and Losi, and Oxy-Metal Finishing Corporation, from which it is apparent that these baths have opened the way for the successful deposition of a wide range of gold alloys. The literature does not reveal the extent to which these baths are finding application. It has been reported,[86] however, that Au-Cu deposits obtained from these electrolytes contain only small percentages of heterogeneously crystallised copper. Almost complete incorporation of the copper in solid-solution form occurs. Amine-sulphite baths are therefore of interest as a source of Au-Cu electrodeposits which have a low susceptibility to tarnishing.

The use of sulphite baths essentially of the Shoushanian[334] or Smith[343] types for the plating of white golds has been described,[140] Konig[344] has claimed the addition of soluble salts of platinum metals to such baths as a means for reducing the concentration of the alloying metals necessary to achieve particular effects and of improving thereby the properties of the deposits.

The compositions and operating conditions of some of the gold sulphite and amine-sulphite baths referred to above are given in Table XLIX.

10.3 Electrodeposition from gold(III) halide complexes

This has been reviewed by Page[4] and discussed briefly by Mason.[3] Although used in the early days in the plating of gold, gold chloride electrolytes are today applied almost exclusively in the electrochemical refining of gold. No recent reports of developments in this area have been noted, and Mason has commented on the general lack of available information concerning the properties of deposits from gold chloride electrolytes.

10.4 Electrodeposition from gold(I) thiomalate complexes

The thiomalate complex of gold which has been used is sodium gold(I) thiomalate, $NaOOC \cdot CH_2 \cdot CH(SAu) \cdot COONa$, which is reported to be highly stable. Its application in the plating of gold and of Au-Sn alloys has been patented.[346,347]

10.5 Electrodeposition from gold(I) thiosulphate complexes

Although the electrodeposition of gold from thiosulphate solutions of fulminating gold was claimed as early as 1842,[348] and although the thiosulphate complexes of gold(I) are well known and stable in the solid state they appear, like the sulphite complexes of gold, to have attracted for over one hundred

Table XLIX. Composition* and Operating Conditions of some Sulphite Gold Baths

	Ref. 327	Ref. 334	Ref. 334	Ref. 329	Ref. 329	Ref. 335	Ref. 340	Ref. 340	Ref. 341
Au (as $Au(SO_3)_2^{3-}$)	3–4	10	6	–	–	10	–	–	8
Au (as En – Au $(SO_3)_2^-$)†	–	–	–	2	8	–	5	10	–
EDTA (Na_2 or K_2 salts)	16	30	–	20	20	–	20	20	–
Alkali sulphite (Na or K)	–		90	90	50	25	25	10	10
K_2HPO_4	–	–	15	–	–	–	–	–	–
As (as As_2O_3)	–	0.004	–	–	–	0.03	–	–	–
Co (as EDTA complex)	–	–	0.2	–	–	–	–	–	–
Hexmethylene tetramine	40	–	–	–	–	–	–	–	–
En sulphate	–	–	–	–	40	–	–	–	–
Potassium citrate	–	–	–	–	–	70	–	–	–
Oxalic acid	–	–	–	–	–	0.3	–	–	–
Cu (as complex)	–	–	–	–	–	–	4	1.5	–
Sb (as tartrate)	–	–	–	–	–	–	0.2	–	–
Pd (as sulphite-amine complex)	–	–	–	–	–	–	–	–	0.3
Na_2SO_4	–	–	–	–	–	–	–	–	20
pH	–	10	9.5	9–10	7–7.5	8–11	6.5–7	6	7
Temperature (°C)	–	54	49	30–40	60	–	50–60	55–60	65
Current density (A/dm^2)	–	0.8	0.6	2–3	0.5	–	0.3–0.5	0.7	0.5
Caratage of deposit	24	24	24	24	24	24	23 (pink)	16 (rose)	20 (grey)

* Concentrations are in g/litre
† En = ethylene diamine

years[349–351] little attention as a basis for gold plating baths. Recent reports relating to their use in this way are relatively few.

Pouradier and Gadet[352] have, however, reported that the standard potential of the system $Au/Au(S_2O_3)_2^{3-}$ is +0.153 V at 25°C, the stability constant of the $Au(S_2O_3)_2^{3-}$ complex being 10^{26}, which is substantially less than that of the $Au(CN)_2^-$ ion. Peschevitskii *et al.* have also reported values for $E°$ (1.68 ± 0.005 V) and for β_2 (3×10^{29}). Moreover, Safarzynski *et al.*[353,354] have described the effects of current density, buffering salts, EDTA and thiosulphate concentrations upon the plating of gold from a bath containing gold as its thiosulphate complex, 10 g/litre; buffering salt, 1–20 g/litre; and EDTA, 0–0.5 M. Current efficiences and plating rates were high, the latter being two to five times higher than that from acid cyanide baths.

Mixed amine-thiosulphate complexes of gold, analogous to the amine-sulphite complexes which have been discussed above, do not appear to have been studied so far.

11. Electroforming of Gold and Gold Alloys

Electroforming of gold and gold alloys is practised commercially to only a small extent, despite development of interest in this field on the part of jewellery manufacturers. Only limited guidance in regard to processes has been published. This reflects a general lack of information concerning the technology of the production of thick gold deposits and of the electrolytes required. A primary requirement in electroforming is that the deposits should not be too stressed and that they should be mechanically strong, so that they do not distort, crack or collapse on removal of the support upon which they are formed. The electrolytes should be stable over the long plating times required and should have good throwing and levelling powers, since bright and uniform deposits over intricate shapes are usually required.

Electroforming may be conducted either on a pattern produced or generated by photography (photoforming) on a substrate or on a mandrel produced mechanically.

The photoforming process is generally used for the production of perforated products such as sieves, filters, electron microscope grids, optical apertures or electronic devices where comparatively thin (10 μm–200 μm) deposits are made on selected areas of a substrate, which are defined accurately by photographic methods. For the latter purpose the substrate metal is coated with a photosensitive lacquer or photoresist which may be either of the positive type, which on exposure to actinic light becomes soluble in a developer, or of the negative type, which is soluble in the developer initially, but made insoluble on exposure to light. Whichever of the two types may be used, the final result is that certain preselected areas of the substrate metal surface are exposed. After the resist over the remainder of the substrate surface has been suitably hardened to make it resistant to the electrolyte, the exposed areas of substrate metal are plated to the desired thickness with gold or gold alloy and the substrate removed, if so required, to give the electroformed product.

Electroforming on mandrels may be used for producing a variety of products. The mandrels are either of base metal or base metal alloys, or made from epoxy resins, waxes, or other material. In the latter case the surface of the mandrel must be made electrically conducting by coating it with graphite or base metal powders or by electroless plating with copper, nickel or silver. In either case, a parting layer must be present on the surface so that the electroform can be separated easily from the mandrel. In some instances base metal substrates may be preferentially dissolved or etched away from the final electroform.

Developments in these areas have been reviewed recently by Mason[3] and by Mohan.[355]

REFERENCES

1. Hunt, L. B., The early history of gold plating: A tangled tale of disputed priorities, *Gold Bulletin*, **6** (1973), 16–27.
2. Weisberg, A. M., (see Ref. 3, pp. 3–11.
3. Reid, F. H. and Goldie, W., "Gold Plating Technology". Electrochemical Publications, Ltd., Ayr, Scotland, 1974.
4. Page, R. T., "Review of Gold Electroplating Solutions". Fuel and Metallurgy Journals, Ltd., London, 1974.
5. Page, R. T., A review of gold electroplating solutions: Parts 1–8, *Met. Finish. J.* **19** (1973), 274–279, 304–308, 338–345 and **20** (1974), 4–9, 29, 33–36, 62–65, 87–90, 119, 122–125.
6. Rapson, W. S., "The Electrodeposition of Gold Alloys: Parts I–IV". Chamber of Mines of South Africa, Research Organisation, Research Reports Nos. 21/75, 25/75, 27/75, 28/75, 1975.
7. Von Krusenstjern, A., "Edelmetall-Galvanotechnik: Dekorative und technische anwendungen". Eugen G. Leuze Verlag, D7968, Saulgau/Wurttemberg.
8. Brenner, A., "Electrodeposition of Alloys: Principles and Practice". Academic Press, New York and London, Vols I and II, 1963.
9. Lowenheim, F. A. (Editor), "Modern Electroplating", 3rd Edition. John Wiley, New York, 1974.
10. Wise, E. M. (Editor), "Gold: Recovery, Properties and Applications". Van Nostrand, Princeton, N.Y., 1964.
11. Raub, E., Theoretical and practical aspects of alloy plating: Part I, *Plating Metal Finish.* **63** (2) (1976), 29–37 and **63** (3) (1976), 30–43.
12. Niehoff, R. T., Faust, C. L., Cady, J. and Freberg, C., "A Bibliography with Abstracts Covering 1940–1965 on the Properties of Electrodeposited Gold and Gold Alloys for Electronic Uses". American Electroplaters' Soc., Research Report No. 58, 1970. Batelle Memorial Institute, Columbus, Ohio.
13. Chemical Society, London, "Stability Constants of Metal Ion Complexes". Chemical Society Special Publication No. 17. The Chemical Society, London, 1964.

14. Finkelstein, N. P. and Hancock, R. D., A new approach to the chemistry of gold, *Gold Bulletin*, **7** (1974), 72–77.
15. Hancock, R. D., Finkelstein, N. P. and Evers, A., Stabilities of the cyanide complexes of the monovalent Group IB metal ions in aqueous solution, *J. Inorg. Nucl. Chem.* **34** (1972), 3747–3751.
16. Finkelstein, N. P., "The Chemistry of the Extraction of Gold" *in* "Gold Metallurgy in South Africa", edited by R. J. Adamson, Chamber of Mines of South Africa, 1972.
17. Peshchevitskii, B. I., Erenburg, A. M., Belevantsev, V. I. and Kazakov, V. P., Stability of gold complexes in aqueous solutions, *Sib. Chem. J.*, No. 4 (1970), 517–523 (English translation).
18. Kazakov, V. P., Lapshin, A. I. and Peshchevitskii, B. I., Redox potential of the gold(I) thiourea complex, *Russ. J. Inorg. Chem.* **9** (1964), 708–709.
19. Groenewald, T., Electrochemical studies on gold electrodes in acidic solutions of thiourea containing gold(I) thiourea complex ions, *J. Appl. Electrochem.* **5** (1975), 71–78.
20. Finkelstein, N. P. and Hancock, R. D., "E°/E° Diagrams. Their Use in the Estimation of Unknown Standard Reduction Potentials, with Particular Reference to the d^{10} Electronic Configuration Ions". National Institute Metallurgy, South Africa, NIM Report No. 1153, 1970.
21. Gedansky, L. M. and Hepler, L. G., Thermochemistry of gold and its compounds, *Engelhard Ind. Tech. Bull.* **10** (1) (1969), 5–9.
22. Charlot, G., Collumeau, A. and Marchon, M. J. C., "Selected Constants: Oxidation-Reduction Potentials of Inorganic Substances in Aqueous Solution". Butterworths, London, 1971.
23. Bockris, J. O'M. and Reddy, A. K. N., "Modern Electrochemistry". Plenum Press, New York, Second printing, 1971.
24. Delahay, P., "Instrument Analysis". MacMillan, London, 1957.
25. Riddiford, A. C., The temperature coefficient of heterogeneous reactions, *J. Phys. Chem.* **56** (1952), 745–748.
26. Schneider, E. B. and Lindell, M. E. (Litton Systems Inc.), "Method of Electroplating and Electroforming Gold in an Ultrasonic Field". U.S. Patent 3,427,231 (1969).
27. Raub, E., Die elektrolytische Abscheidung von Gold und Goldlegierungen, *Metallwissenschaft u. Technik*, **21** (1967), 709–717.
28. Brenner, A., Electrodeposition of alloys, past, present and future, *Plating*, **52** (1965), 1249–1257.
29. Andreeva, G. F., Fedot'ev, N. P. and Vyacheslavov, P. M., Physicochemical properties and structure of electrolytic alloy Au-Cu, *Zh. Priklad. Khim.* **37** (1964), 1465–1471 (English translation).
30. Raub, E. and Sautter, F., Der Aufbau galvanischer Legierungsniederschläge. X. Die Gold–Kupfer–Legierungen, *Metalloberfläche*, **10** (1956), 65–72.
31. Danemark, M. A., A review of the principles of electroplating gold alloys, *Metal Finish.* **10** (1964), 483–489.
32. Yampolsky, A. M., Russian practice in metal plating, *Electroplat. Metal. Fin.* **16** (1963), 76–79.
33. Figuli, E. S., Velocity gold plating of integrated circuits, *West. Electr. Eng.* **18** (1) (1974), 18–24.
34. Hischmann, M., "Verfahren zur Erzeugung galvanischer Niederschläge von Goldlegierungen". German Patent 748,266 (1944).

35. Heilmann, G. (Degussa), "Process for Electrolytic Deposition of Gold–Copper–Cadmium Alloys". U.S. Patent 3,056,733 (1962).
36. Heilmann, G. (Degussa), "Process for Electrolytic Deposition of Gold–Copper–Cadmium Alloys". U.S. Patent 3,586,611 (1971).
37. Deutsche Gold– und Silber–Scheideanstalt, "Gold–Copper–Cadmium Alloy Plating". Dutch Applic. 69.06.666.
38. Hänsel, G., Kusch, W. and Koeditz, H., "Verfahren zur galvanischen Abscheidung von Goldlegierungen". East German Patent 59,022 (1967) (*C.A.* **69** (1968), 112915h).
39. Aliprandini, G., Desthomas, G. and Meyer, A., "Bain pour le dépôt électrolytique d'alliages d'or et son utilisation en galvanoplastie". Swiss Patent 529,843 (1972).
40. Bertorelle, E., The "triangular cell", a new means of checking galvanic baths, *Galvanotechnik*, **4** (6) (1953), 169–175.
41. Winkler, J., "Electrodeposition of Metal Alloys". U.S. Patent 1,951,893 (1934) (cf. German Patent 576,585 (1933)).
42. Winkler, J., German Patent 723,497 (1942).
43. Dettke, M. *et al.*, "Method of Electrodepositing Gold Alloys". U.S. Patent 3,749,650 (1973).
44. Miller, R. C., "Asymmetrical Controlled Current Electroplating". U.S. Patent 3,738,927 (1973).
45. Avila, A. J. and Brown, M. J., Design factors in pulse plating, *Plating*, **57** (1970), 1105–1108.
46. Cheh, H. Y., Electrodeposition of gold by pulsed current, *J. Electrochem. Soc.* **118** (1971), 551–557.
47. Lendvay, J. and Raub, C. J., Mechanical properties of electrolytically separated gold films, *Metalloberfläche*, **29** (1975), 165–167.
48. Brown, R., *Plating*, **63** (1976), 15.
49. Morissey, R. J., Weisberg, A. M. and Shoushanian, H., Methods of reducing porosity in gold baths, *Plating*, **63** (1976), 15.
50. Rehrig, D. L., "The Deposition of Soft Gold by Pulse Plating Techniques". Paper presented to the 63rd Annual Technical Conference of the American Electroplaters' Society, Denver, Colorado, June-July 1976.
51. Rehrig, D. L., Effect of deposition method on porosity in gold thin films, *Plating*, **61** (1974), 43–46.
52. Harrison, J. A. and Thompson, J., The electrodeposition of precious metals: A review of the fundamental electrochemistry, *Electrochim. Acta*, **18** (1973), 829–834.
53. Maja, M., Il comportamento elettrochimico dell'Oro in bagni cianidrici, Atti. Accad. Sci. Torino: *Classe Sci. Fis. Mat. Nat.* **99** (1965), 1111–1122.
54. Cheh, H. Y. and Sard, R., Electrochemical and structural aspects of gold electrodeposition from dilute solutions by direct current, *J. Electrochem. Soc.: Electrochem. Sci. and Technol.* **118** (1971), 1737–1747.
55. Harrison, J. A. and Thompson, J., The reduction of gold cyanide complexes, *J. Electroanal. Chem.* **40** (1972), 113–120.
56. MacArthur, D. M., A study of gold reduction and oxidation in aqueous solutions, *J. Electrochem. Soc.: Electrochem. Sci. and Technol.* **119** (1972), 672–676.
57. Burrows, I. R., Harrison, J. A. and Thompson, J., Gold deposition, *J. Electroanal. Chem.* **53** (1974), 283–291.
58. Bek, Yu. R. and Zelinskii, A. G., *Izv. sib. Otdel. Akad., Akad. Nauk S.S.S.R., Ser. Khim. Nauk*, **3** (1970), 50 (from Ref. 52).

59. Harrison, J. A. and Thompson, J., The kinetics of gold deposition from chloride solution, *J. Electroanal. Chem.* **59** (1975), 273–280.
60. Knoedler, A., Raub, E. and Pfeiffer, W., Gold electrodeposits from cyanaurate(III) solutions, and some characteristics of the cyanoaurate complex, *Plating*, **53** (1966), 765–769.
61. Frank, F. C. *et al.*, Nature, **163** (1949), p. 398, *Phil. Trans. Roy. Soc. London, Ser. A.* **123** (1951), 299.
62. Frank, F. C., *Discuss., Farad. Soc.* **5** (1949), 48–54).
63. Bockris, J. O'M. and Razumney, G. A., "Fundamental Aspects of Electrocrystallisation". Plenum Press, New York, 1967.
64. Despic, A. R. and Popov, K. I. *in* "Modern Aspects of Electrochemistry", edited by B. E. Conway and J. O'M. Bockris, Butterworths, London, 1972.
65. Krichmar, S. I., The theory of the levelling action in the electrodeposition of metals, *Elektrokhimiya*, **1** (1965), 858–861.
66. Kardos, O., Current distribution on microprofiles, Part I, *Plating*, **61** (1974), 129–138. (References to other reviews on macrothrowing power are quoted in this article.)
67. Raub, E. and Müller, K., "Fundamentals of Metal Deposition". Elsevier Publish. Coy, Amsterdam and New York, 1967.
68. Silman, H., Designing for gold plating: Influence on cost and reliability, *Gold Bulletin*, **9** (1976), 38–44.
69. Foulke, D. G. and Johnson, D. C., *Proc. Amer. Electroplat. Soc.* **50** (1963), 107 (cf. Ref. 3).
70. Field, S., *J. Electrodep. Tech. Soc.* **7** (1932), 83.
71. Foulke, D. G. and Kardos, O., Current distribution on microprofiles, *Tech. Proc. Amer. Electroplat. Soc.* **43** (1956), 172–194.
72. Meyer, W. R., *Proc. Amer. Electroplat. Soc.* **23** (1935), 116 and 135, **24** (1936), 135 ff.
73. Schmellenmeier, H., "Smear" effect in the electrolytic deposition of metals, *Korros. Metallschultz,* **21** (1945), 9–11.
74. Gardam, G. E., Smoothing action as a mechanism in bright nickel plating, *J. Electrodepos. Tech. Soc.* **22** (1947), 155–168.
75. Craig, S. E., Harr, R. E., Henry, J. and Turner, P., A comparison of various 24K gold electrodeposits, *J. Electrochem. Soc.: Electrochem. Technol.* **117** (1970), 1450–1456.
76. Vrobel, L., The influence of ultrasonic vibrations on the electrodeposition of gold, *Trans. Inst. Metal Finish.* **44** (1966), 161–164.
77. Huettner, D. J. and Sanwald, R. C., The effect of cyanide electrolysis products on the morphology and ultrasonic bondability of gold, *Plating*, **59** (1972), 750–755.
78. Duva, R. and Foulke, D. G., Properties of some gold electrodeposits, *Plating.* **55** (1968), 1056–1062.
79. Fedot'ev, N. P. and Vyacheslavov, P. M., The phase structure of binary alloys produced by electrodeposition, *Plating*, **57** (1970), 700–706.
80. Wiesner, H. J. and Distler, W. B., Physical and mechanical properties of electroformed gold–copper alloys, *Plating*, **56** (1969), 799–804.
81. Steinemann, S., Renaud, J. P., Hintermann, H. E. and Braun, A., Structure fine et traitement thermique des plaqués or, *Bull. ann. Soc. Suisse de Chronom., Lab. Suisse Rech. Horl.* (1958), 1–14.
82. Fedot'ev, N. P., Vyacheslavov, P. M. and Volyanyuk, G. A., Deposition technology and physicochemical properties of electrolytic gold-antimony alloy, *Zh. Prikl. Khim.* **40** (1967), 1693–1696 (English translation).

83. Vyacheslavov, P. M., Lokshtanova, O. G. and Smorodina, T. P., Phase structure and "structure-sensitive" properties of electrolytic alloys of gold with metals of the iron group, *Zh. Prikl. Khim.* **45** (1972), 1755–1758 (English translation, pp. 1830–1833).
84. Solov'eva, Z. A. and Shishakov, N. A., Structure of electrolytic iron-gold alloys, *C.A.* **65** (1966), 5012.
85. Solov'eva, Z. A. and Lapshina, A. E., Effect of ammonia on the co-deposition of iron and gold, *Electrokhimya*, **3** (1967), 1106–1109 (*C.A.* **68** (1968), 8685a).
86. Grossmann, H., Horn, F. and Bauer, H., Structural investigations of electrodeposited hard gold deposits with various metallic additions, *J. Less-Common Met.* **43** (1975), 291–294.
87. Dettke, M. and Ludwig, R., Zur Abscheidung 18-Karätiger Goldlegierungen: Struktur, Gefüge und Eigenschaften, *Galvanotechnik*, **63** (1972), 729–736.
88. Hischmann, M., "Verfahren zur Erzeugung von Gold und Goldlegierungsniederschlägen aus Dauerbädern mittels des elektrischen Stromes". German Patent 743,955 (1943).
89. Hischmann, M., Neue Vergoldungsverfahren, *Metalloberfläche*, **1** (1949), B57–B59.
90. Ostrow, B. D., "Bright Gold and Gold Alloy Plating Baths". U.S. Patent 2,660,554 (1953).
91. Raub, E., Über die Struktur galvanisch abgeschiedener Metalle und Legierungen, *Z. Elektrochem.* **55** (1951), 146–151.
92. Wullhorst, B., Über die Grundlagen der galvanischen Vergoldung in zyanidischen Lösungen und ihre Anwendung in der Praxis, *Metalloberfläche*, **7** (1953), A49–A58.
93. Loebich, O., Über die galvanische Plattierung mit Gold, *Metalloberfläche*, **7** (1953), B58–B61.
94. Parker, E. A., Electroplating of gold alloys, *Plating*, **38** (1951), 1134–1156 and 1256–1259, **39** (1952), 43–50.
95. Volk, F., "Electrolytic Deposition of Gold and Gold Alloys", U.S. Patent 2,812,299 (1957) (cf. German Patent 801,312 and Swiss Patent 277,997).
96. Spreter, V. and Mermillod, J., "Process for the Deposition of Gold or Gold Alloys". U.S. Patent 2,702,271 (1955).
97. Spreter, V. and Mermillod, J., "Baths for the Deposit of Gold Alloys by Electroplating". U.S. Patent 2,724,687 (1955).
98. Rochat, R., L'effet du traitement thermique sur certains alliages galvaniques, *Bull. Ann. Soc. Suisse de Chronom., Lab. Suisse Rech. Horlog.* **4** (1957), 45–49.
99. Dérobert et Cie, "Étude du plaqué or galvanique 'thermocompact' et contribution à l'étude des plaqués galvanique". Geneva, 1957.
100. Raub, E., Die galvanische abscheidung von Gold und Goldlegierungen, *Galvano*, **2** (1961), 14–15, 119–120, 123–126 and 166–169.
101. Korovin, N. V., Alloy deposition: Theory, practice and current applications, *Electroplat. Metal Finish.* **17** (1964), 117–121, 151–156, 188–192 and 249–250.
102. Rinker, E. C. and Duva, R., "Electroplating smooth ductile gold". U.S. Patent 3,104,212 (1963).
103. Ehrhardt, R. A., Acid gold plating, *Proc. Am. Electropl. Soc.* **47** (1960), 78–82.
104. Cathrein, R. and Danemark, M., "Electrolyte and Method for Coating Articles with a Gold–Copper–Antimony Alloy and Article Thereof". U.S. Patent 3,380,814 (1968).

105. Wiesner, H. J. and Frey, W. P., The electrodeposition of thick deposits of gold–copper alloys: Process development, *Plating*, **56** (1969), 527–532.
106. Bauer, C. L., Notes on the electrodeposition of thick gold deposits, *Plating*, **39** (1952), 1335–1338.
107. Gore, J. K. and Seegmiller, R., "A Cyanide Bath for Heavy Plating". 47th Ann. Tech. Proc. of the Am. Electroplat. Soc., 1960.
108. Edson, G., "Gold Electroplating". U.S. Patent 3,445,352 (1969).
109. Bick, M. and Lochet, J. A. (Auric Corp.), "Gold Plating Baths for Barrel Operations". U.S. Patents 3,791,941 (1974), 3,770,596 (1974) and 3,783,111 (1974) (*C.A.* **80** (1974), 152110a, 43493g and 77638y).
110. Nobel, F. I. and Thomson, D. W., Metal distribution in barrel gold plating: Further studies, *Plating*, **57** (1970), 469–474.
111. Zimmerman, R. and Brennerman, R., Brit. Patent 1,275,386 (1972).
112. Hodgson, R. W. and Szkudlapski, A. H., Gold plating in the manufacture of solid state electronics, *Plating*, **57** (1970), 693–699.
113. Ostrow, B. D. and Nobel, F. I., "Improvements in Electrodeposition of Gold and Gold Alloys". Brit. Patent 1,198,527 (1970).
114. Pokras, D. S., "Method and Solution for Gold Electroplating". U.S. Patent 3,505,182 (1971).
115. Duva, R. and Simonian, A., "Plating of Smooth Semi-bright Gold Deposits". U.S. Patent 3,562,120 (1971).
116. Schumpelt, K., "Acid–gold Electroplating Bath". U.S. Patent 3,367,853 (1968).
117. Duva, R. and Foulke, D. G., "Gold Plating". U.S. Patent 3,156,634 (1964).
118. Camp, E. K., Dietz, G. and Mitchell, N. W., "Acidic Gold Cyanide Electroplating Bath and Process". U.S. Patent 3,303,112 (1967).
119. Greenspan, L., "Gold Plating". U.S. Patent 3,423,295 (1969).
120. Poll, G. H., Gold plating: spots, stripes, strips, *Prod. Finish. (Cincinnatti)*, **39** (1975), 46–54.
121. Anon., Texas Instruments uses spot plating of gold to cut costs of keyboards, *Plating and Surface Finish.* **62** (1975), 846–847.
122. Thews, E. R., The production of colored gold finishes, *Met. Finish.* (1951) (Sept.), 80–85.
123. Korbelak, A., Engineering and industrial gold plating, *Metal Finish. Guidebook*, **27** (1959), 332–343.
124. Gardam, G. E. and Tidswell, N. E., The electrodeposition of gold and other alloys by a new method, *Trans. Inst. Met. Finish.* **31** (1954), 418–425.
125. Yamamura, K., "Glossy Pink Gold–Copper Plating Compositions". Jap. Patent 6932,650 (1969) (*C.A.* **72** (1969), 117224d).
126. Danemark, M., Cathrein, R. and Simonian, A. Y., "Electrolyte and Method for Depositing a Pink Gold Alloy". U.S. Patent 3,380,898 (1968) (cf. U.S. Patent 3,380,814).
127. Bacquias, G., Les techniques de placage or rose, *Galvano*, **40** (1971), 917–921.
128. Bacquias, G., Le dépôt électrolytique d'alliages en milieu cyanuré: l'importance des rapports intermétalliques dans les solutions, *Galvano*, **41** (1972), 289–294.
129. "Auruna 506". Degussa, Frankfurt, 1975.
130. Kawai, S., Kuroda, K. and Ishiguro, I., Studies on bath composition for electroplating of AuCu and Au-Cu-Cd alloys, *J. Met. Finish. Soc., Japan*, **19** (1968), 350–355 *(C.–A.* **70** (1969), 25120x).
131. Bek, E. and Thoma, E., "Process of Finishing Metal Articles and the Product Thereof". U.S. Patent 1,712,244 (1929) (cf. German Patents 528,885 (1931), 545,589 (1932), 556,315 (1932), 563,615 (1932).

132. Parker, E. A., Recent developments in gold alloy plating, *Plating*, **45** (1958), 631–635.
133. Rinker, E. C., "Method and Electrolyte for Producing Bright Gold". U.S. Patent 2,799,633 (1957).
134. Ostrow, B. D. and Nobel, F. I., "Bath for plating Bright Gold". U.S. Patent 2,765,269 (1956).
135. Raub, E., "Verfahren zur Herstellung harter galvanischer Silber– und Goldüberzüge". German Patent 849,787 (1952) (cf. German Patent 843,785).
136. Wilson, R., The control and characteristics of acid alloy gold used as a decorative plate, *Trans. Inst. Met. Finish.* **47** (1969), 42–49.
137. Raub, E. and Bihlmaier, K., Galvanische Weissgoldniederschläge, *Mitt. Forschungsinst. Probieramts. Edelmetalle staatl. höheren Fachschule Schwäb. Gmund*, **11** (1937), 59–65.
138. Menzel, E. (Degussa), "Verfahren zur Herstellung von Weissgold aus Gold einerseits und Nickel, Platin, Platinmetallen oder anderen geeigneten Weissfärbenden Metallen andererseits". German Patent 526,745 (1931).
139. Yamamura, K., "White Gold Alloy Plating Solution". Jap. Patent 70 31,366 (1970) (*C.A.* **75** (1971), 14316t).
140. Greenspan, L. (Engelhard Minerals & Chemical Corp.), "Weissgoldeter Metallgegenstand und seine Herstellung". U.S. Patent, 3,915,814 (1975), German Offen. 2,342,691 (1974) (cf. French Demande 2,196,908 (1973).
141. Olivier, A. (Degussa), "Galvanisches Bad zum Abscheiden von Goldlegierungen". German Offen. 2,334,813 (1974).
142. Vinogradov, S. N. *et al.*, Electrodeposition of a gold–palladium alloy, *Zashch. Met.* **8(i)** (1972), 92–95 (*C.A.* **76** (1972), 107201n).
143. Vinogradov, S. N. *et al.*, Mechanism of the electrodeposition of a gold-palladium alloy, *Zashch. Met.* **10**(1) (1974), 75–77 (*C.A.* **80** (1974), 115371r).
144. Mason, D. R., Problems in the industrial use of electrodeposited alloy golds, *Gold Bulletin* **7** (1974), 104–105.
145. Mason, D. R., Blair, A. and Wilkinson, P., Alloy gold deposits: Have they any industrial use?, *Trans. Inst. Met. Finish.* **52** (1974), 143–147.
146. Nobel, F. I., Thomson, D. W. and Brasch, W. R., A comparative study of metal distribution from 18 karat and 24 karat gold plating solutions, *Plating and Surf. Finish.* **62** (1975), 462–466.
147. Fedot'ev, N. P. *et al.*, Hard gold plating, *Zh. Prikl. Khim.* **29** (1956), 489–492 (English translation, pp. 537–539).
148. Raub, E., Pahlke, S. and Wiehl, H. P., Electrodeposition of alloys of gold and silver with nickel and cobalt, *Metall.* **25** (1971), 735–739.
149. Mermillod, J., "Bain pour le dépôt galvanique d'alliage d'or brillant, procédé de préparation et utilisation dudit bain". Swiss Patent 354,643 (1961).
150. Fedot'ev, N. P. *et al.*, Technology of electrodeposition and physicochemical properties of gold–cobalt alloy, *Zh. Prikl. Khim.* **41** (1968), 296–300 (English translation, pp. 281–285).
151. Atanasyants, A. G. *et al.*, Hard gold plating, *Zh. Prikl. Khim.* **30** (1958), 926–930 (English translation).
152. Raub, E., Die galvanische Abscheidung der Legierungen von Gold mit Nickel und Cobalt, *Metall.* **25** (1971), 735–739.
153. Raub, E., Die elektrolytische Abscheidung von Gold und Goldlegierungen, *Metall.* **21** (1967), 709–717.
154. Ohsuga, A., Iwashita, T. and Matsumura, S., "Electroplating of Gold Nickel

Alloys". Jap. Kokai 74 49,837 (1974) (*C.A.* **82** (1975), 162713j) and Jap. Kokai 74 49,838 (1974) (*C.A.* **81** (1974), 180385h).
155. Fedot'ev, N. P. *et al.*, Electrodeposition technology and physicochemical properties of gold-nickel alloy, *Zh. Prikl. Khim.* **40** (1967), 2253–2258 (English translation, pp. 2167–2171).
156. Yamakazi, T., "Bath for a Black Gold Alloy". Jap. Patent 74 11,978 (1974) (*C.A.* **81** (1974), 85280m).
157. Rinker, E. C. and Duva, R., "Electroplating Bright Gold". U.S. Patent 2,905,601 (1959).
158. Knödler, A., Der einbau von Fremdstoffen in Gold–Cobalt und Gold–Nickel–Niederschlägen aus sauren Bädern, *Metalloberfläche*, **28** (1974), 465–472.
159. Ostrow, B. D. and Nobel, F. I., "Electroplating Baths for Hard Bright Gold Deposits". U.S. Patent 2,967,135 (1961).
160. Parker, E. A. and Powers, J. A., "Bright Gold Plating Process". U.S. Patent 3,149,058 (1964).
161. Foulke, D. G., "Gold Plating". U.S. Patent 3,156,635 (1964).
162. Camp, E. K., Dietz, G. and Mitchell, N. W., "Acidic Gold Cyanide Electroplating Bath and Process". U.S. Patent 3,303,112 (1967).
163. Greenspan, L., "Gold Plating". U.S. Patent 3,466,233 (1969).
164. Parker, E. A. and Powers, J. A., "Acid Gold Plating". U.S. Patent 3,149,057 (1964).
165. Pasciak, A., NASA Report TMX 2290/N 71 24 715 (May, 1971).
166. Vyacheslavov, P. M. *et al.*, Electrodeposition of gold–cobalt alloy from a citric acid electrolyte, *Zh. Prikl. Khim.* **46** (1973), 72–76 (English translation pp. 68–71).
167. Raub, Ch. J., Knödler, A. and Lendvay, J., The properties of gold electrolytes containing carbonaceous material, *Plating and Surf. Finish.* **63** (1976), 35–40.
168. Munier, G. B., Polymer codeposited with gold during electroplating, *Plating*, **56** (1969), 1151–1157.
169. Mason, D. R. and Blair, A., An intermediate investigation into purity of gold deposits, *Trans. Inst. Met. Finish.* **50** (1972), 138–140.
170. Socha, J., Raub, E. and Knödler, A., Über die galvanische Abscheidung von Metallen und Legierungen aus Athylendiamintetraessigsäure und Citronensäure enthaltenden Elektrolyten. Teil I. Die Abscheidung von Gold, Kobalt und Nickel, *Metalloberfläche*, **26** (1972), 125–129.
171. Silver, H. G., Gas content of gold foils deposited from polymer-saturated electroplating solutions, *J. Electrochem. Soc. (Electrochem. Sci,)*, **116** (1969), 591–594.
172. Silver, H. G., Analysis of gold foils and electroplating solution for metallic and anionic impurities, *J. Electrochem. Soc. (Electrochem. Sci.)*, **116** (1969), 741–743.
173. Eisenmann, E. T., Dependence of selected properties of hard gold on eight plating variables, *Plating*, **60** (1973), 1131–1137.
174. Reinheimer, H. A., Carbon in gold electrodeposits, *J. Electrochem. Soc. (Electrochem. Sci.)*, **121** (1974), 490–500.
175. Davies, T. A. and Watson, P., Potassium inclusions in gold electrodeposits: Their influence in reed relay contact behaviour, *Plating*, **60** (1973), 1138–1145.
176. Holt, L. and Stanyer, J., Carbon inclusion in gold electroplate, *Trans. Inst. Met. Finish.* **50** (1972), 24–27.

177. Holt, L., Gold plating from the acid cyanide system: Some aspects of the effect of plating parameters on codeposition, *Trans. Inst. Met. Finish.* **51** (1973), 134–140.
178. Holt, L., Ellis, R. J. and Stanyer, J., The deposition of cobalt/gold alloys from acid gold cyanide electrolytes, *Plating*, **60** (1973), 918–921.
179. Holt, L., Ellis, R. J. and Stanyer, J., Carbon inclusion in gold electroplate: Some aspects of the role of codepositing metals, *Plating*, **60** (1973), 910–917.
180. Kim, C. K. and Vasile, M. J., The simultaneous determination of carbon and oxygen in electroplated gold by helium-3 activation, *Anal. Chim. Acta.* **56** (1971), 339–345.
181. Vasile, M. J. and Malm, D. L., Study of electroplated gold by spark source mass spectrometry, *Anal. Chem.* **44** (1972), 650–655.
182. Malm, D. L. and Vasile, M. J., A study of contamination on electroplated gold, copper, platinum and palladium, *J. Electrochem. Soc. (Electrochem. Sci.)*, **120** (1973), 1484–1487.
183. Huettner, D. J. and Sanwald, R. C., The effect of cyanide electrolysis products on the morphology and ultrasonic bondability of gold, *Plating*, **59** (1972), 750–754.
184. Antler, M., Structure of polymer co-deposited in gold electroplates, *Plating*, **60** (1973), 468–473.
185. Raub, E., Raub, C. J., Knödler, A. and Wiehl, H. P., Incorporation of carbon containing impurities in the electrodeposition of gold and gold alloy coatings in acid and alkaline cyanide baths, *Werkst. und Korros.* **23** (1972), 643–647.
186. Darby, E. C. and Harris, S. J., Structure and mechanical properties of gold–cobalt electrodeposits, *Trans. Inst. Met. Finish.* **53** (1975), 138–144.
187. Aichinger, W. and Buschle, R. (Degussa), "Galvanisches Goldlegierungsbad, insbesondere für die Trommel-galvanisierung". German Offen. 2,220 622 (1973).
188. Smagunova, N. A. and Yudina, A. K., "Electrochemical Deposition of a Gold Based Alloy". U.S.S.R., 287,483 (1970) (*C.A.* **75** (1971), 29284n).
189. Nobel, F. I. and Ostrow, B. D., "Improvements in Electrodeposition of Gold and Gold Alloys". British Patent 1,198,527 (1970) (cf. French Patent 1,539,226 (1968)).
190. Foulke, D. G. (Sel-Rex Corp.), "Bad zum galvanischen Abscheiden glänzender Überzüge aus Goldlegierungen". German Patent 1,621,162 (1972) (cf. German Patent 1,111,897 (1958)).
191. Philippi & Co. K. G., "Bad zum Abscheiden harter und glänzender Goldlegierungsüberzüge". German Auslegeschrift 1,236,897 (1967).
192. Sel-Rex Corporation, "Electroplating Bright Gold". British Patents 851,973 (1960), 928,088 (1963), 1,177,672 (1970).
193. Technic Inc., "Acid Gold Plating". British Patent 880,656 (1961).
194. Technic Inc., "Improvements in or Relating to Electrodepositing Bright Hard Gold". British Patent 921,960 (1963).
195. Il'in, V. A. and Fortovova, L. S., "Electrolyte for Deposition of a Gold Base Alloy". U.S.S.R. 273,614 (197) (*C.A.* **74** (1971), 27534s).
196. Bick, M. and Lochet, J. A., "Bright Gold Alloy Electroplating: Bath and Process". U.S. Patent 3,716,463 (1973).
197. Dalton, I. M. and Jones, D., "Gold Electroplating Electrolytes". U.S. Patent 3,551,305 (1970).
198. Katz, M., "Method for Electrolytic Gold—Silver Plating". British Patent 1,244,095 (1971).
199. Dettke, M., Fuchs, K. H. and Ludwig, R., "Bad zur galvanischen Abscheidung

von Gold und Goldlegierungen". German Offen. 2,244,434 (1974) (cf. U.S. Patent 3,749,650 (1973)).
200. Nobel, F. I. and Ostrow, B. D., "Saures Gold– oder Goldlegierungsbad". German Auslegeschrift 1,280,015 (1968).
201. P.M.D. Chemicals Ltd., "Bain de dorure électrolytique". French Patent 1,388,541 (1965).
202. Weisberg, A. M. and Kroll, H., "Verfahren zum galvanischen Abscheiden glänzender Goldüberzüge mit hohem Goldgehalt". German Offen. 2,236,493 (1973).
203. Scholze, G., "Bath for Electroplating Bright Gold Alloys". East German Patent 96,261 (1973) (*C.A.* **79** (1973), 152323b).
204. Luebke, H. J., "Galvanisches Goldlegierungsbad zur Erzielung definierter hoch- und niedriglegierter Goldüberzüge". German Offen. 1,941,822 (1971).
205. Bruk, E. S. and Kovaleva, R. G., Electrodeposition of gold–nickel and gold–cobalt alloys from citrate-cyanide electrolytes, *C.A.* **74** (1971), 8927f.
206. Dettke, M. and Todt, G., "Elektrolyt zum Abscheiden glänzender Goldlegierungs-niederschläge". German Offen. 1,909,144 (1970).
207. Nobel, F. I. and Ostrow, B. D., "Gold Alloy Electroplating Baths". U.S. Patents 3,642,589 (1972) (cf. U.S. Patent 2,660,554 (1953), 2,967,135 (1961)).
208. Taormina, S. C. and Marinaro, A. T., "Gold Alloy Plating". U.S. Patent 2,754,258 (1956).
209. Campana, C. R., "Gold Alloy Plating Bath". U.S. Patent 2,719,821 (1955).
210. Greenspan, L. (Engelhard Minerals & Chemicals Corp.), "Alliages d'or blanc brillant électrodéposés". French Applic. 73 30,549 (1974).
211. Dirat, F., "Procédé de placage d'or". French Patent 1,038,224 (1953).
212. Yamamoto, S. and Onota, Y., "Acid Gold–Nickel Alloy Plating System". Japan. Patent 71 09,921 (1971) (*C.A.* **76** (1972), 30198g).
213. Foulke, D. G. (Sel-Rex Corp.), "Gold Plating". British Patent 1,156,186 (1969).
214. Yamamoto, S. and Miyazaki, S., "Gold Alloy Plating Solution". Jap. Patent 70 31,367 (1970) (*C.A.* **75** (1971), 14358h).
215. Yamamura, K. *et al.*, "Bright Gold Plating Solution and Process". U.S. Patent 3,475,290 (1969).
216. Moore, T. R. and Butler, F. P., "Gold and Gold Alloy Plating Solutions". U.S. Patent 3,533,923 (1970) (cf. U.S. Patent 3,458,542 (1969)).
217. Moore, T. R. and Butler, F. P., "Heavy Metal–Diamine–Gold Cyanide Complexes". U.S. Patent 3,458,542 (1969).
218. Philippi and Co. K. G., "Gold or Gold Alloy Plating". German Patent 1,262,723 (1968) (cf. German Patent 1,236,073 (1967)).
219. Vrobel, L., "Bath for Electrodepositing of Hard Glossy Gold Coatings". Czech. Patent 107,385 (1963) (*C.A.* **60** (1964), 5079).
220. Kuronova, N. V., Deposition of multicomponent alloys of gold, *Vses. Nauch. Issled. Instit. "Goznaka"*, (1971), (6) 3–10.
221. Nobel, F. I. and Ostrow, B. D., "Improvements in Electrodeposition of Gold Alloys". British Patent 1,065,308 (1967).
222. Culbert, D. P. A., "Improvements in or Relating to Electrodeposition of Metals and Alloys". British Patent 1,218,732 (1971).
223. Imai, H., "bath for Electroplating a Thick Bright Gold Alloy Layer". Jap. Patent 74 23,459 (1974) (*C.A.* **82** (1975), 23753e) (cf. Japan. Patent 74 45,248 (1972) (*C.A.* **79** (1973), 152321z).
224. Kojo, H., "Solutions for Electrodeposition of Thick Bright Gold–Nickel Alloys". Jap. Patent 69 20,844 (1969) (*C.A.* **71** (1969), 119019u).

225. Parker, E. A. and Powers, J. A., "Acid Gold Plating". U.S. Patent 3,149,057 (1964).
226. Fedot'ev, N. P. *et al.*, Increase of the hardness of gold and silver galvanic platings, *Legk. Prom.* **17** (3) (1957), 43–44 (*C.A.* **52** (1958), 925).
227. Fedot'ev, N. P. *et al.*, Hard electroplated gold, *Zh. Prikl. Khim.* **29** (1956), 489–492 (*C.A.* **50** (1956), 14409).
228. Willcox, P. and Cady, J. R., Structure and properties of a gold–0.4 weight per cent nickel electrodeposit, *Plating* **61** (1974), 1117–1124.
229. Solov'eva, Z. A., "Protective Metallic and Oxide Coatings, Metallic Corrosion and Electrochemistry", edited by N. P. Fedot'ev. Translation into English by the Israel Program for Scientific Translations, 1968, p. 217 (*C.A.* **65** (1966), 5004f).
230. Solov'eva, Z. A. and Lapshina, A. E., Effect of ammonia on the codeposition of iron and gold, *Elektrokhimya*, **3** (1967), 1106–1109 (*C.A.* **68** (1968), 8685b).
231. Lokshtanova, O. G., Fedot'ev, N. P. and Vyacheslavov, P. M., Technology of the electrodeposition and some properties of a physical-chemical nature of gold-iron alloys, *Zashch. Met.* **4** (1968), 548–552 (*C.A.* **70** (1969), 25122z).
232. Solov'eva, Z. A. and Shishakov, N. A., Ref. 212, p. 229 (*C.A.* **65** (1966), 5012b).
233. Vyacheslavov, P. M. *et al.*, Phase structure and structive sensitive properties of electrolytic alloys of gold with iron group metals, *Zh. Prikl. Khim.* **45** (1972), 1755–1758.
234. Fedot'ev, N. P. *et al.*, "Electrodeposition of Alloys of Gold with Iron Group Metals". U.S.S.R. Patent 212,693 (1968) (*C.A.* **69** (1968), 15400d).
235. Field, S., The deposition of gold–silver alloys, *Trans. Faraday, Soc.* **16** (1920), 502–513.
236. Grube, G., "Die Elektrolytische Abscheidung von Goldlegierungen". Heraeus Festschr. Z. 70 Geburtst. Wilhelm Heraeus, pp. 34–44, 1930.
237. Raub, E., Galvanische Legierungsniederschläge, *Metalloberfläche*, **7** (1953), A17–A27.
238. Fedot'ev, N. P. *et al.*, Electrodeposition of gold–silver alloys and their properties, *C.A.* **65** (1966), 5011h.
239. Harr, R. E. and Cafferty, A. G., Gold–silver alloy plating, *Proc. Am. Electroplat. Soc.* **43** (1956), 67–69.
240. Harr, R. E. and Cafferty, A. G., Electroplating 22-karat gold–silver alloy, *Met. Finish.* **56** (1958), 55–57.
241. Raub, E. and Disam, A., Elektrolytische Zinc–Gold– und Cadmium–Gold–Legierungen, *Metalloberfläche*, **17** (1963), 17–22.
242. Fletcher, A. and Smith, P. T., "Electrolytic Gold Plating Solutions and Methods of Using Such Solutions". British Patent 1,349,796 (1974).
243. Hensel, F. R., "Bearing Alloys". U.S. Patent 2,393,905 (1946) (*C.A.* **40** (1946), 210z).
244. Sel-Rex Corporation, "Method for the Electrolytic Deposition of Gold Alloys and Aqueous Plating Bath Therefor". British Patent 1,293,356 (1972).
245. Bick, M. and Lochet, J. A. (Auric Corporation), "Gold Electroplating Baths and Process". U.S. Patent 3,833,488 (1974).
246. Cathro, K. J. and Koch, O. F. A., The dissolution of gold in cyanide solutions—an electrochemical study, *Austral. Inst. Min. Met., Proc.*, No. (210) (1964), 111–117.
247. Cathro, K. J. and Koch, O. F. A., The effect of thallium on the rate of extraction of gold from pyrites calcine, *Austral. Inst. Min. Met., Proc.*, No. (210) (1964), 127–137.

248. Ellsworth, H. D., "Indium–Gold Article and Method". U.S. Patent 2,438,967 (1948).
249. Foulke, D. G., The effect of addition agents on the structure and physical properties of gold electrodeposits, *Plating*, **50** (1963), 39–44.
250. Kersten, H. J., "Electrodeposition of Tin–Gold Alloys". U.S. Patent 1,905,105 (1933) (cf. U.S. Patent 1,924,439 (1933).
251. Raub, E. and Disam, A., Elektrolytische Wismut–Gold– und Blei–Gold–Legierungen, *Metalloberfläche*, **16** (1962), 317–321.
252. Serizawa, S., "Gold Plating Bath". Jap. Kokai 74 72,140 (1974) (*C.A.* **82** (1975), 36676g).
253. Reinheimer, H. A. (Western Electric Co. Inc.), "Electrodeposition of Soft Gold". German Offen. 2,363,462 (1974) (*C.A.* **81** (1974), 180399r).
254. Waltz, M. C., "Method of Plating Silicon". U.S. Patent 2,814,589 (1957).
255. Raub, E., German Patents 843,785 (1949) and 849,787 (1952).
256. Raub, E., Dehoust, G. and Ramcke, K., Die galvanische Abscheidung von Legierungen des Antimons mit Kupfer, Silber und Gold, *Metall.* **22** (1968), 573–576.
257. Fedot'ev, N. P., Vyacheslavov, P. M. and Volyanyuk, G. A., Deposition technology and physicochemical properties of electrolytic gold–antimony alloy, *Zh. Prikl. Khim.* **40** (1967), 1759–1763 (English translation pp. 1693–1696).
258. Societé Continentale Parker, "Verfahren zur galvanischen Vergoldung und Lösungen zu ihren Durchführung". German Offen. 2,317,028 (1972).
259. Domnikov, L., Deposition of a ternary gold–silver–antimony alloy, *Met. Finish.* **68** (1970), 54–56.
260. Vyacheslavov, P. M. *et al.*, Coating of electrical contacts with gold–silver–antimony alloys, *Zh. Prikl. Khim.* **43** (1970), 79–82 (English translation pp. 72–74).
261. Yamomoto, T. and Homma, R., "Bright Gold Plating". Jap. Patent 3954 (1954) (*C.A.* **49** (1955), 8014).
262. Sel-Rex Corporation, U.S. Patent 3,020,217 (1972).
263. Lainer, V I., Velichko, Y. A. and Zuikova, V. S., Electrolytic deposition of gold–antimony alloy, *Zh. Prikl. Khim.* **39** (1966), 2505–2509 (*C.A.* **66** (1967), 34179t).
264. Vasileva, G. S., Gold plating cyanide bath, *Med. Prom. S.S.S.R. (Moscow),* (1959), 48 (*C.A.* **54** (1960), 12828) (cf. U.S.S.R. Patent 123,828 (1959).
265. Mel'nikov, P. S., Saifullin, R. S. and Vozdvizhenskii, G. S., Gold-antimony alloy coatings from a self-regulating cyanide electrolyte, *Zashch. Met.* **7** (1971), 354–357 (*C.A.* **75** (1971), 44196w).
266. Mori, H., "Gold–Antimony Alloy Plating". Jap. Patent 69 00,123 (1969).
267. Depierre, J., "Obtaining Bismuth–Gold Alloys by Electrodeposition". French Patent 2,053,770 (1971) (*C.A.* **76** (1972), 67455q).
268. Foulke, D. G., Ref. 3, p. 36.
269. Graham, A. K., Heiman, S. and Pinkerton, H. L., Deposition of precious metal alloys, *Plating*, **35** (1948), 1217–1219.
270. Technic Inc., "Deposition of Gold–Palladium Alloys". British Patent 1,051,383 (*C.A.* **66** (1967), 34299g, **72** (1970), 92926u).
271. Technic Inc., "Bad und Verfahren zum galvanischen Abscheidung von Gold–Palladium–Legierungen". German Patent 1,521,043 (1971).
272. Vinogradov, S. N. *et al.*, Electrodeposition of a gold–palladium alloy, *Zashch. Met.* **8** (1) (1972), 92–95 and **10** (1) (1974), 75–77 (*C.A.* **76** (1972), 107201w and **80** (1974), 115371r).

273. Ohkubo, K. and Mukoujima, H., "Solutions for Rhodium and Rhodium Alloy Plating". U.S. Patent 3,515,651 (1970).
274. Scherzer, J. and Weisberg, A. M., "Gold and Ruthenium Plating Baths". U.S. Patent 3,530,049 (1970).
275. Nazarova, E. S. *et al.*, "Electrochemical Deposition of a Gold Based Alloy". U.S.S.R. Patent 322,411 (1971) (*C.A.* **76** (1972), 80298k).
276. Benham, R. B. and Melrose, S P. G. (Johnson, Matthey & Company), "Improvements in and Relating to the Electrodeposition of Gold". British Patent 928,083 (1963).
277. Schering, A. G., "Procédé de dépôt galvanique d'alliages d'or". French Applics. 2,134,401 (1972) and 2,202,954 (1974) (cf. German Offen. 2,251,285 (1974)).
278. Fischer, J. and Schwarze, W., "Bright Metal Plating". U.S. Patent 2,800,439 (1957).
279. Schmid, E. and Schnur, K., "Verfahren zum Herstellen von hochfrequenzleitenden Oberflächenschichten und nach diesem Verfahren hergestelltes elektrisches Bauelement". German Offen. 1,948,734 (1971).
280. Menzel, T. J., "Electroplating Baths and Methods for Electroplating Gold Alloys and a Product Thereof". U.S. Patent 3,607,682 (1971) (cf. German Offen. 2,033,970 (1971)).
281. Smagunova, N. A. *et al.*, Effect of Trilon B in the electrodeposition of a gold–silver alloy, *Zashch. Met.* **4** (3) (1968), 336–338 (*C.A.* **69** (1968), 73385p).
282. Il'in, V. A. and Fortovova, L. S., "Electrolytic Deposition of Alloys Based on Gold with Copper, Silver and Antimony". U.S.S.R. Patent 188,254 (1966) (*C.A.* **67** (1967), 7535v).
283. Rinker, E. C. and Foulke, D. G. (Sel-Rex Corporation), "Bath for Electroplating Bright 14–23.5 Karat Gold–Silver Alloys". German Patent 1,213,196 (1966) (*C.A.* **64** (1966), 17048).
284. Foulke, D. G. and Duva, R., Electroplating of 60–80% gold alloys for the contact industry, *in* 4th "Plat. Electron. Ind., Symp"., (1973), 131–140 (*C.A.* **81** (1974), 20116t).
285. Foster, A. J., A practical approach to gold electroplating, *Electroplat. Met. Finish.* **27** (1) (1974), 17–23 and **27** (2) (1974), 22–25 and 27.
286. Marakhtanova, Z. H., Electrodeposition of gold–silver–cadmium alloy, *Zashch. Met.* **5** (4) (1969), 437–440 (*C.A.* **71** (1969), 108302e).
287. Degussa, "Electrolytic Deposition of Gold–Copper–Cadmium Alloys". U.S. Patent 3,056,733 (1962) (*C.A.* **58** (1963), 246b).
288. Degussa, "Procedure for the Electrolytic Deposition of Coatings of Gold–Copper–Cadmium Alloys". U.S. Patent 3,586,611 (1971) (cf. French Applic. 2,011,755 (1970), Dutch Applic. 6,906,666 (1966)).
289. Heilmann, G. (Degussa), "Ergänzungslösung zur Verringerung der Zunahme der Alkalität in Cyanidischen Bädern zur galvanischen Abscheidung von Goldlegierungen". German Patent 1,032,636 (1958).
290. Melnikov, P. S., Protective galvanic coatings of gold alloys of increased hardness and wear resistance, *Zashch. Met.* **6** (3) (1970), 365–367 (*C.A.* **73** (1970), 51476z).
291. Aliprandini, G., Desthomas, G. and Meyer, A., "Bad zur elektrolytischen Auflage von Goldlegierungen und seiner Anwendung in der Galvanotechnik". German Offen. 2,233,783 (1973) (cf. French Applic. 72.26091 (1973)).
292. Young, C. B. F. and Herschlag, V. E., An investigation of electrodeposited gold alloys, *Met. Ind. (N.Y.)*, **38** (1940), 194–196.

293. Smagunova, N. A. *et al.*, Electroplating of Gold–Base Alloys". U.S.S.R. patent 280,154 (1970) (*C.A.* **74** (1971), 49093z).
294. Komura, K., "Gold Alloy Electroplating Bath for Producing Thick Crack Free Deposits From Electrolyte Containing Triethylenetetramine and Hydroxycarboxylic Acids". Jap. Patent 74 27,934 (1974) (*C.A.* **82** (1975), 117875y).
295. Yamaguchi, H. and Atabe, Y., "Electroplating Bath for Gold Alloy Including Zinc Group Elements Except for Mercury". Jap. Patent 74 20,705 (1974) (*C.A.* **82** (1975), 91773m).
296. Schering, A. G., "Werkwijze ter bereiding van een bad voor de galvanische afscheiding van Au-Cu-Cd-legeringen ens.". Dutch Applic. 73 14,000 (1974).
297. Dettke, M., Ludwig, R., Martin, K. U. and Riedel, W., Zur Abscheidung 18-Karätiger Goldlegierungen, *Galvano*, **62** (1971), 773–778.
298. Lea-Ronal Inc., "Procédé de bain de placage d'alliages d'or". French Applic. 2,176,638 (1973).
299. Kuroda, K. and Ishiguro, I., "Electrolyte for Electrodeposition of Gold Alloys". Belgian Patent 651,976 (1964) (*C.A.* **64** (1966), 12199).
300. Kanai, M., "Bath for Electroplating Hard Bright Gold Alloy". Jap. Patent 74 25,532 (1974) (*C.A.* **82** (1975), 91768p).
301. Schickel, M. and Eichhorst, I., "Verfahren zur Herstellung der Belegung von Relaiskontakten". German Offen. 2,042,700 (1972).
302. Bick, M. and Lochet, J. A. (Auric Corp.), "Bright Gold Electroplating Bath". U.S. Patent 3,856,638 (1974) (*C.A.* **82** (1975), 161843k).
303. Weisberg, A. M. and Butler, F. P. (Technic Inc.), "Bright Electrodeposits of Gold and its Alloys". U.S. Patent 3,870,619 (1975) (*C.A.* **82** (1975), 177222t).
304. Edson, G I. and Reynolds, K. R., "Gold Alloy Electroplating". U.S. Patent 3,809,628 (1974) (*C.A.* **81** (1974), 44731r).
305. Coniglio, A. and Eichhorst, I., "Verfahren zum galvanischen Abscheiden von chromhaltigen Uberzügen". German Offen. 2,048,928 (1972).
306. Dole, M., Diffusion theory of the co-deposition of gold and copper, *Trans. Electrochem. Soc.* **82** (1942), 241–255.
307. Krasikov, B. S., Electrodeposition of gold–copper alloys, *Zh. Prikl. Khim.* **30** (1957), 799–801 (English translation pp. 843–845).
308. Fedot'ev, N. P., Kruglova, E. G. and Vyacheslavov, P. M., Electrochemical deposition of gold–copper alloys, *Zh. Prikl. Khim.* **32** (1959), 2014–2021 (English translation pp. 2063–2070).
309. Zvolner, H., "Electrodeposition of Gold Alloys". Thesis submitted to Graduate School for M.S. degree, Evanston, Illinois, 1941.
310. Krasikov, B. S. and Grin, Yu. D., Production of bright coating in electrodeposition of copper–gold alloys, *Zh. Prikl. Khim.* **39** (1959), 837–841.
311. Redomger, V. (Firma Robert Hellerich), "Gold Alloy Electroplating of Eyeglass Frames". German Offen. 2,256,641 (1974) (*C.A.* **81** (1974), 44728v).
312. Anon., New gold bath doubles plating production, *Plating* (1974), 994–995 and 998.
313. Zaitsev, V. N. *et al.*, Bright gold plating of jewellery with gold alloys, *C.A.* **80** (1974), 33185u.
314. Vasileva, G. S., Electrochemical gold coating of high wear resistance, *Med. Prom., S.S.S.R. (Moscow)* (5) (1959), 48–52 (cf. U.S.S.R. Patent 123,828 (1959)).
315. Sel-Rex Corporation, "Werkwijze voor het elektrisch afzetten van glanzende, harde, legeringen van goud en koper". Dutch Patent 68 03,173 (1969).

316. Thornton, H. R. and Wolanski, Z. R., Lubrication with electrodeposited films of silver–rhenium and gold–molybdenum, *J. Am. Soc. Lubric. Eng.* **23** (1967), 271–277.
317. Kopchikin, D. S. and Nazarova, E. S., Electrodeposition of gold–rhodium alloys, *C.A.* **80** (1974), 33186v.
318. Noguchi, T., Mohri, H. and Akebi, M., "Electroplating of Gold–Antimony Alloy". Jap. Patent 74 05,097 (1974) (*C.A.* **81** (1974), 144757x).
319. Lerner, L. B., "Electrodeposition of Gold–Antimony Alloys". U.S. Patent 3,926,748 (1975).
320. Lacal, R. (R.T.C. La Radiotechnique--Compelec), "Procédé de dépôt simultané d'or et d'étain par la voie galvanique". French Patent 1,566,760 (1969).
321. Ohsuga, A., Iwashita, T. and Shifo, S., "Plating Bath for a Gold–tin Alloy". Jap. Kokai 74 65,950 (*C.A.* **81** (1974), 9345s).
322. Kersten, H. J., "Electrodeposition of Zinc–Gold Alloys". U.S. Patent 1,905,106 (1933).
323. Yamada, H. *et al.*, "Electroplating an Alloy of Gold and Zinc". Jap. Patent 74 05,820 (1974).
324. Woolrich, J. S., British Patent 9,431 (1842). (The use of sulphite baths was also patented in 1842 by W. Siemens—see *Z. Elektrochem.* **19** (1913), 151–152.)
325. Johnson, P. R., "Complexes of Gold with Ligands Having Sulphur Donor Atoms". Report to the Chamber of Mines of South Africa, 1975.
326. Oddo, B. and Mingoia, Q., *Gazz. Chim. Ital.* **57** (1927), 820–826.
327. Smith, P. T., "Gold Plating Bath and Process". U.S. Patent 3,057,789 (1962).
328. Socha, J., Safarzynski, S. and Zak, T., Electrolytic deposition of gold from sulphite electrolyte, *J. Less-Common Met.* **43** (1975), 283–290.
329. Zuntini, F., Aliprandini, G., Gioria, J. M., Meyer, A. and Losi, S., "Amine Gold Complex Useful for the Electrodeposition of Gold and its Alloys". U.S. Patent 3,787,463 (1974).
330. Dunand, A. and Gardil, R., Crystal and molecular structure of ethylene diammonium bis (cis [ethylenediamine disulphitoaurate III]), *Acta. Cryst.* **B31** (1975), 370–374.
331. Kazakov, V. P. and Konovalova, M. V., Kinetics of the reduction of $AuBr_4^-$ by sulphite and thiocyanate ions, *Zh. Neorg. Khim.* **13** (1968), 447–453.
332. Peshchevitskii, B. I. and Erenburg, A. M., Stability of some sulphur containing compounds of gold in aqueous solution, *Izv. Sib. Otd. Akad. Nauk. S.S.S.R.* (4) (1970), 83–87 (*C.A* **74** (1971), 80424g).
333. Bradford, C. W. and Brentford, H. (Johnson Matthey), "Verfahren zur Elektroplattierung einer Unterlage mit Gold und Goldkomplex zur Verwendung bei diesem Verfahren". German Offen. 2,338,176 (1974).
334. Shoushanian, H. H. (Technic Inc.), "Gold Plating Bath and Process". U.S. Process 3,475,292 (1969).
335. Baker, K. D., "Gold Plating Electrolyte". U.S. Patent 3,776,822 (1973).
336. Smagunova, N. A., Gavrilova, J. P. and Balashova, N. N., "Electrochemical Gold Plating". U.S.S.R. Patent 231,991 (1968) (*C.A.* **70** (1969), 63569h).
337. Smagunova, N. A., Gavrilova, J. P. and Yudina, A. K., Gold plating in a noncyanide electrolyte, *Zashch. Met.* **6** (1970), 716–719 (*C.A.* **74** (1971), 37613f).
338. Losi, S. A., Zuntini, F. L. and Meyer, A. R., Dépôt électrolytique d'alliages or-cuivre en milieu non-cyanuré, *Electrodepos. Surface Treatment*, **1** (1972/73), 3–19.
339. Losi, S. A., Zuntini, F. L. and Meyer, A. R., Dépôt électrolytique d'alliages or-cuivre en milieu non-cyanuré, *Surfaces (Paris)*, **12** (1973), 67–75.

340. Zuntini, F., Aliprandini, G., Gioria, J. M. and Meyer, A., "Bain aqueux pour le placage électrolytique d'or ou d'alliage d'or sur un article conducteur, procédé de fabrication dudit bain aqueux et utilisation de celui-ci". Swiss Patent 506,628 (1971) (cf. U.S. Patent 3,092,559, French Applic. 2,060,092 (1971), German Offen. 2,039,157 (1970), Netherlands Applic. 70 12171 (1971)).
341. Meyer, A. and Antony, P., "Elektrolytisches Bad zur elektrochemischen Abscheidung von Goldlegierungen und dessen Anwendung". German Offen. 2,244,437 (1972).
342. Oxy Metal Finishing Corporation, "Werkwijze voor de galvanische afzetting van goud en goudlegeringen uit sulfietbaden alsmede bad voor het uitvoeren van de werkwijze". Dutch Applic. 73 11108.
343. Smith, P. T. (Sel-Rex Corporation), "Bad und Verfahren zur Goldplattierung". German Offen. 2,042,127 (1971).
344. Konig, W., "Composition de bain pour la production galvanique de revêtements en alliage d'or". French Applic. 73 14874 (1973) (cf. German Offen. 2,222,310 (1973)).
345. Sasaki, T. *et al.*, "Gold Electroplating Bath". Jap. Patent 74 41,255 (1974) (*C.A.* **83** (1975), 177219x).
346. Duva, R. and Raleigh, J. P., "Electroplating Gold and Thiomalate Electrolyte Therefor". U.S. Patent 3,520,785 (1970).
347. Sel-Rex Corporation, "Gold Alloy Electroplating Bath". British Patent 1,221,862 (1971).
348. von Siemens, W., Lebenserinnerungen. Springer Verlag, Berlin, 1922 (cf. *Z. Elektrochem.* **19** (1913), 151–152).
349. Kushner, J. B., *Prod. Finish. (Cincinnati)*, **6** (3) (1941), 22–30 (*C.A.* **43** (1942), 1243).
350. Fischer, J., Deutsch. Goldschmiede-Ztg. **45** (1942), 43–44.
351. Fischer, H., Neuzeitliche Probleme und Verfahren der Elektrochemie, *Electrotech. Z.* **61** (1940), 121–125.
352. Pouradier, J. and Gadet, M. C., Electrochemistry of gold salts. VII. Thiosulphato-aurate, *J. Chim. Phys. Phys-Chim. Biol.* **66** (1969), 109–112 (*C.A* **70** (1969), 111142n).
353. Safarzynski, S., Socha, J. and Zak, T., Thiosulphate bath for electroplating gold, *C.A.* **82** (1975), 177082x.
354. Anon., Gold plating from a thiosulphate bath, *Gold Bulletin*, *8* (1975), 126.
355. Mohan, A., The electroforming of gold: A manufacturing technique for intricate components, *Gold Bulletin,* **8** (1975), 66–69.
356. Raub, Ch. J., Khan, H. R. and Lendvay, J., Temperature sensitive properties of gold and gold alloy electrodeposits, *Gold Bulletin,* **9** (1976), 123–128.

9 Gold Coatings of Other Types

1. Liquid Golds

Gold has been used in the form of powder for the surface decoration of ceramics for hundreds of years. The films obtained were thick (about 25 microns), however, and it was only with the development of liquid golds by the Royal Porcelain Factory of Meissen in the period 1800–1830, and the publication in 1851 of details of their preparation and use, that the formation of thinner, bright and less costly films on ceramic ware became generally possible. These early liquid golds were prepared by reacting selected turpentines with sulphur to form sulphur derivatives, which were of undefined chemical constitution, but which reacted with gold chloride to yield solutions containing up to 24% of gold, presumably as terpene gold mercaptides.[1] When such solutions are applied to ceramic substrates and fired, the organic matter and sulphur are volatilized and/or burned off and thin bright coatings of gold are left bonded to the ceramic surface.

Liquid golds of this early type have been improved greatly in recent years. This has been made possible as a result of the commercial availability of relatively pure terpene fractions which have been reacted with sulphur under controlled conditions. The products of such processes are of higher quality and more constant composition than the earlier products obtained by boiling crude terpene mixtures with rock or flowers of sulphur. Terpenes rich in Δ–3–carene have been reported[2] to give particularly good liquid golds. The separate preparation of gold terpene mercaptides from a variety of terpenes and their subsequent use in the compounding of liquid gold preparations has been described.[3] The gold terpene marcaptides as isolated were white solids which were stable in storage.

In addition to much improved liquid golds of the "terpene" type, however, the decorator has available today liquid golds of a more specifically synthetic character which contain gold mercaptides prepared by the interaction of synthetic alkyl mercaptans of known structure with gold chloride.[4–7] Not all such mercaptides have the high solubility in organic solvents which is necessary if they are to be used in liquid gold formulations; the tertiary isomers, however, are highly soluble. Thus a gold tertiary dodecyl mercaptide of the type.

$$
\begin{array}{c}
\mathrm{R} \\
| \\
\mathrm{R{-}C{-}S{-}Au} \\
| \\
\mathrm{R}
\end{array}
$$

although a solid containing 49% of gold, dissolves to the extent of 35% in toluene. Moreover it is very sensitive to heat, and decomposes with liberation of metallic gold at 150°C. Materials which are relatively sensitive to high temperatures can therefore be coated with gold using solutions of this material.

Silver can replace gold in tertiary alkyl mercaptide formulations of this type, and liquid silvers are also known[8] in which are exploited the solubility in organic solvents and ease of pyrolysis of silver tertiary alkyl carboxylates of the type.

$$
\begin{array}{c}
\mathrm{R} \\
| \\
\mathrm{R{-}C{-}CO{-}OAg} \\
| \\
\mathrm{R}
\end{array}
$$

By the use of mixtures of liquid golds and liquid silvers of these types, gold–silver alloys of varying compositions can be formed on various substrates at relatively low temperatures.

Gold mercaptides can also be complexed with both silver mercaptides and silver carboxylates to give compounds of the structures $R_3C\text{-}S\text{-}Au{\cdot}Ag\text{-}S\text{-}CR_3$ and $R_3C\text{-}S\text{-}Au{\cdot}Ag\text{-}O\text{-}CO\text{-}CR_3$, in each of which both gold and silver are present in the same thermally sensitive molecule.[9] Other types of liquid golds[10] which have been patented do not appear to have found application.

Liquid gold formulations, whether of the terpene mercaptide or synthetic mercaptide type, are normally not simply solutions of the gold compound in a solvent. Organic resins and related materials may also be incorporated in order to adjust the viscosity and handling properties of the preparation to meet requirements for brushing, spraying, screen printing, etc. and soluble organic compounds of other metals such as bismuth, lead, tin, nickel, chromium, iron and silicon may be added, which decompose on firing to leave a residue of the metal oxide dispersed in the gold.

Of most importance in the latter respect is the addition of a small amount (up to 0.5%) of an organic rhodium compound, since gold films which contain dispersed rhodium oxide show very considerable resistance to agglomeration when heated, and are smooth and lustrous. Important also is the incorporation of organic compounds of such metals as bismuth, tin and vanadium which form oxides on pyrolysis which promote bonding of the gold film to the substrate. It will be recalled also that, suitably chosen, metal oxides co-deposited with gold can affect greatly its emissivity and absorptivity to peak solar radiation and that liquid golds have been studied and developed from this point of view in the design of selective absorber coatings.

Liquid gold coatings can be as thin as 0.19 μm, but thicker films can be formed either by the application of multiple coatings or, alternatively, by incorporating fine gold powder in the liquid gold. In the latter case, however, the coatings may need burnishing if a bright coating is desired. Their high infrared reflectivity and low emissivity have led to extensive use of certain liquid gold coatings as heat shields.

2. Mechanically Clad Golds

2.1 Introduction

Base metals may be clad mechanically or inlaid with gold or gold alloys for either decorative or technical applications. In the former the nature of the coating is usually described simply by reference to its caratage and colour while in the latter not only the full composition of the coating but also properties such as hardness, contact resistance and resistance to tarnish may be relevant and specified. In addition, the proportion of gold or gold alloy on the final clad strip or plate is normally specified in such a manner that the relative dimensions of the substrate and the clad coating or inlay are defined.

In the jewellery trade metals, for example, mechanically-clad with carat golds, which are referred to as rolled gold or gold filled or doublé, may be described[11] in the following ways:

(a) By the proportion by weight of carat gold to the whole expressed as a fraction, e.g. $\frac{1}{10}$ 12 carat yellow gold on gilding metal.
(b) By the proportion by weight of fine gold in the whole expressed in parts per 1000, e.g. 50/1000 12-carat yellow gold on gilding metal.
(c) By the thickness of carat gold alloy required and the overall thickness of the finished material, e.g. 20 μm 12 carat yellow gold on gilding metal, overall thickness 0.60mm.

Conversion tables are available for determining the dimensional implications of specifications of types (a) and (b) which are in proportion by weight.

2.2 Manufacture

In manufacture gold filled sheet is made[12] by bonding, by brazing if necessary, a gold or gold alloy bar of the appropriate quality and calculated dimensions to a base metal bar, which may be nickel coated in advance to prevent excessive diffusion of the final coating into the substrate metal during brazing and subsequent annealing treatments. The bonded product is then rolled to the desired thickness. Gold filled wire or tubing, on the other hand, is made by inserting a base metal rod or tube into a carat gold tube of appropriate dimensions, and drawing the two together with anneals at appropriate stages. Effective bonding usually results from the pressure and heat treatments. Clad metal inlays, made primarily for the fabrication of electrical contacts[13,14] are also pressure bonded, with a nickel interliner between the gold or gold alloy and the substrate metal. A 6% nickel-gold alloy has recently shown considerable promise when used in inlay form for the contacts in a telephone handset, the inlayed alloy having acceptable porosity, wear, arc erosion and ease of forming. It was comparable in performance to cobalt-hardened electroplated gold which has been used for this purpose, but lower in cost.[14] A gold inlaid Cu-Ni alloy has also been found to be cheaper than an electroplated product in the production of a computer connector.[14a]

The basis metals used in decorative applications are for the greater part nickel silver or commercial bronze though nickel may be used, for example, in the manufacture of gold filled spectacle frames because of its resistance to tarnish where exposed by cutting. In technical applications a wider range of basis metals may be used.

Most manufacturers supply information relating to the further processing of their products, e.g. in respect of cold working, annealing, soldering and polishing. In general, however, gold-filled materials are best purchased in as near to final form as possible, so that properties which are most easily controlled in production are retained.

2.3 Properties

Both the manufacture and properties of mechanically clad golds have received considerable attention because of their electrical applications. As alternatives to electroplated golds they offer advantages[15,15] in respect of lower porosity, better formability, and the relative ease with which a variety of alloy coatings of lower gold content can be applied by cladding as compared with electroplating.

They have disadvantages, however, in situations where a decorative finish has to be applied by engraving or diamond milling, since the final article may then not comply with specifications or requirements in respect of the thickness of the clad layer. In the case of plated golds, where the plating is applied at the end of the production cycle, such difficulties are avoided.

Russell[17,18] has compared a range of sixteen clad inlay wrought gold alloy coatings in regard to properties such as porosity, formability, hardness, wear composition, diffusion effects, tarnish resistance, contact resistance, etc. Only certain aspects of the findings call for mention here.

Porosity. The most serious cause of porosity in clad gold coatings are puncturing of the clad coatings by adventitious dirt, and scratches or chafe marks arising during handling, so that scrupulous cleanliness and care in manufacture and use are necessary. At comparable coating thicknesses, however, porosities tend to be less with clad than with plated coatings (Figure 52).

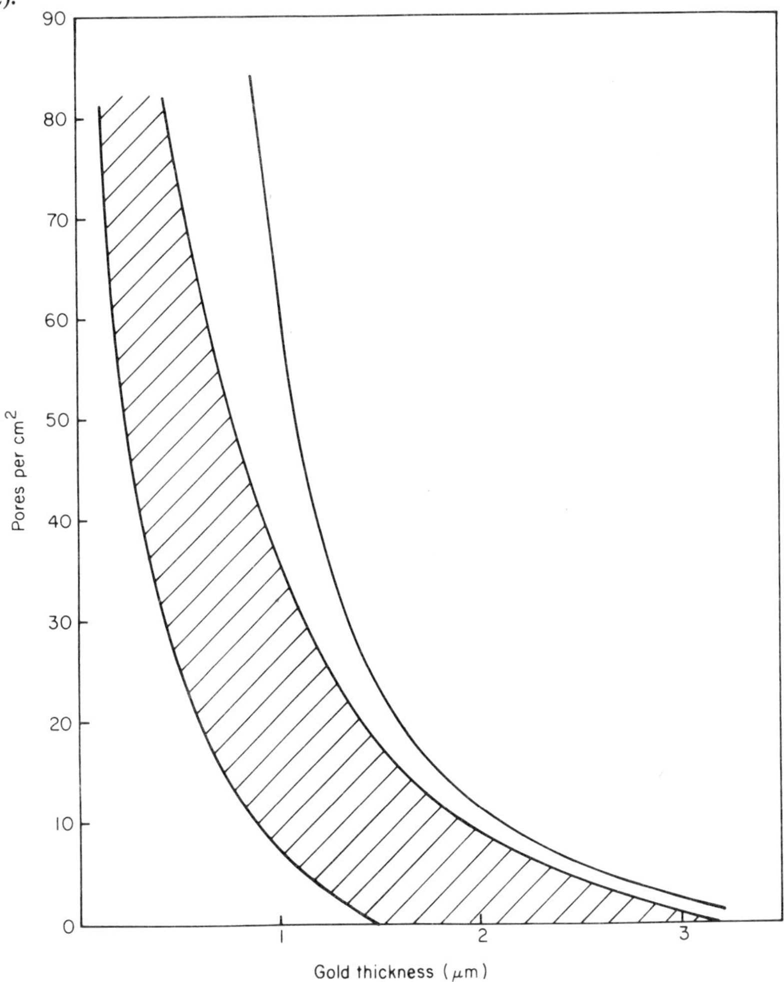

FIG. 52. The range of porosity counts (corrosion counts from the nitric acid vapour test) as a function of gold thickness. The cross-hatched band shows the range of points for a number of gold alloys in inlay-clad form. The single curve shows average values for electrodeposited gold. (After Russell.[18])

Formability. While there is no advantage of clad over plated coatings on the concave sides of bends, such advantage is obtained on convex bends.

Hardness and wear resistance: Quite apart from the work hardening which results from rolling, hardness can be achieved in clad coatings by choice of the appropriate alloy for cladding. Knoop hardnesses (KNH50) of up to 281 are recorded by Russell.

Diffusion effects. During production of clad metals, annealing is necessary, and diffusion occurs across the gold alloy – nickel interface to an extent which depends on the number of anneals and the times and temperatures involved. Auger thin film analysis combined with ion sputter etching of samples from production runs revealed[17] nickel as penetrating to depths of up to about 0.9 μm into the gold or gold alloy coating during the process. While the effects of such penetration must clearly depend on a variety of factors, and are apparently not significant in most connector applications of clad golds, they are obviously an important factor in determining the properties of clad golds.

Tarnish and contact resistance: The data presented[17] in respect of changes in contȧct resistance during exposure in both hor air (100°C) and a flowers of sulphur test indicate special advantages in general of higher gold content cladding alloys, though a 55Au-39Ag-3Cd-3In alloy[19] was confirmed as having high resistance to development of sulphide tarnish.

3. Gold Coatings Formed by Thick Film Processes

3.1 The nature of thick film pastes

Of the various technologies developed for the interconnection of active devices in electronic circuits, such as printed circuit board, thin film and thick film technologies, the latter is perhaps the most flexible.[20,21] Moreover, of the thick film conductor materials which are available, gold containing conductors provide the widest selection of properties to meet the needs of various situations. These conductors are generated on ceramic (alumina) substrates by printing, using gold or gold–alloy containing thick film pastes. After being dried at 120°C, the substrates are fired in air at temperatures which do not exceed 1000°C. During the firing the organic components of the printed paste are volatilised and/or burned off, and the gold or gold alloy components of the paste become sintered and bonded to the substrate.

The pastes are of two types: firstly, thick film conductor pastes which contain glass frit and, secondly, thick film conductor pastes which are glass-free. In each, noble metal (gold, silver, platinum, palladium) powders are blended in the desired proportions with an organic vehicle and a bonding agent. The organic vehicle must be such as to facilitate the printing of the paste to form the extremely fine circuit patterns that are required.

The bonding agent in the first class of pastes is the glass frit.[22–25] During firing, this softens and reacts with the ceramic surface. Further glass becomes attached to this anchored glass surface, so that on cooling the gold tends to be largely concentrated at the surface. The use of vanadium oxide[26] and bismuth oxide [27] in metallising compositions of this type has been claimed to improve the bonding and other properties of the films.

In the second class of thick film pastes, bonding of the conducting metal or alloy to the substrate surface is achieved by reaction bonding. This development is essentially the application in the electronics field of technology previously known in the dental and decorative ceramic industries. It has recently been reviewed from various points of view.[28–32] In thick film pastes of this type, direct reaction bonding of the gold or gold alloys to the alumina substrate is promoted by incorporating in them small quantities of certain base metal oxides. One proprietary paste of this type has been reported to contain oxides of both copper and cadmium as reactive bonding materials. The compositions of the pastes are illustrated by the claims of Smith,[33–35] whose metallising compositions contain copper oxides (0.1–7.0% cuprous and cupric), cadmium oxide (0.01–15.0% CdO) and gold powder (78.0–99.89%). For application purposes, solid mixtures of the above compositions are dispersed in an organic bonder, which may constitute 10–95% of the final mixture applied.

For both types of thick film paste, the quality of the alumina substrate is important, and although 96% alumina substrates have apparently been used widely, it has been demonstrated that 99% alumina substrates as used for thin film vacuum processes give stronger and better bonded coatings.[32a]

The metal powders employed in these pastes are usually prepared by precipitation in order to obtain particles of suitably small size. In the case of gold, this should not exceed about 5 μm diameter. Larger particles tend to be deformed into flakes during preparation of the pastes and these flakes may be large enough to block the screens used for printing. Particle shapes vary, however, in different commercial products, of which a large number are now available.[36]

The metal powder composition used depends upon the specifications which have to be met. For thick film circuitry for the consumer market where cost is a major consideration, silver-containing conductors are used widely. For precision and high reliability applications in computers and in aerospace equipment, on the other hand, conductors of gold, or of Au-Pd or Au-Pt alloys are usually specified, though certain ternary alloys in which the gold in Au-Pd and Au-Pt compositions is replaced partly by silver are available. The extent to which palldium or platinum may be used with gold in thick paste formulations depends on the conductivity and other specified properties of the final fired film. Some degree of alloying may be desirable in certain situations where it is desired to limit interdiffusion effects between conductor and resistor films in contact with one another. A major consideration, however, is the nature of the bonding operations which may have to be made in the final conductor circuit.

3.2 Gold conductors generated from thick film pastes

Even the most highly conductive thick film golds have conductivities which are significantly below that of bulk gold. While this may in some measure be due to the presence of bonding agents, it is primarily a result of the presence of porous regions in the fired conductor films, as revealed by scanning electron microscopy. Nevertheless, thick film golds are, by their very nature, more highly conducting than thick film gold alloys.

Thick film gold conductors are also superior to their gold alloy counterparts where they have to be bonded either by thermocompression or ultrasonic methods to gold wires. In this connection, however, reaction bonded gold conductors are superior to those bonded with glass frit. In the latter, not only is the gold surface never completely free of glass frit, but bonding pressures may cause cracking of the glass joining the film to the ceramic with the result that the metallisation may easily become detached from the substrate. Glass frit bonding also has disadvantages where high conductivity is called for in the peripheral metal layers. This tends to limit the use of this type of thick film conductor in microwave applications.

The greatest disadvantage, apart from their cost, of thick film gold conductors, however, arises from the fact that they cannot be soldered easily with Sn-Pb solders. This is because of the high rate of dissolution of pure gold in such solders. Where soldering to such conductors is necessary, it is normally effected with Au-Sn or indium-based solders.

Thick film gold conductors find widest application where conductivity must be high, where the width of the conductors is small (about 3 mils or 750 μm), and where thermocompression bonding techniques have to be used. They are used for metallisation of beam leads, in multilayer circuitry and increasingly for microwave strip lines.

3.3 Gold alloy conductors generated from thick film pastes

The substitution of a proportion of the gold in a thick film conductor by palladium currently reduces the cost of thick film paste. Perhaps the greatest advantage of Au-Pd alloy conductors, however, is the fact that they can be soldered with Sn-Pb solders, though the formation of intermetallics, particularly on ageing, may lead to a weakening of the bond. Although Au-Pd alloys can be soldered with Sn-Pb solders, beam lead and lead frame bonding with such conductors is not feasible, and die bonding can be achieved satisfactorily only with Au-Si preforms placed between the die and the Au-Pd conductor. Moreover, wire bonds are less strong. Their conductivity is less than that of gold conductors. The Au-Pd conductors are also reported to be less compatible with certain resistor compositions.

Au-Pt thick film conductors are apparently superior to Au-Pd conductors in certain respects but they do not have the cost advantages of the latter.

Ternary alloy conductors do not appear to have been reported upon in any detail.

A treatise on thick film technology has recently been published.[36a]

3.4 Gold-epoxy conductors

Gold powders are used, like silver and copper powders, in making up conductive epoxy pastes. The epoxy components of the paste set by curing, which is accelerated by heating. These pastes have been found to be useful in making low temperature electrical connections and in replacing fired conductors in crossovers and capacitor counter electrodes. They can be applied to most plastics, ceramics, glasses and metals by brush, hypodermic, dipping or screening.

Early gold epoxy compositions suffered from outgassing problems on heating, but the latest compositions are described as free from this defect, and a recent one is described[37] as a soft smooth thixotropic paste, which can be screen printed with accurate definition, has high electrical conductivity and a short curing cycle, and is suitable for bonding of silicon semiconductor chips to various substrates.

4. Vacuum Coated Gold and Gold Alloys: Thin Film Processes

4.1 Evaporated and sputtered golds

There are two methods which are used for the vacuum deposition of gold coatings on various substrates, namely, sputtering and evaporation.

At moderately low gas pressures (1–100 μ) discharge between two electrodes in a gas results in a high velocity bombardment of the cathode by positive ions, and the release of cathode material which condenses upon and coats nearby surfaces. This sputtering process was discovered in 1852 by Sir William Grove and was first employed about fifty years later by Edison who used sputtered gold to generate a conductive surface, for subsequent plating by electrodepostion, upon his phonograph masters. The fact that other metals cannot be applied in this manner in air media because of their susceptibility to oxidation was a signficant factor in this early development of the sputtering of gold.

At very low gas pressures ($< 1\ \mu$), on the other hand, it is possible to evaporate gold by heating it to appropriate temperatures. The evaporated gold condenses in the form of bright coatings on nearby surfaces. With the development of high capacity diffusion pumps and the use of very high vacuum chambers on an industrial scale, the use of evaporative procedures for deposition of vacuum gold coatings tended to outstrip that of sputtering. Nevertheless, the fact that most of the gold charged to an evaporation chamber condenses on the walls of the chamber and the greater degree of control which is possible with sputtering, the improved physical characteristics of sputtered coatings, and the fact that sputtering can be applied effectively for the deposition of gold alloy coatings, is reflected in greatly increased interest in it in recent years.

Sputtered metal leaves the cathode as a result of transfer of energy from the bombarding ions and not as a result of thermal energy vibrations. It therefore reaches the substrate surface in a higher energy state (about 10 eV) than evaporated gold (less than 1 eV). Sputtered golds therefore tend to be better bonded to substrates than evaporated golds.

The rate of transfer of a metal during sputtering varies with the number, velocity and mass of the impinging ions and to some extent with the temperature of the cathode. For noble metals such as gold it is higher than with other metals because of the freedom of their surfaces from oxide and other films. A fairly heavy gas such as argon is usually preferred when working with gold as one means for increasing bombardment energies and transfer rates. There is a limit, however, to the extent to which ion densities can be increased by increasing gas pressures since as soon as a limiting current of a few amperes is exceeded the discharge transforms into an uncontrollable arc which wanders about randomly over the target surface.

It has been found preferable, therefore, to increase ion densities by other methods, such as the use of an auxiliary thermionic cathode or the application of a radio frequency alternating current discharge. The latter is preferable as it is more versatile and permits higher energy densities in the discharge. In principle, ionised plasma can be produced by an electrodeless circular discharge, the induction coil being situated outside the vacuum chamber. In a variation of this technique the plasma generating coil is fabricated from the target metal and placed inside the chamber. By such methods high energy outputs can be obtained in small volumes and the bombardment rates are such that efficient cooling is necessary.

Sputtered gold coatings have been used extensively in the production of heat reflective coatings of bismuth oxide and gold on glass for windows of aircraft and buildings. They are also used extensively with the aid of suitable masking procedures, as are evaporated gold coatings, for the application of the patterned coatings which are such a feature of microelectronic circuitry.[38] As mentioned above, however, thin film techniques of evaporation and sputtering are not as flexible as thick film techniques, which can be used not only for producing conductors, but also resistors and dielectrics. Gold leaf is also produced by continuous coating of Mylar film or other supports with sputtered gold.[39] Sputtering is also applied in a variety of other situations, such as the application of gold contacts to the ends of pyrolytic carbon resistors,[39] and the application of heat conductive gold coatings to beryllium oxide insulators.[40]

4.2 Sputtered gold alloys

Of great interest is the development of the sputtering of gold alloys.[41] Although vacuum evaporation using, for example, electron beam heating, produces excellent gold films, it is less suitable for the production of gold alloy films, as the rate of evaporation of each constituent of the alloy is different. This difficulty can be overcome by various procedures which, however, complicate the operation. Moreover, a problem exists in that the alloy components are deposited as single atoms and energy is necessary to ensure the transformation of the coating as formed to the equilibrium alloy structure. The latent heat of condensation of the gold is not adequate to supply this energy, so that the substrate must be heated during the deposition and/or be subjected subsequently to heat treatment.

For reasons such as these, radio-frequency sputtering is perhaps the most elegant and least complicated method of depositing gold alloy films. The sputtering rates of the metals of the alloy are determined not by their inherent properties but by bombardment rates and energies, and as a result the energies with which the metals reach the target can be made adequate to ensure that the coating assumes the equilibrium alloy structure as it is deposited. Just as a wide range of gold alloys can be applied selectively to other metals by inlay cladding, so they can now be applied by sputtering.

The problems which are involved in the development of thin film conductor systems which employed reduced amounts of gold and other precious metals are well illustrated by an account by Morabito *et al.*[42] of studies of a Ti-Cu-Ni-Au system developed to replace a Ti-Pd-Au system. Conditions have also been described [42a] for the deposition by sputtering of SiPt-Mo-Au or SiPt-Ti-Pt-Au films for devices having microwave bipolar transistors.

A study of evaporated Au-In films which have been used as metallisation on various microelectronic devices is also revealing.[42b]

5. Electroless Gold Coatings

5.1 Introduction

So-called "electroless" metal coatings are of three types:

(a) Coatings formed on other metals by electrochemical displacements from a solution of a salt of the metal. This is accompanied by dissolution of the substrate metal which must therefore be less noble than the metal being plated. For this reason, the only metal which can be plated in this way on gold is silver, which is nobler than gold in cyanide solutions. Moreover, deposition ceases immediately access of plating metal ions to the basis metal is not possible, so that only thin coatings can be formed by this method.

(b) Coatings formed on surfaces, both metallic and not metallic, by homogeneous reduction processes such as silvering. In such processes, metal liberated from a solution of one of its salts by the action of a reducing agent is deposited indiscriminately over all objects in contact with the solution. The metal deposited initially may, however, catalyse further deposition, in which case plating may thereafter be largely autocatalytic in character.

(c) Coatings formed by autocatalytic plating processes. In such processes, metal initially deposited on a surface catalyses further deposition of the metal on to that surface from a solution containing one of its salts together with an appropriate reducing agent.

In the case of metallic substrates of an active metal, an initial deposit of the catalytically active metal being plated is formed by electrochemical displacement, after which autocatalytic deposition takes over. In other instances, a brief cathodic pulse may be required to deposit sufficient catalytic metal from the solution to initiate the reaction. This can be achieved either by an external power source, or by contact, below solution level, with an active metal such as aluminium which forms a local galvanic cell with the basis metal.

With non-metallic and non-conducting articles, such as plastics, initiation of reaction may be achieved by adsorption of a reducing agent on the substrate, followed by immersion of the substrate in a solution of some other metal such as palladium or gold, which is easily reduced, and which initiates the plating of the desired metal. This procedure, the mechanism of which is still imperfectly understood, is the basis of the considerable industry based today upon the plating of plastics. The pre-treatment of the surface of the plastic by "sensitization" and "activation" is an important step in the procedure in each instance. Such pre-treatment must not only promote the adsorption of the reducing agent (usually a stannous salt) on the surface, and the deposition of the activating palladium or gold catalyst, but must also etch the plastic surface so that the metal plating is "keyed" to it. The pre-treatment will therefore vary not only from plastic to plastic, but may also vary in its effectiveness according to the method of production of even the one type of plastic.

The basic reaction in autocatalytic plating processes can be represented in the simple case where the oxidation state of the metal in the plating solution is one, as

$$M^+ + Red \xrightarrow[\text{catalyst}]{M} M + Ox$$

or, more generally, in the case of metals in higher oxidation states n, by two half redox reactions,

$$n\text{Red} \longrightarrow n\text{Ox} + ne^- \quad (1)$$

$$M^{n+} + ne^- \longrightarrow M \quad (2)$$

Where M is the metal being plated and Red represents a reductant, which for the sake of simplicity is represented as being transformed in the process into Ox, the oxidised state, by transfer of a single electron.

In practice, the metal M is usually present in the plating solution in complexed form, and the concentration of M^{n+} ions in the plating solution is determined by the stability constant of the complex, and the concentration of the complexant.

Redox reactions of the above type are often catalysed by metallic surfaces and it has been suggested[43] that the role of the metal in such instances is simply that of a conductor of electrons, accepting the electrons released in the first stage of the reaction from the reductant at one point on its surface, and releasing them to the oxidant in the second stage at another point thus:

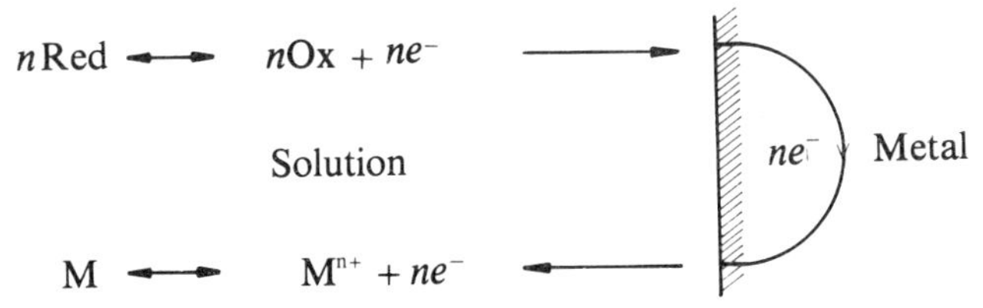

In brief, the situation in an autocatalytic plating reaction is that two separate redox reactions occur at the conducting metal surface, each of which strives to establish its own equilibrium potential on the metal.

Electrochemical theory[44–49] predicts that in a situation such as this, a steady state will be established with the metal at what is termed a steady state mixed potential, the exact value of which will be determined by the shapes of the current density potential and Tafel curves for the two electrode processes which are involved. At this steady state mixed potential both redox systems are removed from their own characteristic equilibrium potentials, so that net electrochemical reaction therefore occurs in each redox system.

5.2 Systems for "electroless" plating of gold

Systems which appear to be not truly autocatalytic: A number of "electroless" gold plating solutions, some of them proprietary, have been described and/or are on the market. They have been reviewed by Okinaka,[50] who has stressed that in a number of instances it is by no means clear whether the processes are autocatalytic or not. At the one extreme there are solutions which clearly effect plating by electrochemical displacement. They give rise to thin coatings only and will not plate gold itself. In between such solutions, however, and the borohydride baths which are truly autocatalytic, there are "electroless" gold baths which deposit gold on basis metals such as copper and nickel rapidly and to considerable thicknesses. Nevertheless, they do not plate gold itself, so that these plating processes cannot be truly autocatalytic.

Okinaka notes three possibilities which could explain the behaviour of such baths:

(i) the reaction is an electrochemical displacement and produces a porous deposit;
(ii) the basis metal acts as a catalyst, but gold does not, and the deposit is porous;
(iii) the reaction is autocatalytic, but only the freshly deposited gold is active as a catalyst.

The studies of the processes in this category which have been reported do not, however, provide a basis for testing these possibilities. That the latter possibility is a real one is indicated by a recent observation by Groenwald[51] of a catalytic effect of gold which is operative only when the gold is freshly deposited.

The types of "electroless" gold baths discussed by Okinaka[50] and by Goldie[52] include hypophosphite baths,[53–59] hydrazine baths,[60,61] diethylglycine-citric acid and diethyglycine-tartaric acid baths,[62,63] tartrate baths,[64] ammonia[65] and thiourea[66,67] baths, borohydride baths,[68–76] dimethylamine borane,[77] an aldehyde-amine borane bath[78] and a formaldehyde bath.[79] Others have since been patented. Of them all, the borohydride baths have been investigated and reported upon in greatest detail.

Information relating to the composition of the various proprietary electroless gold solutions which are marketed today is lacking.

5.3 The use of gold as an activator in the plating of plastics and other non-conductors

The sensitization and activation processes as developed and applied up to 1968 in the autocatalytic plating of plastics have been described by Goldie.[52]

In brief, the position is that a number of metals, including palladium, silver, gold and platinum have been found to initiate the autocatalytic deposition of other metals on plastics. Although at the present time palladium is used exclusively for this purpose, several authors[80–82] have claimed gold to be technically superior to palladium in particular applications, and one investigator[83] reporting on the size of the microdeposits from gold and palladium systems found that gold gave by far the finest deposits. The use of gold colloids for the activation of ABS plastic has also been investigated.[84]

Although there are other references[85–95] in the patent literature to the use of gold as an activator for the electroless plating of plastics, it does not appear so far to have found application in practice. It is possible, however, that the recent developments[84] in regard to the use of gold colloids for the activation of plastics may make these attractive for some operations in this field. One report[96] has been noted of the activation of non-conducting surfaces, for subsequent electroless plating with cobalt or copper, in which gold is applied to the surfaces by vacuum evaporation.

5.4 Electroless aerosol gold plating systems

Aerosol systems for electroless plating of gold have been reviewed by Goldie.[97] They involve the simultaneous spraying of the substrate surface with two solutions, one of which is a solution of a gold complex, and the other a solution of a reducing agent. Since the droplets of the two solutions must coalesce and react by the time they impact on the substrate surface, the systems must, in contrast to the solutions which are used for electroless plating by immersion, be highly reactive. The use of weak complexing agents, strong reducing agents and relatively high solution concentrations is therefore indicated.

As described by Levy and Momyer[98] the substrate, whether metallic or plastic is first coated with an epoxy paint. It is then sprayed with a conditioner to aid wetting and sensitized by a spray of a dilute solution of stannous chloride. The two plating sprays are then applied, and depending upon the methods of atomization used, the spraying time, the concentrations of the solutions and the distance between the atomizers and the substrate surface, different rates of deposition and different efficiencies of gold deposition are achieved. The spray parameters have been found[99,100] to be similar to those established[101] in respect of the aerosol plating of silver.

The solutions used by Levy and his co-workers in their studies[94-96,101-104] were proprietary in character,[105] but a patent[106] assigned to the Lockheed Aircraft Corporation describes the use of hydrazine as a reducing agent in aerosol plating, with the gold being supplied as its chloride or bromide, in solution with a complexing agent such as ammonium hydroxide or an aliphatic amine, preferably ethylene diamine.

Typical solutions are:

Solution “A”	Gold chloride (0.2M)	50 ml-litre
	Ethylene diamine	1.5 ml/litre
Solution “B”	Hydrazine (20M)	100 ml/litre
	Sodium hydroxide (1M)	50 ml/litre

The solutions are applied at ambient temperatures and the pH of the resultant medium is 6 – 8.

Earlier media proposed[107,108] for the aerosol deposition of gold are referred to by Goldie,[97] but the extent, if any, to which they have found application is not known.

6. Fire-Gilded Gold Coatings

Fire-gliding or mercury-gliding no longer competes with more modern methods for applying coatings of gold to metal substrates, but is worthy of mention because of the excellent properties of the coatings obtained by earlier craftsmen using this process, of which there are many outstanding examples of considerable age. In the process the cleaned, scratched and degreased substrate metal is first "quicked" or surface-amalgamated by immersion in an alkaline solution made up from mercuric nitrate and potassium cyanide, and then smeared over with the necessary quantity of a pasty amalgam, prepared from mercury (6 parts) and gold (1 part), using a copper spatula, the amalgam being distributed evenly over the surface by means of a brass wire brush. The mercury is then evaporated off by slow heating, during which the surface becomes first matt-white, and then matt-golden. On cooling the surface is burnished to a brilliant finish.

The health hazards of these operations are such that mercury-gilding is to be undertaken only under most stringently controlled working conditions.

REFERENCES

1. Chemnitius, F., Zur Kenntnis des Glanzgoldes, *J. Prakt. Chem.* **117** (1927), 245–261.
2. Levin, P. A. and Komarov, V. P., Production of liquid gold preparations based on individual terpenes, *Zh. Prik. Khim.* **45** (1972), 1235–1242.
3. Ballard, K. H. (E. I. du Pont de Nemours and Co.), "Gold Compounds and Ceramic Decorating Compositions Containing Same". U.S. Patent 2,490,399 (1949).
4. Fitch, H. M. (Engelhard Industries Inc.), "Gold Tertiary Mercaptides and Method for the Preparation Thereof". U.S. Patent 2,984,575 (1961).
5. Fitch, H. M. (Engelhard Industries Inc.), "Gold Decorating Compositions Containing Gold Primary Mercaptides". U.S. Patent 2,994,614 (1961).
6. Fitch, H. M. (Engelhard Industries Inc.), "Gold Secondary Mercaptides". U.S. Patent 3,163,665 (1964).
7. Fitch, H. M. (Engelhard Industries Inc.), "Gold Aryl Mercaptides". U.S. Patent 3,268,568 (1966).
8. Fitch, H. M. (Engelhard Industries Inc.), "Decorating Compositions Containing Silver Carboxylate-amine Co-ordination Compounds and Method of Applying Same". U.S. Patent 3,262,790 (1966).

9. Langley, R. C. and Fitch, H. M. (Engelhard Industries Inc.), "Gold–Silver Co-ordination Compounds and Decorating Compositions Containing Same". U.S. Patent 3,313,632 (1967).
10. Gibson, C. S., "Preparation of Gold Films". U.S. Patent 2,171,086 (1939).
11. Johnson Matthey Metals Limited, "Products and Services for the Manufacturer in Precious Metals: Rolled Gold and Rolled Gold Workshop Practice". London, 1973.
12. Technical Materials Inc., "Handbook for Clad Metals". Lincoln, Rhode Island, 1974.
13. Tembe, G., Gold materials for connectors: factors affecting their selection and design, *Gold Bulletin*, **5** (1972), 14–18.
14. Rao, M. U., Molchan, A. J. and Russell, R. J., "An Application of Inlay Materials for Electrical Contacts in Telephones". Paper presented to the 8th Annual Connector Symposium, Cherry Hill, N.J., U.S.A., Oct. 22–23, 1975 (cf. *Gold Bulletin*, **9** (1976), 49).
14a. Barry, R. F., Use of cupro–nickel alloy and gold inlay reduces cost of computer connector by 25%, *Insul./Circuits*, **21** (1975), 23–24.
15. Bacquias, G., Plaqué laminé et plaqué galvanique, *Galvano*, **42** (1973), 321–325.
16. Anon., Gold inlaid contacts in new edge connectors, *Gold Bulletin*, **8** (1975), 110.
17. Russell, R. J., "Properties of Inlay Clad Wrought Gold Alloys". Paper presented to the 7th Annual Connector Symposium, Cherry Hill, N.J., U.S.A., October 1974.
18. Russell, R. J., Inlay-clad gold alloys, *Gold Bulletin*, **9** (1976), 2–6.
19. Abbott, W. H., "Low Energy Electrical Contacts of Gold-Base Alloys". U.S. Patent 3,661,569 (1972).
20. Finch, R. G., Gold in thick film hybrid microelectronics, *Gold Bulletin*, **5** (1972), 26–30.
21. Caley, R. H., Gold in thick film conductors: the relative merits of all-gold and alloyed systems, *Gold Bulletin*, *9* (1976), 70–75.
22. Holmes, C. L. *et al.*, "Electrically Conductive Composition Element and method of Making the Same". U.S. Patent 3,832,308 (1974).
23. McMunn, *et al.*, "Seal Ring Compositions and Electronic Packages Made Therewith". U.S. Patent 3,809,797 (1974).
24. Hoffman, L. C., "Metallizing Compositions Containing Critical Proportions of Metal (Pt-Au or Pd-Au), and a Specific High Density Frit". U.S. Patent 3,440,062 (1969).
25. Hoffman, L. C., "Metallizing Compositions". U.S. Patent 3,385,799 (1968).
26. Hoffman, L. C. (E. I. du Pont de Nemours and Company), "Copper/Vanadium Oxide Metallizing Compositions, Noble Metal Metallizing Compositions Containing Vanadium Oxide Additives, and Electrical Conductor Elements made Thereof". U.S. Patent 3,440,182 (1969).
27. E. I. du Pont de Nemours and Company, "Werkwijze voor het bereiden van metalliseermengsels op basis van edele metalen". Dutch Applic. 67 07540 (1967).
28. Loasby, R. G., Davey, N. and Barlow, H., Enhanced property thick film conductor pastes, *Solid State Technol.* **15** (1972), 46–50 and 72.
29. Anon., A new generation of thick film gold conductor pastes, *Gold Bulletin*, **8** (1975), 13–15.
30. Sayers, P., Molecular bonding conductive films, *Solid State Technol.* **17** (1974), 66–69.

31. Hitch, T. T., Adhesion, phase morphology, and bondability of reactively-bonded and frit-bonded gold and silver thick film conductors, *J. Electron. Mater.* **3** (1974), 553–577.
32. Zeien, R. H., "Characterisation of Thick Film Fritless Metallization", *in* Proc. Int. Hybrid Electronics Symposium, 1974, pp. 7–15.
32a. Hof, J., The influence of substrate purity and manufacture on thick film conductor adhesion, *Plating*, **63** (1976), 14.
33. Smith, B. R., "An Innovation in Gold Paste", *in* Proc. Int. Microelectronics Symposium, 1972, pp. 2-A-5-1 to 2-A-5-8.
34. Smith, B. R., "Composition and Method of Bonding Gold to a Ceramic Substrate and a Bonded Gold Article". U.S. Patent 3,799,890 (1974).
35. Smith, B. R., "Gold Composition for Bonding Gold to a Ceramic Substrate Utilizing Copper Oxide and Cadmium Oxide". U.S. Patent 3,799,891 (1974).
36. Anon., "Thick-film Conductor Survey". ERA Report 75–43, Electrical Research Association, Leatherhead, Surrey, England, 1975.
36a. Holmes, P. J. and Loasby, R. G., Editors, "Handbook of Thick Film Technology". Electrochemical Publications, Ayr, Scotland, 1976.
37. Anon., Gold epoxy compositions for semiconductor bonding, *Gold Bulletin*, **6** (1973), 47.
38. Harper, C. A., Editor, "Handbook of Materials and Processes for Electronics". McGraw Hill, New York, 1970.
39. Wise, E. M., "Gold: Recovery, Properties and Applications". Van Nostrand, Princeton, N.J., 1964.
40. Anon., Gold metallisation of beryllia: Heat dissipation from electronic components, *Gold Bulletin*, **4** (1971), 69.
41. Reichelt, W., Vapour deposition of gold alloys: developments in radio frequency sputtering of electrical contact surfaces, *Gold Bulletin*, **5** (1972), 55–57.
42. Morabito, J. M., Thomas, J. H. and Lesh, N. G., Material characterisation of Ti-Cu-Ni-Au (TCNA)—a new low cost thin film conductor system, *I.E.E.E. Trans.* **PHP-11** (1975), 253–262.
42a. Pestie, J. P., Les couches métalliques pour dispositifs à transistors bipolairs hyperfréquence, *Ann. Chim., Paris,* **10** (1975), 267–273.
42b. Finstad, T. G., Andreassen, T. and Olsen, T., Characterisation of evaporated gold–indium films on semiconductors, *Thin Solid Films*, **29** (1975), 145–154.
43. Spiro, M., The heterogeneous catalysis by metals of electron transfer reactions in solution, *J. Chem. Soc.* (1960), 3678–3679.
44. Rapson, W. S. and Groenwald, T. E., The use of gold in autocatalytic plating processes, *Gold Bulletin*, **8** (1975), 119–126.
45. Wagner, C. and Traud, W., Über die Deutung von Korrosionsvorgängen durch Überlagerung von elektrochemischen Teilvorgängen und über die Potentialbildung an Mischelektroden, *Z. Elektrochem.* **44** (1938), 391–402.
46. Bockris, J. O'M., "Modern Aspects of Electrochemistry", edited by J. O'M. Bockris and B. E. Conway. Butterworth, London, 1954, p. 253.
47. Stern, M. and Geary, A. L., Electrochemical polarization. I. A theoretical analysis of the shape of polarization curves, *J. Electrochem. Soc.* **104** (1957), 56–63.
48. Paunovic, M., Electrochemical aspects of electroless deposition of metals, *Plating*, **55** (1968), 1161–1167.
49. Spiro, M. and Ravnö, A. B., Heterogeneous catalysis in solution. Part II. The effect of platinum on oxidation-reduction reactions, *J. Chem. Soc.* (1965), 78–96.

50. Okinaka, Y., in "Gold Plating Technology", edited by F. H. Reid and W. Goldie. Electrochemical Publications, Ayr, Scotland, 1974, pp. 82–102.
51. Groenewald, T., Electrochemical studies on gold electrodes in acidic solutions of thiourea containing gold(I) thiourea complex ions, *J. Appl. Electrochem.* **5** (1975), 71–78.
52. Goldie, W. (Editor), "Metallic Coating of Plastics. Vols. I and II". Electrochemical Publications, Ayr, Scotland, 1969.
53. Swan, S. D. and Gostin, E. L., Electroless gold plating, *Met. Finish.* **59** (1961), 52–53.
54. Jostan, J. L. and Bogenschütz, A. F., Electroforming of extremely smooth metallic substrates for application in microelectronics, *Plating*, **56** (1969), 399–404.
55. Tanabe, Y. and Matsubayashi, H., Formation mechanism of electroless plated gold films, *J. Met. Finish. Soc., Japan*, **21** (1970), 335–340.
56. Brenner, A., "Modern Electroplating", edited by F. A. Lowenheim, 2nd Edition. Wiley and Sons, New York, 1963.
57. Brookshire, R. R., "Method of Gold Plating by Chemical Reduction". U.S. Patent 2,976,181 (1961).
58. Ezawa, T. and Ito, H., "Plating Metals by Non-electrolytic Methods". Jap. Patents 70 24841 (1970) and 40 1081 (1965) (via reference 50).
59. Ezawa, T. and Ito, H., Jap Patent 40 25881 (1968) (via Ref. 50).
60. Gostin, E. L. and Swan, S. D., "Method and Composition for Plating by Chemical Reduction". U.S. Patent 3,032,436 (1962).
61. Luce, B. M., "Electroless Plating of Gold". U.S. Patent 3,300,328 (1967) (cf. French Patent 1,564,064 (1968)) (*C.A.* **72** (1970), 35129n).
62. Walton, R. F., Electroless deposition of gold from aqueous solution on base metals of nickel and iron–nickel–cobalt alloys, *J. Electrochem. Soc.* **108** (1961), 767–774.
63. Pokras, D. S., Sullens, T. L. and Walton, R. F., "Electroless Gold Plating Solution". U.S. Patent 3,123,484 (1964).
64. Trueblood, R. K., "Electroless Gold Plating on Refractory Metals". U.S. Patent 3,862,850 (1975) (*C.A.* **82** (1975), 144004y).
65. Schneble, F. W., McCormack, J. F. and Zeblisky, R. J., "Electroless Gold Plating". U.S. Patent 3,468,676 (1969) (cf. Jap. Kokai 74 67838) (*C.A.* **82** (1975), 37940a).
66. Oda, T. and Hayashi, K., "Electroless Gold Plating Solutions". U.S. Patent 3,506,462 (1970).
67. Baker, K. D., "Chemical Plating from Aqueous Solutions of Thiourea Complex Compounds". British Patent 1,193,823 (1970).
68. Sard, R., Okinaka, Y. and Rushton, J. R., Some properties of electroless gold deposits, *Plating*, **58** (1971), 893–900.
69. Okinaka, Y. and Wolowodiuk, C., Electroless gold deposition: replenishment of bath constituents, *Plating*, **58** (1971), 1080–1084.
70. Okinaka, Y., An electrochemical study of electroless gold-deposition reaction, *J. Electrochem. Soc.* **120** (1973), 739–744.
71. Okinaka, Y., Electroless gold deposition using borohydride or dimethylamine borane as reducing agent, *Plating*, **57** (1970), 914–920.
72. Okinaka, Y., Sard, R., Wolowodiuk, C., Craft, W. H. and Retajczyk, T. F., Some practical aspects of electroless gold plating, *J. Electrochem. Soc.* **121** (1974), 56–62.

73. Okinaka, Y., Sard, R., Craft, W. H. and Wolowodiuk, C., Paper presented at Full Meeting of Electrochem. Soc., Miami Beach, 1972.
74. Sard, R., Chemical deposition of single crystal gold films, *J. Electrochem. Soc.* **117** (1970), 1156–1157.
75. Gardiner, J. A. and Collat, J. W., Kinetics of the stepwise hydrolysis of tetrahydroborate ion, *J. Am. Chem. Soc.* **87** (1965), 1692.
76. Gardiner, J. A. and Collat, J. W., Polarography of the tetrahydroborate ion. The effect of hydrolysis on the system, *Inorg. Chem.* **4** (1965), 1208–1212.
77. McCormack, J. F., "Autocatalytic Gold Plating Solutions". U.S. Patent 3,589,916 (1971).
78. Rich, D. W., see Ref. 50, pp. 100–101.
79. Wein, S., U.S. Rept. PB 111,332, Office of Technical Services, U.S. Dept. of Commerce, 1953 (cf. *Met. Finish.* **46** (1948), 8 and 58).
80. Saubestre, E. B., "Copper Sensitizers". U.S. Patent 2,872,359 (1959).
81. Narcus, H., Improved techniques for electroless nickel plating on non-conductors, *Proc. Am. Electropl. Soc.* **53** (1956), 157–161.
82. Pearlstein, F., Electroless nickel deposition: activation of non-metallic surfaces, *Met. Finish.* **53** (1955), 59–61.
83. Matsunaga, M., Hagiuda, Y. and Ito, K., Adhesion of electrodeposits to plastics: an electron microscopic investigation, *Met. Finish.* **66** (1968), 80–84.
84. Fulmer Research Institute Ltd., "An Improved Method for the Preparation of Activators for Use in the Electroless Plating of Plastics". U.K. Prov. Patent No. 27,122/74.
85. Bergstrom, A. E., "Method of Catalytically Depositing Metal Layers". German Offen., 1,915,252 (1969) (*C.A.* **72** (1970), 15213g).
86. Hepfer, I. C., "Electroless Plating of Nickel or Copper on Non-conductors". U.S. Patent 3,370,974 (1968) (*C.A.* **68** (1968), 80980m).
87. Young, W. H., Csuthy, B. and Guidess, J., "Metallised Plastic Part". U.S. Patent 3,565,770 (1971) (*C.A.* **74** (1971), 143434h).
88. Heymann, K. and Woldt, G., "Etching of Plastic Surfaces Before Plating with Metals". German Offen. 1,949,278 (1971) (*C.A.* **75** (1971), 7004m).
89. Hentzschel, H. P. K., "Electroless Nickel Plating of Semiconductor Parts". German Offen., 1,949,754 (1970) (*C.A.* **75** (1971), 27006s).
90. Foulke, D. G. and Simonian, A., "Sensitizing Substrates for Electroless Plating". U.S. Patent 3,516,848 (1970) (*C.A.* **73** (1970), 58796n).
91. Sel-Rex Corp., "Process and Solution for Sensitizing Substrates for Electroless Plating". British Patent 1,101,848 (1968) (*C.A.* **68** (1968), 71734g).
92. Rajchenbaum, N. B., "Metallising Insulating Surfaces". British Patent 1,122,235 (1970) (*C.A.* **69** (1968), 82950w).
93. Schering, A. G., "Treatment of Plastic Surfaces Before Galvanizing". French Patent 1,553,827 (1969) (*C.A.* **72** (1970), 13939z).
94. Okuno, G. and Yamauchi, C., "Metallization of Plastics". Dutch App. 66 10976 (1967) (*C.A.* **67** (1967), 12247g).
95. Baker, K. D., British Patent 1,193,823 (1970).
96. Fusayama, T. *et al.*, "Electroless Plating". German Offen. 2,158,239 (1972) (*C.A.* **77** (1972), 91725u).
97. Reid, F. H. and Goldie, W., "Gold Plating Technology". Electrochemical Publications, Ayr, Scotland, 1974, pp. 171–180.
98. Levy, D. J. and Momyer, W. R., Spray deposition of low emittance gold thermal control coatings, *SAMPE J.*, April/May, 1969.

99. Levy, D. J., The technology of aerosol plating, *Proc. Am. Electropl. Soc.* **51** (1964), 139–149.
100. Levy, D. J. and Delgado, E. F., The effect of atomization conditions on the aerosol plating of gold, *Plating*, **52** (1965), 1127–1132.
101. Upton, P. B., Soundy, G. W. and Busby, G. E., Some factors in spray-silvering, *J. Electroplat. Depositors' Tech. Soc.* **28** (1952), 103–113.
102. Gomes, G. S. and Levy, D. J., Gold plating with spray guns, *Prod. Finish., Cincinnati*, **27** (1963), 36–44.
103. Levy, D. J., "Materials for Space Use", *in* Soc. of Mater. Process Engineers, Symposium, Seattle, 1963.
104. Levy, D. J., Gold plating by chemical reduction, *Met. Finish. J.* **10** (1964), 157–160 and 164.
105. Lockheed Missiles & Space Co., "Lockspray Gold Process". Technical Data Bulletin 6-76-66-9, Palo Alto, California, 1966.
106. Lockheed Aircraft Corp., "Deposition of Gold Films". British Patent 1,027,652 (1966).
107. Andres, F., "Means for Coating with Metal". U.S. Patent 1,953,330 (1934).
108. Schneider, H., "Process and Apparatus for Producing Metallic Coatings on Various Articles". U.S. Patent 2,136,024 (1938).

10
Gold and Gold Alloys in Miscellaneous Forms and Uses

1. Gold and Gold Alloy Wire

In jewellery fabrication there is significant consumption of carat gold wires for chain making, for which sophisticated equipment is available. At each caratage alloys must be selected which are malleable and ductile. Red gold alloys and nickel white gold alloys may present problems in wire and chain making, because of inadequate malleability and ductility.

In the electronics industry, on the other hand, particular interest attaches to the extensive use of pure gold in the form of fine wire in transistor and integrated circuits.

Gold bonding wire for these applications is cold drawn from high purity gold ingots through successive diamond dies to final diameters of 13–37 μm. The operation is difficult because of the low tensile strength of the gold and because of recovery and even crystallisation effects in the highly cold-worked pure gold.[1] Scrupulous cleanliness and dust-free conditions are required during wire drawing.

The main objective in drawing is to achieve a wire with an appropriate combination of elongation and break load. This requirement, and electrical specifications, call for high purity and much gold wire is sold as 99.975% gold. Since gas impurities cause difficulty as a result of the formation of outgassing blow holes, many manufacturers use vacuum processed gold for wire production. An ASTM specification for gold wire for semiconductor lead bonding has been issued.[2]

Both mechanical and metallurgical problems can arise during hot bonding. Mechanical defects arise from too severe indentation, while metallurgical problems can occur, for example, in ball bonding, because of the heating of the wire in the capillary feed and because of the flame-off operation when the molten ball of gold is formed. Crystallisation in the molten ball results in a coarse-grained and weaker structure but this is at least partly offset during the bonding opertion by work hardening. Recrystallisation with grain growth which occurs in the wire above the ball, however, reduces the strength of the wire, and this results in failures particularly in plastic packages exposed to temperature cycles which stress the wires. Such failures can be materially reduced by the addition of grain refining agents such as beryllium which is effective in concentrations of 0.0025–0.0075% by weight,[3] or yttrium and the rare earth metals, which are effective in concentrations of 0.005–0.05% by weight.[4] The data presented by Raub and Thiede[4] in respect of the latter types of addition indicate that they increase the recrystallisation temperature of the gold to such an extent that it continues to exhibit resistance to creep up to 500°C.

Apart from the above, special considerations arise[5] when gold wires are bonded to aluminium surfaces, or aluminium wires are bonded to gold surfaces, because of the formation of intermetallic compounds and of "Kirkendall voids" which develop when the joints are exposed to temperatures in excess of 200°C. Philofsky[6] has recently investigated these phenomena and reviewed earlier literature upon them. He concluded that if Au-Al bonds are exposed to temperatures above 250°C for long enough and there are sufficient supplies of gold and aluminium, then all five of the known intermetallic phases will form, namely, $AuAl_2$, AuAl, Au_2Al, Au_5Al_2 and Au_4Al. The intermetallic compound formed in major amount is always Au_5Al_2. If, however, there is a shortage of gold or aluminium, as is often the case in wire bonding, then only aluminium-rich or gold-rich intermetallics will form. When $AuAl_2$ is present, it lends its purple colour ("purple plague") to the area concerned, though where silicon is present (as may be the case in making bonds to gold-plated silicon dies) another purple phase thought to be a ternary compound containing 6% Si also contributes to the effect. In contrast to certain earlier investigators, Philofsky concluded that the presence of the intermetallic compounds did not in itself weaken the Au-Al bonding. He associated such weakening with the occurrence of Kirkendall voids which tend to develop in near continuous lines as a consequence of the greater rates of diffusion of aluminium as compared with gold. This results in a piling up on the aluminium side of the bond of lattice vacancies which coalesce to form microcracks and voids, which reduce the strength of the bond. The effect is an insidious one, since as mentioned above, it can develop in a joint which was sound when established, if the joint is exposed subsequently to temperatures at which diffusion rates are significant.

Similar problems arise in flip chip bonding where, for example, aluminium pads on a silicon chip may be bonded to gold-plated pillars on the substrate. Titanium has been reported[7] to be effective as an alternative to aluminium for metallisation of silicon chips at points at which it is desired to bond gold wire.

Where it is desired to have both gold and aluminium metallisation in a hybrid microcircuit, however, steps are desirable to avoid the necessity for making connections which involve Au-Al interfaces. Hampy[8] has described one way in which this may be done in a two-level metallisation system.

2. Gold Leaf

The ancient art of gold beating to produce gold leaf has been reviewed by Theobald,[9] by Lübke,[10] and by Brüche and Schulze,[11] and summarised by Wise.[12] The extreme thinness (0.1 μm) to which gold can be beaten, and therefore the considerable area of leaf which can be derived from a small weight of gold, has made possible the extensive use of gold leaf for decorative purposes for altars, statuary, buildings, mosaics, printing and bookbinding. In all these applications its tarnish resistance is of great significance.

Although pure gold may be used as starting material, the ease with which it welds together if accidentally folded can cause trouble in producing thin (0.1 μm) leaf. It is normally used therefore for the production of heavier foils. The ductility of a range of Au-Ag-Cu alloys, however, is still sufficiently high for them to be beaten to thin leaf. The fact that all three metals have an f.c.c. lattice and form solid solutions with one another is significant in this connection.

For the greater part, gold leaf has been made from alloy containing 92.5% Au, but alloys with lower gold percentages have been used, with a wide range of colours depending upon the relative proportions of copper and silver present in the alloy. Lübke[10] lists the range of colours and compositions of alloys from which gold leaf has been produced. Some contain as little as 30% Au.

In application, beaten gold leaf is bonded to the substrate in each instance by an appropriate adhesive. Where it is applied by stamping, for example in printing or embossing, beaten leaf may be replaced by gold deposited on the surface of a polymer film[13] by evaporation, sputtering or other means. The gold surface is then coated with a thermoplastic adhesive, so that when the composite material is pressed on to the workpiece with a hot stamp, the gold will adhere to the workpiece and become detached from its film support. Leaf in this form is more suited to factory type applications.

The structure of beaten gold leaf and the processes occurring in its formation have been studied in particular by Brüche and Schulze;[11,14] and Hirsch, Kelly and Menter.[15] Under the mechanical stresses during beating, the latter authors concluded that deformation of the gold crystals by slip along the glide planes occurs, with the generation of very thin (0.01 μm) gold platelets which lie partly beside and partly overlapping one another in a final gold leaf of about 0.1 μm thickness.

Hutting and Nuttall[15a] have pointed out in an article on the malleability of gold, however, that this model does not adequately explain all the known facts relating to the deformation of thin gold foil by rolling and beating. A significant difference between the behaviour of gold and even such closely related metals as copper and silver is that gold continues to undergo plastic deformation even when the thickness of the beaten foil is less than the size of the sub-grains produced by the plastic strain, whereas with copper, silver and other metals fragmentation occurs when the foil or leaf thickness reaches the sub-grain diameter. Moreover Nutting and Nuttall did not observe the expected increase in the maximum sub-grain size diameter as the total strain in gold leaf was increased. They concluded that "even when the thickness of gold foil becomes less than the maximum sub-grain diameter, new sub-grains can be formed by dislocation movements and interactions totally within the volume of the foil".

They associate this unusual behaviour of gold with its nobility and with the fact that there is no oxide film on a gold surface which can prevent the escape at the metal surface of dislocations formed within the gold. The escape of such dislocations from the surfaces of thin gold leaf or foil is seen as enabling it to accommodate strain without preferential deformation occurring at the sub-grain boundaries.

3. Colloidal Golds

Although certain gold sols have been shown[19] to be effective as activators of the autocatalytic plating of plastics with metals such as copper and nickel, they do not appear to be used for this purpose. Nevertheless radioactive gold sols are used in medicine, and studies of the reactions by which gold sols are formed have contributed materially to knowledge of the processes by which metals are precipitated from solutions of their salts by reducing agents. Their formation and their properties are therefore of significance for an understanding of the processes which occur in the production of gold powders by such reactions and merit some mention from this point of view. They are also of interest in relation to the production of Purple of Cassius.

Electron microscope studies of a wide range of procedures for the preparation of gold sols have demonstrated that the particle shapes, particle sizes and particle size distribution curves of the dispersed gold phases vary greatly and depend upon the reducing agents used and the conditions under which they are applied.

Particle shapes can vary from irregular to spherical, with in some instances triangular and hexagonal platelets being formed as well as pentagonal bipyramidal forms.[16–18] Particle size distribution curves on the other hand appear in any one type of reduction to be dependent largely upon the relative rates of nucleation on the one hand and of particle growth on the other.

Nucleation does not occur with all reducing agents and for example does not occur with hydroxylamine under acid conditions. It has been suggested[17] that the capacity to produce gold nuclei is an attribute of only those reducing agents which at some intermediate stage of the reduction give rise to complexes containing a number of gold atoms, from which free gold atoms are ultimately released in clusters or nuclei. Evidence for interaction between individual nuclei has nevertheless been presented.[20] Particle growth on the other hand, such as can be demonstrated when the hydroxylamine reagent is added to a solution in which gold nuclei have already been generated, could well take place by an autocatalytic mechanism.

The significance of the above principles is well illustrated in the application by Frens[21,22] of controlled nucleation in the production of monodisperse aqueous suspensions of spherical gold particles. Frens found that in the reduction of chloroauric acid by sodium citrate, the number of nuclei generated was dependent on the amount (concentration) of sodium citrate, and the extent to which these nuclei grow, and therefore the ultimate size of the gold particles, was dependent upon the amount of gold available for reduction.

The gold sols which have been employed[19] for activating autocatalytic plating of plastics have not been studied in a similar manner to the above. They have been produced by reduction with stannous chloride in acid media, and it is possible, in the light of the results of studies[23–27] of palladium sols analogously prepared, that they are not simple gold sols.

Concentrated stable gold sols have recently been claimed as resulting from the reduction of gold salts with a polyelectrolyte such as polyacrylic hydrazide in the presence of a strong electrolyte such as 2N°KNO_3.[28]

One of the best known forms of colloidal or finely divided gold, however, is Purple of Cassius. This is an amorphous flocculant purple-violet precipitate which is formed when stannous chloride, which contains some stannic chloride, is added to a dilute solution of gold chloride. The presence of stannic chloride is essential, since pure stannous chloride produces only a brown precipitate. The intensity of the colour can be gauged from the fact that its formation is capable of revealing the presence of one part of gold in 10^8 parts of water when the reaction or test is carried out under appropriate conditions.

Although modern texts and encyclopedias attribute the discovery of Purple of Cassius to Andreas Cassius Junior in 1685, Hunt's investigations[29] have recently revealed that its preparation was described by Johann Rudolph Glauber 26 years earlier in 1659. Various procedures have been recorded for its production. Zsigmondy[30] for example has described its preparation by mixing together of gold chloride solution (200 ml, 3 g Au/litre), stannous chloride solution (250 ml, 3 g Sn/litre), a little HCl and water (4 litres). After three days the purple was deposited and the supernatant liquid was free from gold and tin. In this and analogous procedures, the precipitate is apparently formed by flocculation of micellar particles containing small particles of gold protected by colloidal stannic acid. In accord with this, the precipitate can be peptised by the action of such reagents as ammonia and alkalis. The Zsigmondy preparation after ignition contained ca. 40% Au and 60% SnO_2.

Andreas Cassius (junior) employed Purple of Cassius in the production of purple enamels. It is still used in the preparation of a range of enamels, ranging from pink to maroon in colour, and this use is growing.

Similar coloured dispersions of fine gold particles in other oxides can be prepared without use of tin, for example, by reduction of gold chloride solution by glucose in a suspension of aluminium hydroxide.

4. Gold in Glass

Although it had been known for many centuries that compounds of gold could impart a red colour to glass, present-day knowledge of gold ruby glasses had its beginnings in the discovery in the latter half of the seventeenth century that if a little Purple of Cassius was introduced into glass, the product though colourless initially, turned ruby red on reheating.[29] This was the basis of the production of ruby glasses by Johann Knunckel at Potsdam between 1678 and 1689, but the results were not predictable, and an understanding of the process adequate to ensure its control was not achieved until relatively recently. The background investigations are summarised by Weyl,[31] but the significant contributions were those of Stookey[32,33] and of Maurer,[34] whose work has been summarised by Wise.[35]

When an oxidising glass is melted with a gold salt or with finely divided gold, the gold dissolves in colourless ionic form, and the glass remains colourless even at concentrations of 0.02–0.10% Au. However, if small amounts of polyvalent elements such as tin, antimony, bismuth, lead, selenium or tellurium are present, which tend to act as reducing agents at lower temperatures, the glass may become red either upon cooling or upon reheating for some hours at 600–700°C, as a result of the precipitation of gold in the glass in finely divided form. The exact behaviour depends, however, upon the concentration of reducing agents. Where this is low, development of ruby colour or "striking" may not occur, unless the solid glass is exposed to X-rays (or to near U.V. radiation if cerium is present). Such exposure induces photochemical reduction of gold ions to form minute nuclei of metallic gold upon which additional gold will then precipitate on re-heating to above 500°C to accelerate diffusion processes in the glass

At a low concentration of reducing agent, no further nucleation occurs during the heat treatment, but if the concentrations are higher or a more powerful reducing agent is present, further nucleation will occur on heating.

The photochemical effect has been exploited as the basis of a "photo process",[36] using glasses in which the gold (or silver or copper) particles formed by photo-nucleation through a screen act as nuclei for the precipitation of phases such as lithium metasilicate which are soluble in dilute hydrofluoric acid and can be selectively etched to create a pattern on the glass.

Both copper and silver can be used similarly in glass, though at higher concentrations, to produce copper ruby and silver yellow glasses.

Ultra violet absorbing glasses for use as filters in photography with colour films are claimed[37] which contain gold. Gold has also been described[38] as a component of certain glasses which change colour reversibly in light, and a red hard glass composition containing gold has been claimed[39] as suitable for use in infrared lamps because of its transparency to i.r. radiation (0.7–3.5 μ).

5. Gold Powders

Gold, along with palladium and platinum, has been used in significant amount in the form of powder in recent years as a component of pastes which are screen printed on ceramic wafers in the production of micro electronic devices by the thick film process.[40] After being screen printed, the wafers are fired, organic components of the pastes destroyed and a residual film left bonded to the wafers. This film may be conductive, resistive or insulating, depending on the nature of the paste used. Pastes to produce conductive films may contain gold, either unalloyed or alloyed with platinum and/or palladium.

Since these films must be precisely printed and finely defined, pastes must be so formulated that fine pattern detail can be achieved using them. Whilst the nature of the organic vehicle for the paste is all important in this connection, the sizes and shapes of the particles of the metal powders are also significant. Thus gold particles which are greater than 5 μm in size tend to be deformed into laminar flakes during the preparation of the pastes. The gold flakes can then block the screen in the printing process and lead to poor definition.

Gold and gold alloy powders are also of interest in connection with the production of gold composites and gold composite coatings (see below), and the decoration of ceramic ware.[41] They may also be used in dental restorations, and in the manufacture of postage stamps with relief gilded decoration,[42] in soldering and brazing pastes, and in conductive metal powder connectors.[43]

As indicative of the forms in which gold powder is marketed, data concerning products sold by one company[44] are reproduced in Table L.

Gold powders of other types are produced and used by various manufacturers, however, and a wide variety of procedures have been claimed for the production of gold and gold alloy powders to conform with various specifications in regard to particle sizes and shapes.

Thus Short[45,46] and Short and Weaver[47] describe conditions for the production of gold powders for controlled particle shapes and sizes by reducing gold salts with various reagents in the presence of a protective colloid such as gum arabic to prevent agglomeration of the gold particles. Reduction with K_2SO_3 and with hydroquinone and its derivatives, for example, yields spherical or approximately spherical particles, while reduction with oxalic acid or oxalates gives flaky powders. Unsaturated alcohols such as allyl and propargyl

Table L. Properties of Gold Powder Marketed by One Company

	Particle structure	Particle size distribution	
		Approximate percentage	Particle size (μm)
1.	Dendritic	10	1–2
		20	2–5
		50	5–10
		20	10–20
2.	Dendritic	50	1–2
		35	2–5
		15	5–10
3.	Platelets	10	5–10
		40	10–20
		30	20–30
		20	40–40
4.	Platelets	30	5–10
		40	10–15
		20	20–30
		10	40–50

alcohols when used as reductants in the presence of protective colloids also yield platelets.[48] The use of an emulsifier (preferably butyl stearate) during the precipitation of gold powder has been claimed by Daiga[49,50] as giving rise to gold particle sizes less than 20μm, each particle being coated with emulsifier, which acts as a lubricant when the powder is used in printing formulations.

Precipitation processes such as the above can apparently be applied[51,52] to the production of certain gold alloy powders by the addition of salts of metals such as palladium, platinum, silver and rhodium to the original gold solution, and Owens Illinois Inc.[53] have claimed that extremely homogeneous powders for electronic applications are produced by suspending gold particles in a solution of a salt of silver plus at least one other metal such as palladium, and then precipitating the silver and the other metal or metals on the gold particles. No chloride or cyanide must be present. The production of noble metal powders for use in thick film pastes and other metallising compositions by evaporation of the metals in inert atmospheres,[54,55] by milling[56,57] and by freeze drying of colloidal dispersions of metals in various media[58] has also been claimed. A process for the direct production of high purity gold powder from Merrill slimes obtained in the extraction of gold has been described by the Anglo American Corporation of South Africa, Limited.[59]

6. Gold Composites

Gold composites have been produced in "bulk" by powder metallurgical techniques, and in the form of surface coatings by electrodeposition and vacuum techniques. Composites containing Al_2O_3 as dispersed phase have also been produced by internal oxidation of Au-Al alloys.[60,61]

The objectives have been to obtain dispersion-hardened golds which would be superior to gold or its alloys in respect of wear resistance in jewellery and in sliding contact applications, in respect of welding and metal transfer characteristics in make-and-break contact applications, or in respect of creep in higher temperature uses.

As disperse phases a number of metallic oxides, carbides, nitrides, silicides and nitrides have been used. Organic polymers such as poly-tetrafluorethylene have also been electrodeposited[62] with gold to give composite coatings with polymer as the disperse phase, but no uses for the products appear to have been developed.

Sautter and his co-workers,[63-67] for example, demonstrated that composite coatings formed by co-deposition of alumina with gold from gold electrolytes containing alumina in suspension possessed considerably improved creep resistance as compared with gold itself. Peiffer *et al.,*[68] moreover, have reported that the incorporation of tungsten carbide in plated gold decreased welding and metal transfer in reed switch and certain relay applications. Further, a French patent application[69] claims benefits in sliding applications from the incorporation of sulphides of molybdenum, niobium and tungsten in plated gold, and a method for the electrodeposition of abrasion-resistant gold-graphite coatings has been described.[70] Wear-resistant gold-diamond composite coatings have also been produced.[71]

More recently Larson[72] has reported in some detail on the electrodeposition of gold-tungsten carbide and gold–titanium carbide composites, and on the properties of the deposits themselves. Substantial increases in hardness and tensile strength were found with increasing volume per cent of carbide, and in tune with these, wear resistance increased markedly. The wear resistance was measured against a hard gold alloy wire (JM 625) in a continuous sliding contact with which it was found that the contact resistance was only marginally higher than it was for pure gold surfaces. The hardness of the deposits was retained during annealing at temperatures up to almost 300°C. Larson concluded that the observed mechanical properties of the composites seemed to indicate that hardening occurred in the composites not only by restriction of dislocation movements in the metal matrix, but also by mechanical restraint of metal deformation by the hard disperse phase. It was felt that deposits of this type could find extensive use in certain sliding contact applications, and that they might also prove of interest in decorative jewellery applications.

Composite coatings of gold with easily oxidisable metals such as aluminium[73] and titanium[74] have also been produced by vacuum deposition of gold and such metals in alternate layers, the oxidisable metal in each alternate layer being converted to its oxide by heat treatment. Magnesia-gold composite coatings have also been produced by sputtering.[75]

The production of gold composites by powder metallurgy techniques, and their properties have been reported upon by Gimpl and Fuschillo[76-78] who found that with both alumina and thoria as disperse phases, recrystallisation and grain growth of the dispersion-hardened gold were retarded and its mechanical properties at elevated temperatures improved. Poniatowski and Clasing,[79] on the other hand, used titanium dioxide as the disperse phase, and found similar enhancement of mechanical properties at higher temperatures. They did not observe significant improvement, however, in the hardness of the composite at ambient temperatures as compared with gold, so that wear properties were presumably unimproved.

The most extensive studies of gold composites of this general type appear to have been carried out by Engelhard Industries[80] who have found,[81] *inter alia,* that small quantities of certain oxides such as cerium oxide in gold substantially improve its performance in make-and-break contacts. The low hardness and low recrystallisation temperature of gold lead to excessive mechanical wear, and a high tendency for welding and galling when it is used for make-and-break contacts at higher current values. Moreover, excessive loss of gold by arc erosion occurs. It is for this reason that most commercial gold contact materials to date have been alloys containing other noble metals such as platinum or silver, or base metals such as nickel or copper. The extent of these alloy additions is limited by adverse effects on conductivity and/or tarnish resistance. Additions of 0.1 to 4.0% of cerium oxide by volume to substantially pure gold, however, yield outstanding contact materials with substantially zero arc erosion, a reduced tendency to welding and only slightly higher contact resistance.

Villot and Lingenberg[82] have claimed the use for jewellery purposes of a wide range of metal-hard metal composites, which include composites of gold and gold alloys with boride, nitride, carbide or silicide compounds of Group IVa, Va and VIa metals, formed by pressing and sintering to a pore-free condition.

Metal friction couples which operate satisfactorily at 550°C in air without lubricant have been described as being pressed composites of Cr_2O_3 with gold made by a proprietary method.[83] Metal surfaces have also been described as lubricated by a polymeric matrix, such as poly-tetrafluoroethylene, containing ultrafine gold (or certain other metal) particles in flake form.[84] Composites of molybdenum and gold,[85] and of ZrO_2 and gold[86] have also been patented.

Preliminary studies have been reported[87] of the combined effects of dispersion strengthening and order strengthening in Cu_3Au single crystals.

7. Gold "Skeletal" Foams and "Expanded" Gold

In flexible foams made from rubber or rubber-like materials the air cells may be completely separate, in which case the foam is impermeable to liquids and gases, or they may be partly connected to one another, in which case they are permeable. From the latter type a third kind of structure, a skeletal foam, can be made by knocking out the "windows" between adjacent cells.

These skeletal foams which may be as much as 97% void, can be coated with gold, and if desired the substrate polymer removed to yield a gold skeletal foam.[88,89] Whilst no record has been noted of applications of such gold foams, it seems appropriate that they be mentioned in the present context.

"Expanded" gold has recently been introduced as a reinforcement for dental plates.[90] Although such reinforcement has traditionally been fabricated in a 75% gold alloy in the form of gauze woven from 127 μm diameter wire, the expanded metal produced from sheet by continuous shear and press operation is stronger, and makes possible both thinner and less costly construction. It seems likely that gold and gold alloys in "expanded" form will find other applications.

8. Gold Alloys with Shape Memory

Reversible shape change or shape memory is the property which certain alloys exhibit, whereby after deformation at one temperature into a new shape, they return spontaneously to their original shape after heating to above a critical temperature. It was observed first in single crystals of Au-Cd alloys, which apart from being expensive to produce, were brittle and therefore not capable of being worked and used. The same property was, however, later found in poly-crystalline and ductile alloys such as the compound TiNi (which has found applications) and certain other gold alloys.[91,92] In respect of the latter, studies have been reported[93] of selected Au-Cu-Zn alloys falling in the pseudo-binary system between the ductile Cu-Zn memory alloys, the beta brasses, and the less ductile Au-Zn alloys based on AuZn.

The beta phase forms a continuous single phase field between the Au-Zn and Cu-Zn systems and is characterised by Heusler ordering at approximately 25Au-25Cu-50Zn atomic composition. The temperature M_s at which the memory phase phenomenon is exhibited is determined by the martensitic transformation of the β-phase. Brook and Iles[93] were able on the basis of known data[94-99] to plot (see Figure 53) estimated constant M_s contours within

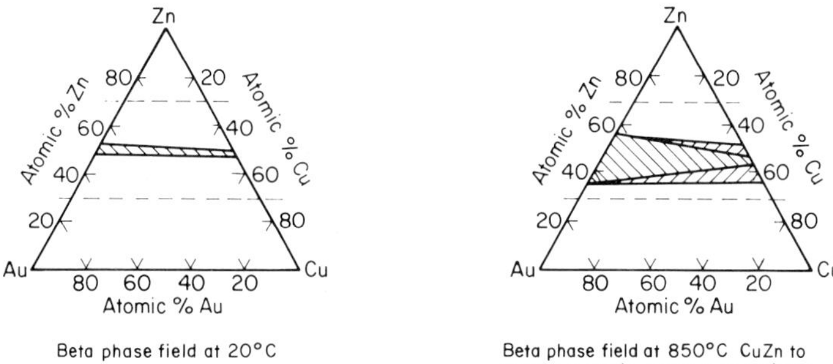

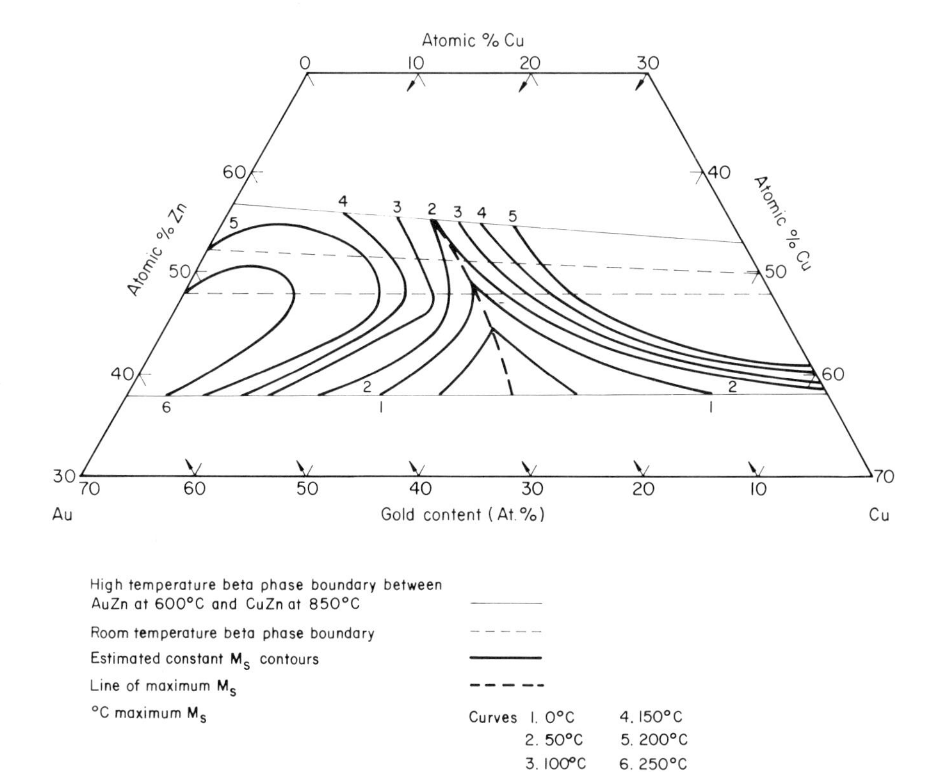

FIG. 53. Section of the Au-Cu-Zn ternary phase diagram, showing beta-phase field at high temperature between AuZn at 600°C and CuZn at 850°C), and contours of constant M_s temperature estimated from known data. (From Brook and Iles.[93])

the β-phase field at high temperature, between Au-Zn at 600°C and Cu-Zn at 850°C, and on the basis of these contours to select five alloys for study.

The behaviour of these alloys indicated the diagram (Figure 53) of estimated constant M_s contours to be roughly true, so that within the Au-Cu-Zn system a locus of maximum M_s temperatures forms a ridge at roughly equiatomic concentrations of gold and copper, extending in the general direction of increasing zinc content. This means that for any particular M_s temperature, which defines the temperature below which the alloy has shape memory, there is a choice of a gold-rich or a copper-rich alloy. There is thus within this system a considerable degree of flexibility for choosing alloy compositions with shape memory, so that other important properties of memory alloys of this class such as hot workability, mechanical strength and ductility at ambient temperature can also be optimised.

Although various applications for these alloys have been mooted[91] they do not appear as yet to have been used in practice.

9. Gold in Photography

The use of gold in photography for toning and stabilisation of images, and for sensitisation of silver chloride emulsions, has recently been reviewed by Ellis.[100]

9.1 Gold toning, image stabilisation and latensification

Gold toning is rarely if ever used today. It was first applied to the silver-mercury amalgam images of daguerrotypes, the beauty and permanence of which were found to be greatly increased by treatment with gold chloride. Then later, when the silver print-out images produced by the Fox-Talbot process were found to be unstable, weak and unattractive in colour, the same technique of gold toning was applied to them and found to improve them also. Gold toning was therefore in use for many years until print-out images were replaced by developed silver images. These were blacker and more stable than those previously obtainable, so that with their introduction gold toning survived mainly as a treatment for special effects.

Gold toning has been reviewed by Blay[101] and gold toning bath formulae have been the subject of many patents. Typically[102] a toning bath consisted of a solution containing 0.1% gold chloride and 1% potassium or ammonium thiocyanate, and different colour effects were produced according to the nature of the original print, and its degree of exposure. Toning produced not only changes of colour, however, but also an image that was much more permanent than the original image. Formstecher[103] found that the mechanism of the toning process was electrochemical, silver of the image being replaced by gold:

$$Ag + Au^{+} \rightarrow Ag^{+} + Au$$

$$\text{or} \quad 3Ag + Au^{3+} \rightarrow 3Ag^{+} + Au$$

In accord with the equations for the relevant replacement reactions, monovalent gold was found much more effective than trivalent gold in toning. The replacement of the silver by gold was limited to 40–50%, however, in the absence of thiocyanate which increased it to 100%.[104]

In present day practice, such replacement of the silver of conventional images by gold may be practiced for image stabilisation, for example, in the case of microforms which may be kept in archives for many years. Crabtree, Eaton and Mueller[105] have used thiocyanate complexes of gold for the silver replacement reaction whilst Henn and Mack[106] have used gold thiourea complexes for the purpose. The gold images which result are not subject to deleterious effects as a result of exposure to air pollutants such as SO_2, H_2S and ozone, as is the case with silver images.

Quite apart from its effect on *developed* silver images, treatment with gold has been found to have a marked effect on *latent* silver images before development. Thus if an exposed emulsion is bathed for a few minutes in a gold salt solution, the silver of the latent image is replaced at least partly by gold, and subsequent development of the image is greatly accelerated.[107-109] This effect is known as latent-image intensification or latensification. The effect is not limited to simple replacement of silver by gold, however, since if gold treatment of the latent image is prolonged, a gold image is formed which is visible in the electron microscope. Gold thus also exercises a development effect with an increase in the size of the latent image centres. The size of the development centres apparently increases as a result of autocatalytic deposition of metal, but the number of the centres remains unaltered.[110]

9.2 Emulsion sensitisation by gold

The most important use of gold in photography today is for sensitising emulsions.

In brief,[100] photographic emulsions are prepared by precipitation of silver-bromide in an aqueous solution of gelatin, which prevents the silver-bromide from settling out, so that a suspension of fine silver halide crystals results. The size distribution of these crystals determines the photographic properties of the emulsion and is carefully controlled. Thus, in order to allow the size of the silver halide particles to come to equilibrium and in order to induce an increase in the sensitivity of the final emulsion, the reaction mixture is next digested or ripened for a period at 50°–80°C. The gel obtained on cooling or the precipitate obtained by coagulation of the gelatin is then washed free of soluble salts.

At this stage a second ripening treatment may be given in order further to increase the sensitivity of the final emulsion. In this, gelatin with sensitising properties may be added to the melted emulsion and the mixture held at a steady temperature for a period. In the process the size of the silver halide particles as unaffected, and the increase in sensitivity is thought to result from the presence of "sensitising" gelatins of sulphur-containing sensitisers such as allyl thiourea and allyl isothiocyanate, which form sensitivity centres of silver sulphide or metallic silver on the silver halide crystals.[111]

The first suggestion that gold salts might produce a much greater sensitisation than sulphur compounds such as these came in 1925.[112] Despite the efforts of a number of workers,[113-117] however, a successful method of gold sensitisation was not developed until 1936, when Koslowsky[118] in the Agfa Filmfabrik laboratories discovered that a practical process could be based upon the addition of complex aurothiocyanates such as $NH_4Au(CNS)_2$ to the emulsion after the silver, or during the first or second ripening period. Information concerning the process was not released by Agfa, however, either in scientific publication or patent form, and became available later through patents assigned to Kodak and other photographic firms[119-121] and through scientific publications such as those of Faelens[122] and Mueller.[123,124]

Even today, the mechanism of gold sensitisation is apparently not entirely clear. It was early suggested[125] that the basis of the sensitisation was the replacement of silver in silver sensitivity specks by gold from the sensitising gold complex and this is supported by a variety of observations. Very different effects are produced by different gold complexes and $KAu(CN)_2$, for example, is quite inactive. It has been found that gold complexes which effect solution of silver halide, silver sulphide or silver are not effective sensitisers and it has been suggested that with such complexes the ripening sensitivity speck is dissolved before replacement by gold can take place. Because of the dissolving effect of free cyanide ions associated with $KAu(CN)_2$, this complex is therefore not active, but because thiosulphate and thiocyanate ions do not dissolve silver or gold, thiosulphate and thiocyanate complexes of gold are active. The phenomenon is very complex,[126-128] however, and there is other evidence which suggests that more than one mechanism may operate.

9.3 Photosensitivity of gold compounds

Apart from the effects of gold in developed silver images and in the sensitisation of silver emulsions, gold can itself, like silver, be used in the making of photo-sensitive materials. Thus Ishitani and Gakula[129] have claimed that if a film of a polymer such as Nylon 6 is immersed in a gold solution, then gold complexes are formed on the surface of the polymer which are photo-sensitive, so that the film can be used for photocopying.

10. Radioactive Gold

10.1 Introduction

Although the only stable isotope of gold is $^{197}_{79}Au$, some twenty-six isotopes of gold have been identified, of which some exhibit metastable as well as ground states. Only one, namely, ${}^{198}_{79}Au$ with a half life of 2.7 days, has found significant uses, though ${}^{195}_{79}Au$ and ${}^{199}_{79}Au$ with half lives of 183 days and 3.15 days respectively have been used in scientific studies. Their production and applications have been reviewed by Myerscough.[130]

10.2 Production and properties of ${}^{198}_{79}Au$, ${}^{199}_{79}Au$ and of ${}^{195}_{79}Au$

The neutron heavy ${}^{198}_{79}Au$ is produced at low cost by a neutron capture reaction by exposure of metallic gold in a nuclear reactor

$${}^{197}_{79}Au + {}^{1}_{0}n \rightarrow {}^{198}_{79}Au$$

The high cross section of the reaction (98.8 barns) enables a useful specific activity to be achieved for clinical purposes in a nuclear reactor of moderate power. Thus with a neutron flux of 2×10^{13} $n/cm^2/sec$ the specific activity reaches 1 curie per gram in 6.5 hours. A secondary reaction

$${}^{198}_{79}Au + {}^{1}_{0}n \rightarrow {}^{199}_{79}Au$$

has a very high cross section and introduces a small proportion of Au^{199}_{79}, which has a half life of 3.15 days and rather less energetic beta and gamma emissions.

Production of carrier free Au^{199}_{79}, i.e. Au^{199}_{79} free of non-radioactive gold, is possible by neutron irradiation of natural platinum and subsequent beta decay. Natural platinum contains 7.21% of ${}^{198}_{78}Pt$ which gives rise to ${}^{198}_{79}Au$ as follows:

$${}^{198}_{78}Pt + {}^{1}_{0}n \rightarrow {}^{199}_{78}Pt \rightarrow {}^{199}_{79}Au + {}^{0}_{-1}\beta$$

The neutron deficient ${}^{195}_{79}Au$, on the other hand, is produced by bombardment of natural platinum, which contains 33.8% of ${}^{195}_{78}Pt$, with protons in a cyclotron

$${}^{195}_{78}Pt + {}^{1}_{1}p \rightarrow {}^{195}_{79}Au + {}^{1}_{0}n$$

Radioisotopes of gold formed from the other four stable isotopes of the natural platinum have very much shorter half lives than ${}^{195}_{79}Au$, so that subsequent separation of the latter from the target is possible. The cost of this isotope is relatively high.

The nuclear properties of ${}^{195}_{79}Au$, ${}^{198}_{79}Au$ and ${}^{199}_{79}Au$ are set out in Table LI, and the tenth value thicknesses for gamma ray absorption from ${}^{198}_{79}Au$ and ${}^{195}_{79}Au$ for a number of materials are listed in Table LII.

10.3 Therapeutic and diagnostic applications of ${}^{198}_{79}Au$

${}^{198}_{79}Au$ is made routinely in most developed countries for clinical use, world production being of the order of 150 kilocuries per annum. It is used either in the form of implants known as "gold grains", as a source of gamma radiation for the treatment of malignancies, or in the form of its colloidal solutions where it is used, in effect, as a liquid beta emitter.

Gold grains have almost completely replaced so-called radon seeds previously used for the same purpose. They normally consist of small cylinders of pure gold, 2.2 mm long and 0.5 mm in diameter, enclosed in platinum of 0.15 mm thickness, which is designed to absorb the beta radiation. Implants are made via a cannula using an implantation gun which accepts aluminium magazines holding a number of grains. The magazines are irradiated in a reactor to generate the $^{198}_{79}Au$, and after allowing short-lived isotopes generated in the aluminium magazine and the platinum sheathing to decay, the activity of the gold is measured and the gold grains issued with a certificate. In Great Britain the output of activated grains is about 30,000 per annum and a large number of inactive grains is exported for activation and use elsewhere.

Table LI. Nuclear Properties of $^{195}_{79}Au$, $^{198}_{79}Au$ and $^{199}_{79}Au$*

Mass number	Half life (days)	Type of decay and particle energies (MeV)			γ-energies and % of total number of transformations of isotope	
195	183	EC		100%	0.031	1.2%
					0.099	11.5%
					0.13	0.8%
					0.067	Pt X-rays
					0.009	
198	2.70	ß–	0.29	1.2%	0.412	95.8%
			0.96	98.8%	0.68	1.0%
			1.37	0.025%	1.09	0.2%
199	3.15	ß–	0.25	23%	0.050	0.4%
			0.30	70%	0.158	38.2%
			0.46	7%	0.208	9.4%

*From Radiochemical Manual, The Radiochemical Centre, Amersham, England, 1966.

Table LII. Tenth Value Thicknesses (cm) For Gamma Ray Absorption From $^{195}_{79}Au$ and $^{198}_{79}Au$

Material	$^{195}_{79}Au$	$^{198}_{79}Au$
Water ($\rho = 1$)	13	22
Carbon ($\rho = 1.67$)	9	15
Concrete ($\rho = 2.3$)	6	10
Al ($\rho = 2.7$)	5	9
Fe ($\rho = 7.85$)	0.76	3.3
Pb ($\rho = 11.34$)	0.035	1

Colloidal $^{198}_{79}Au$ for injection is produced by dissolution of an irradiated gold target to yield gold chloride, which is reduced with a glucose and sodium hydroxide in the presence of gelatin as a protective colloid. Blood plasma substitute has been claimed as a stabilising agent also.[131] The particle size[132] of the resulting colloid, and its activity are controlled, and preparations must be sufficiently stable to withstand sterilisation at 120°C. The chemical inertness of metallic gold, combined with the short half life of $^{198}_{79}Au$, make it particularly suitable as a liquid source of beta radiation. It may be used intravenously, interstitially, and in particular in the pleural and peritoneal cavities. It may also be used in treatment of chronic effusion of the knee joint resulting from rheumatoid or degenerative arthritis. Although it is not selectively absorbed, colloidal $^{198}_{79}Au$ also finds considerable use in delineating the lungs and liver by scintigraphy, in which radiation from these organs after treatment is scanned so as to give two dimensional representations of them, as an aid to identifying areas of malfunction.

10.4 Other applications of $^{198}_{79}Au$, $^{199}_{79}Au$ and of $^{195}_{79}Au$

The high cross section of the neutron capture reaction by which $^{198}_{79}Au$ is formed, and the gamma emission of the latter, make possible the determination of gold by activation analysis. This technique has been applied for the determination of gold in trace (p.p.b.) concentrations in a variety of materials such as meteorites, marine organisms and sea water. Radio-chemical techniques are sometimes used to remove interfering radioisotopes as an intermediate step.

$^{198}_{79}Au$ has also been used as a tracer in studies[133-135] of the continuous casting of steel, and in following the movement of sea sediments.[130]

In contrast applications of $^{195}_{79}Au$ and $^{199}_{79}Au$ to date have been limited mainly to scientific studies.

11. The Use of Gold in Medicine

Although the empirical use of gold in medicine is recorded far back in history,[136-138] more scientifically based applications of gold in medicine can be said to have begun in 1890 with the description by Robert Koch[139] of the bacteriostatic action of gold cyanide and gold chloride on tubercle bacilli, and with a series of studies by other workers of the effects of gold in various forms upon other pathogenic organisms. The finding in 1913 that a number of thio-compounds of gold could be applied by injection led subsequently to the use of such compounds clinically in the treatment of a variety of conditions in addition to tuberculosis.

Although chrysotherapy in this sense has not survived the advent of antibiotics and other chemotherapeutic agents, it was noted by Landé[140] in 1927 during trials of the use of gold thioglucose in the treatment of bacterial endocarditis that joint pain was relieved in his patients. This led him to conclude that gold compounds might be beneficial in the treatment of arthritis, and it is in the treatment of rheumatoid arthritis with sodium aurothiomalate and aurothioglucose that gold finds its major therapeutic use today. The extent of their application can be judged by the fact that in 1970[141] some 17,000 prescriptions for sodium aurothiomalate were issued in Great Britain.

Although there were conflicting reports over a number of years concerning the efficacy of gold therapy in the treatment of rheumatoid arthritis, such efficacy has been established in a series[142-145] of controlled trials. It brings about a moderate improvement in the condition which has been observed to persist after treatment. There appears to be little doubt[146] that gold is one of the few drugs which can alter the course of the disease. Despite a very considerable volume of research, in which the biochemical and related effects of a wide range of gold compounds have been studied, the mechanism by which it acts still remains obscure. A review by Walz, Dimartino and Sutton[147] of the design and laboratory evaluation of gold compounds as anti-inflammatory agents is significant in this field, as is also a review of the biological chemistry of gold by Sadler.[147a]

An adverse and serious feature of gold therapy is the high rate of incidence of toxic effects.[136,147,148]

Medical applications of radioactive gold have been discussed above, and the use of gold foil in the treatment of decubitus ulcers which develop in areas of pressure in patients confined to certain positions has been referred to by Konof and by Smit.[136]

12. Gold as a Catalyst

Extensive experimental investigations have shown that gold, alone or alloyed with other metals or supported on other materials, exhibits catalytic activity in a number of reactions. Moreover, when alloyed with catalytically active metals, gold has been found capable of modifying their activity. In no instance, however, are the catalytic effects of gold strong, and despite the existence of a number of patents covering catalysts containing it, gold cannot be regarded as an industrial catalyst.[149]

12.1 Heterogeneous catalysis by gold

In heterogeneous hydrogenation and oxidation processes, the weak catalytic properties of gold are associated with its inertness and its inability to adsorb hydrogen[150] or oxygen,[151] and to make them available for reaction in an activated form. The absence of unpaired d-electrons in gold is to be noted in this connection.

Thus it is significant that molecular hydrogen is not chemisorbed on gold at room temperature, and that only above 200°C do dissociation of molecular hydrogen and adsorption of hydrogen atoms occur on the surface of the gold. These latter effects are limited in the sense that surface coverages are probably quite small, and are interpreted as resulting from promotion of electrons from the *5d* to the *6s* level at higher temperatures with creation of unpaired *5d* electrons. They are too small to maintain hydrogenation reactions at a useful rate.

Gold can nevertheless be caused to catalyse hydrogenation reactions if hydrogen atoms are either supplied to, or generated at, its surface. Thus it has been shown that if a Pd-Ag diffusion thimble is plated with gold and molecular hydrogen supplied to the inside of the thimble, hydrogen atoms which diffuse through the Pd-Ag alloy and the gold plating and emerge at the surface of the gold can hydrogenate alkenes to alkanes.[152] That it is its capacity to generate hydrogen atoms from hydrogen molecules that is lacking in gold, and not its capacity to chemisorb such atoms, even if weakly, is demonstrated by the fact that hydrogen atoms generated by electrical discharge combine to form molecular hydrogen at gold surfaces,[153] and by the fact that hydrogen atoms liberated by electrochemical discharge of protons at a gold electrode can be used to hydrogenate olefins.[154] Interest also attaches to recent observations[155] of the activity for hydrogenation and isomerisation reactions of gold catalysts prepared by thermal decomposition of chloroauric acid supported on silica or alumina. This activity resembles that of platinum in such reactions and it has been postulated that it may result from electron deficiency in the very small gold crystallites resulting from donation of electrons to the support. In this way the catalytic properties of the gold may be altered to resemble those of platinum.

In this latter connection the relationships which have been demonstrated between structural disorder and catalytic activity of metals must be borne in mind. Thus Kishimoto and Nishioka[156] have found that there is a convincing relationship for gold foil between the increase in hardness as a result of cold working, the thermoelectric force, and its catalytic activity in decomposing hydrogen peroxide. The catalytic activity of Au-Cd alloys in the decomposition of formic acid has also been correlated with their Brinell hardness.[157]

In at least one instance, namely, the hydrogenation of buta-1,3-diene to butane, a supported gold catalyst has been found to be highly selective, in that the hydrogenation of butenes to butane is promoted by the catalyst.[158]

Oxidation dehydrogenation processes, which can be represented by reactions of the types

$$AH_2 \rightleftarrows A + H_2$$
$$2H_2 + O_2 \rightarrow 2H_2O$$

in which A represents the dehydrogenated molecule, depend on the ability of the catalyst to abstract hydrogen atoms from a hydrogen containing molecule, and subsequently to facilitate the combination of these hydrogen atoms with oxygen, so that the equilibrium is displaced. Despite the weak adsorption of oxygen by gold, gold has been found to catalyse a number of reactions of this type, such as

cyclohexanone	$+ O_2$	$\rightarrow$	cyclohexenone + H_2O[159]
ethylene glycol	$+ O_2$	$\rightarrow$	glyoxal + H_2O[160]
ethylpyridine	$+ O_2$	$\rightarrow$	vinylpyridine + H_2O[161]
methyl alcohol	$+ O_2$	$\rightarrow$	formaldehyde + H_2O[162]

The weakness of adsorption of oxygen by the gold is probably significant in determining that no extensive oxidation of the organic structures to carbon dioxide and water takes place in the above reactions. Similar effects are to be noted in catalysis by gold of the oxidation of certain olefins by oxygen. Thus gold, generated by decomposition of gold(I) ketenide[163] on a glass support, catalyses the direct addition of oxygen to propene to form epoxy-propene. In this, it resembles silver catalysts prepared similarly from silver ketenide, which are active in promoting the conversion of ethylene to ethylene oxide. It seems possible that a template effect, connected with the unique metal layer structures of the ketenide precursors, may be a factor in these reactions.[164]

An observation of potential industrial significance is that gold shares with iridium and platinum their capacity for promoting the isomerisation of hydrocarbons, in particular the isomerisation of straight chain hydrocarbons to branched-chain hydrocarbons, which is such an important aspect of petroleum processing. Its activity is considerably less than that of platinum and iridium, however.[165]

Figar and Haidinger[166] have shown that in two processes, namely the decomposition of formic acid and the oxidation of propene, the activity of a gold catalyst can be increased by generating hot electrons in it. The effect is interpreted as resulting from the transfer of their excess energy (above the Fermi level) by hot electrons at the metal surface to chemisorbed complexes, which are thus activated to a higher energetic state and made more reactive. In the oxidation of propene, the formation of propylene oxide was enhanced and that of acrolein diminished, indicating that as might be expected there was selectivity in the enhancement of activity.

It has also been observed that the oxidation of certain organic compounds can be effected by means of gold anodes.[167]

12.2 Heterogeneous catalysis by gold alloys

From the few reactions which have been discussed above it will be apparent that the inherent catalytic activity of pure gold is low, though it can be enhanced or modified by special means.

As a result the greatest potential for the use of gold in catalysts would appear to be its use either alloyed or physically mixed with other inherently more active metal catalysts in order to modify their activity, and to enhance their selectivity.[168] Fundamental interest naturally focuses on the electronic structures of alloy catalysts which are critical for their activity.[150,168] In particular a great deal of work has been done with Au-Pd alloys in which the effect of gold is usually to decrease catalytic activity. In addition, gold forms a continuous series of solid solutions with palladium which favours investigation of Au-Pd alloys of various composition. This work has been carried out largely with the alloys in the form of wire, foil or evaporated films. More attention is being given, however, to supported catalysts of gold mixed or alloyed with other metals.

Examples of the use of gold with other metals, which have received attention, include its use alloyed with silver for catalytic oxidation of ethylene to ethylene oxide[169] and of cumene to cumene hydroperoxide,[170] its use with palladium in the synthesis of vinyl acetate from ethylene, oxygen and acetic acid[171] and in the oxidation of ethylene to acetic acid, acetaldehyde and acetic anhydride,[172] its use with palladium and platinum and iridium in the oxidative dehydrogenation of alkanes to alkenes and of n-butenes to butadiene[173] and of methanol to formaldehyde;[174] its use with iridium for hydroforming of paraffins and cycloalkanes,[175] and it use along with rhenium[176] and palladium[177] as a petroleum reforming catalyst. Gold in a form obtained by leaching of Al from an Au-Al alloy has also been claimed as a strong catalyst with good thermal conductivity for the treatment of exhaust gases from internal combustion engines.[178]

The excellent review by Clarke[168] as well as those by Allison and Bond[179] and by Moss and Whalley[180] should be consulted for further information on the catalytic activity of gold alloys.

12.3 Homogeneous catalysis by gold

A recent development in this area has been described by Muller.[181]

The platinum(II) complexes of the $SnCl_3^-$ ligand (e.g. $[Ph_3P]_2Pt[Cl]SnCl_3$ and $PtCl_2[SnCl_3]_2^{2-}$) catalyse the homegeneous hydrogenation ofolefins. This is associated with the π-accepting properties of the $SnCl_3^-$ ligand, which help the Pt(II) to form a sufficiently labile 5-co-ordinate complex containing both a hydrogen atom and the olefin co-ordinated to the metal, thus allowing reaction to take place at the platinum via an alkyl intermediate.

Gold(III) has the same $5d^8$ electronic configuration as platinum(II), but the formation of equivalent Au(III)-$SnCl_3$ complexes is not possible, because gold(III) is reduced by stannous chloride. The $SbCl_3$ ligand is isoelectronic with the $SnCl_3^-$ ligand but the preparation of complexes from $SbCl_3$ and $AuCl_4^-$ could not be achieved, probably because of the relative inertness of the electronic lone pair of Sb(III). The preparation of gold(III) to antimony bonded complexes was achieved, however, by replacing the chlorine atoms attached to the antimony by the pseudo-halide group X

F F
F — S—
F F
X

which is known for its strongly electron withdrawing properties. It was hoped that these would strengthen the gold-antimony bond by π-back bonding to the extent that formation of gold(III) to antimony bonded complexes would be possible. This hope was realised and a series of three gold-antimony complexes was prepared with the structures

$[Ph_4As]_3\,[AuX_4(SbX_4)_2]$

$[Ph_4As]_3\,[AuX_4(SbX_3Cl)_2]$

and $[Ph_4As]_3\,[AuX_4(SbX_2Cl_2)_2]$

These contained 6-co-ordinated gold(III) in the solid state, with two Au-Sb bonds, one of which was strong and one weak.

Attempts were made to hydrogenate ethylene in a solution in alcohol of $HAuCl_4$ or $HAuX_4$ containing a large excess of SbX_3, under which conditions these complexes had been found to be formed. These attempts were successful, indicating that gold(III) to antimony bonded complexes possess the same capacity to catalyse the homogeneous hydrogenation of ethylene as the platinum(II)-tin chloride complexes.

In the light of the very special circumstances under which this catalysis was achieved, it seems improbable that gold will easily find application in homogeneous catalysis.

REFERENCES

1. Ramsay, T. H., Metallurgical behaviour of gold wire in thermal compression binding, *Solid State Technol.* **16** (1973), 43–47.
2. American Society for Testing Materials. "Standard Specification for Gold Wire for Semiconductor Lead-bonding". Annual Book of ASTM Standards, Part 8 (Nov.), 1971, pp. 638–643.
3. Brenner, B., "Beryllium–Gold Alloy and Article Made Therefrom". U.S. Patent 3,272,625 (1966).
4. Raub, C., Thiede, M. and Thiede, H., "Verwendung einer Goldlegierung für warm- und kriechfest Gegenstände". German Patent 1,608,161 (1971).
5. Harman, G. G., Metallurgical failure modes of wire bonds, *in* Ann. Proc. Reliab. Phys. (Symp.), Vol. 12, 1974, pp. 134–141.
6. Philofsky, E., Intermetallic formation in gold–aluminium systems, *Solid-State Electron.* **13** (1970), 1391–1399.
7. Anon., Titanium–gold metallisation of semiconductor devices, *Gold Bulletin*, **8** (1975), 55.

8. Hampy, R. E., A versatile Ta_2N-W-Au/SiO_2/Al/SiO_2 thin film hybrid microcircuit metallisation system, *I.E.E.E. Trans.* **PHP-11** (1975), 263–272.
9. Theobald, W., "Funf Jahrtausende Goldschläger Kunst". Denkschrift im Auftrage des Vereins deutscher Blattgoldfabrikanten in Nürmberg, 1919.
10. Lübke, A., Herstellung von Blattgold und seine Verwendung, *Z. Metallk.* **46** (1955), 818–822.
11. Brüche, E. and Schulze, K. J., Electronenmikroskopische Studie zur Blattgoldherstellung, *Metall.* **12** (1958), 21–27.
12. Wise, E. M., "Gold: Recovery, Properties and Applications". Van Nostrand, 1964, pp. 224–226.
13. Oike & Co. Ltd., "Metal Leaf and Method for Producing the Same". British Patent 1,373,790 (1974).
14. Brüche, E. and Schulze, K. J., Bildung, Gefügeaufbau und Zerfall von Blattgold, *Z. Phys.* **153** (1959), 571–590.
15. Hirsch, P. B., Kelly, A. and Menter, J. W., The structure of cold worked gold. I.: A study by electron diffraction, *Proc. Phys. Soc. London*, **B68** (1955), 1132–1145.
15a. Nutting, J. and Nuttall, J. L., The malleability of gold: An explanation of its unique mode of deformation, *Gold Bulletin*, **10** (1977), 2–8.
16. Suito, E. and Uyeda, N., A study of flaky single micro-crystals of gold with a three-stage electron microscope, *in Proc. Int. Confer. Electron Microscopy, London*, **1954,** pp. 223–230.
17. Turkevich, J., Stevenson, P. C. and Hillier, J., A study of the nucleation and growth processes in the synthesis of colloidal gold, *Faraday Soc. Discussions*, **11** (1951), 55–75.
18. Smart, D. C., Boswell, F. W. and Corbett, J. M., Lattice spacings of very thin gold platelets, *J. Appl. Phys.* **43** (1972), 4461–4465.
19. Fulmer Research Institute Ltd., "Improved Method of Preparation of Activators for Use in the Electroless Plating of Plastics". British Prov. Patent 27,122/74.
20. Uyeda, N., Nishino, M. and Suito, E., Nucleus interaction and fine structures of colloidal gold particles, *J. Colloid Interface Sci.* **241** (1973), 20–23.
21. Frens, G., Particle size and sol stability in metal colloids, *Kolloid Z. u. Z. Polymere*, **250** (1972), 736–741.
22. Frens, G., Controlled nucleation for the regulation of the particle size in monodisperse gold suspensions, *Nature (London), Phys. Sci.* **241** (1973), 20–23.
23. Cohen, R. L. and West, R. W., Generative and stabilising processes in tin–palladium sols and palladium sol sensitizers, *J. Electrochem. Soc.* **120** (1973), 502–508.
24. Rantell, A. and Holtzman, A., Mixed $SnCl_2$/$PdCl_2$ activators—a study of their formation and nature, *Plating*, **61** (1974), 320–331 and 1051–1054.
25. Rantell, A. and Holtzman, A., The role of accelerators prior to electroless plating of ABS plastic, *Trans. Inst. Met. Finish.* **52** (1974), 31–38.
26. Rantell, A. and Holtzman, A., Mechanism of activation of polymer surfaces by mixed stannous chloride palladium chloride catalysts, *Trans. Inst. Met. Finish.* **51** (1973), 62–68.
27. Matijevic, E., Poskanzer, A. M. and Zuman, P., The characterisation of the stannous chloride/palladium chloride catalysts for electroless plating, *Plating*, **62** (1975), 958–965.
28. Thiele, H., "Verfahren zur Herstellung von feinteiligen, stabilen, kolloiden Goldhydrosolen und Radiogold". German Offen. 1,533,132 (1969).

29. Hunt, L. B., The true story of Purple of Cassius: The birth of gold-based glass and enamel colours, *Gold Bulletin*, **9** (1976), 134–139.
30. Zsigmondy, R., Die chemische Natur des Cassius'schen Goldpurpurs, *Annalen*, **301** (1898), 361–387.
31. Weyl, W. A., Coloured glasses, *J. Soc. Glass Technol.* London, 1951 (reprinted by Dawson's (London) in 1959).
32. Stookey, S. D., Coloration of glass by gold, silver and copper, *J. Am. Ceram. Soc.* **32** (1949), 246–249.
33. Stookey, S. D., Catalysed crystallization of glass in theory and practice, *Ind. Eng. Chem.* **51** (1959), 805–808.
34. Maurer, R. D., Nucleation and growth in a photosensitive glass, *J. Appl. Phys.* **29** (1958), 1–4.
35. Ref. 12, pp. 274–279.
36. Stookey, S. D., Controlled nucleation and crystallization lead to versatile new glass ceramics, *Chem. Eng. News*, **39** (1961), 116–125.
37. Tokunaga, S., Maruda, K. and Nagai, H., "Ultraviolet Absorbing Glass". Jap. Kokai 74 116,110 (*C.A.* **83** (1975), 47220a).
38. Murakami, Y. and Kume, M., "Glass Reversibly Changing in Color by Light". Jap. Patent 72-21998 (1972).
39. Hayashi, J. *et al.*, "Red Hard Glass Composition". Jap. Patent 73-10047 (1973) (*C.A.* **80** (1974), 40469t).
40. Finch, R. G., Gold in thick film hybrid microelectronics, *Gold Bulletin* **5** (1972), 26–30.
41. Heraeus, G.m.b.H., "Gold Alloy Decoration". British Patent 1,403,481 (1974).
42. Bletry, P. C., "Improvements in or Relating to Methods of Manufacture of Postage Stamps with Relief Gilded Decoration". British Patent 1,369,450 (1974).
43. Forster, K. R. and Wheeler, W. J. (International Business Machines Corp.), "Conductive Metal Powder Connection". British Patent 1,374,889 (1972).
44. Doduco, K. G., "Datenbuch". Doduco K.G., Pforzheim, 1974.
45. Short, O. A. (E. I. du Pont de Nemours & Co.), "Gold Metallizing Compositions". U.S. Patent 3,717,481 (1973) (cf. British Patent 1,313,826).
46. Short, O. A. (E. I. du Pont de Nemours & Co.),"Process for Manufacturing Gold Powder". U.S. Patent 3,771,996 (1973).
47. Short, O. A. and Weaver, R. V., "Gold Powder". U.S. Patent 3,811,906 (1974) (cf. U.S. Patent 3,725,035 (1973)) (*C.A.* **82** (1975), 6889m).
48. Golla, M. and Lutz, K. (Demetron Ges. für Elektronikwerkstoffe m. b. H.), "Verfahren zur Herstellung plättchenförmigem Goldpulver". German Offen. 2,330,413 (1975) (cf. German Offen. 2,329,352).
49. Daiga, V. R., "Gold Powder". U.S. Patent 3,843,379 (1974) (*C.A.* **82** (1975), 34332t) and U.S. Patent 3,768,994 (1973) (*C.A.* **80** (1974) 51596h).
50. Brug, J. E. and Heidelberg, E. X. (Owens Illinois Inc.), "Recovery of Gold from Solution in Aqua Regia". U.S. Patent 3,856,507 (1974).
51. E. I. Du Pont de Nemours & Co., "Method for the Production of Finely Divided Metals". U.S. Patent 3,390,981 (1965) (cf. U.S. Patent 3,885,955).
52. Short, O. A. (E. I. du Pont de Nemours & Co.), "Gold Alloy Metallizations for Capacitor Electrodes". U.S. Patent 3,817,758 (1974).
53. Daiga, V. (Owens Illinois Inc.), "Powders of Metal, Silver and Gold". U.S. Patent 3,816,097 (1974) (cf. British Patent 1,401,081).
54. Tazaki, A. and Hariada, A., "Noble Metal Pastes for Printed Circuits". Jap. Kokai 74 122,513 (1974) (*C.A.* **82** (1975), 141789u).

55. Roberts, C. B. (Dow Chemical Co.), "Metal Particulate Production". U.S. Patent 3,839,012 (1974).
56. Short, O. A. (E. I. Du Pont de Nemours & Co.), "Milling Process for Preparing Flake Gold". U.S. Patent 3,539,114 (1968).
57. Dietz, J. W. (E. I. Du Pont de Nemours & Co.), "Milling of Metal Powders in the Presence of Iodine". U.S. Patent 3,540,663 (1968).
58. Landsberg, A. (U.S. Secretary of the Interior), "Preparation of Homogeneous Powders Composed of Ultrafine Particles". U.S. Patent 3,357,819 (1967).
59. Anglo American Corporation of South Africa Ltd., "Producing High Purity Gold Powder". British Patent 1,387,373 (1975).
60. Gragg, J. E., Jr., Theoretical sheer strength of locally ordered alloys, *Monogr. Rep. Ser. Inst. Met. London*, **36** (1973), 55–59.
61. Sastry, S. M. and Ramaswami, B., Dispersion and order strengthening of copper–gold (Cu_3Au), *Monogr. Rep. Ser. Inst. Met. London,* **36** (1973), 51–54 *(C.A.* **82** (1975), 128301j).
62. Akzo, N. V., "Electrodeposition of Metals with Occlusion of Plastic Particles". Neth. Applic. 72 03,718 (1973) (*C.A.* **79** (1973), 38052n).
63. Sautter, F. K., "Electrodeposition of Dispersion Strengthened Au-Al_2O_3 Alloys". Watervliet Arsenal, N.Y., Tech. Report WVT-RR-6321 (1963).
64. Hill, K. A. and Sautter, F. K., "The Electrodeposition of Single Crystal Gold–Aluminium Oxide Alloys". Watervliet Arsenal, N.Y., Tech. Report WVT-6510 (1965).
65. Sautter, F. K., Die elektrolytische Abscheidung von Dispersionsüberzügen und oxydstabilisierten Legierungen, *Metall.* **18** (1964), 596–600.
66. Sautter, F. K. and Chen, E. S., Oxide dispersion strengthening, *Metal. Soc. Confer.* **47** (1968), 495 (Published by Gordon and Breach).
67. Chen, E. S. and Sautter, F. K., Porosity in electrodeposited gold–alumina alloys, *J. Electrochem. Soc.* **117** (1970), 726–728.
68. Peiffer, H. R., Marley, J. E., Kobler, R. J., Jacobs, W., *in* 17th Ann. National Relay Conference, Oklahoma State Univ., 1969, p. 1.
69. Centre Stéphanois de Recherche Mécanique Hydromécanique, "Electrolytic Method for Producing Wear Resistant Lubricant Metal Surfaces". French Demande 2,081,283 (1972).
70. Filatov, A. G. *et al.*, "Electroplating of a Gold Alloy". U.S.S.R. Patent 301,373 (1971) *(C.A.* **75** (1971), 835839).
71. Ernst Winter and Sohn., "Diamond Wear Resistant Coating and Method for its Electrodeposition". British Patent 1,391,001 (1975).
72. Larson, C., Electrodeposited gold composites, *Gold Bulletin* **8** (1975), 127–130 (cf. *Composites,* 1976, January, pp. 9–11).
73. Alexander, J. A., "Distributing Sub-Micron Oxide Dispersions Through Alloys". U.S. Patent 3,518,106 (1970) (*C.A.* **73** (1970), 58802m).
74. Imai, F. *et al.*, "Resistors for Semi-conductor Integrated Circuits". Jap. Kokai 74 74,396 (1974) (*C.A.* **82** (1975), 25098u).
75. Fan, J. C. C. and Henrich, V. E., Preparation and properties of sputtered magnesia–gold, magnesia–silver and magnesia–nickel cermet films, *J. Appl. Phys.* **45** (1974), 3742–3748 (*C.A.* **81** (1974), 142897u).
76. Gimpl, M. L. and Fuschillo, N., Dispersion hardened alloys made by vapor plating and chemical precipitation techniques, *Met. Soc. Confer.,* **47** (1946), 719–739 (Published by Gordon and Breach).

77. Fuschillo, N. and Gimpl, M. L., Electrical or tensile properties of Cu-ThO_2, Au-ThO_2, Pt-ThO_2 and Au-Al_2O_3, Pt-Al_2O_3 alloys, *J. Mater. Sci.* **5** (1970), 1078–1086.
for electrical applications, *J. Met.* **23** (1971), 39–44.
79. Poniatowski, M. and Clasing, M., Dispersion hardened gold: a new material of improved strength at high temperatures, *Gold Bulletin*, **5** (1972), 34–36.
80. Engelhard Industries, U.S. Patents 2,947,114 (1960), 2,984,894 (1961), 3,049,577 (1962), 3,201,236 (1965), 3,606,766 (1971).
81. Engelhard Minerals and Chemicals Corp. U.S. Patent Applic. 610,526 (5th Sept., 1975).
82. Villot, M. and Klingenberg, H. U., "Metal-hard Metal Composites for Jewellery Purposes". German Offen. 1,944,773 (1970).
83. Popakul, J. and Polti, J. L., Metal friction couples for dry operation in air, *Mem. Tech. CETIM*, **11** (1972), 107–118 (*C.A.* **83** (1975), 83597n).
84. Andrews, G. I., Groszek, A. J. and Witheridge, R. E. (British Petroleum Co. Ltd.), "Surface Modified Metals in Composites and Bearings". U.S. Patent 3,516,933 (1970).
85. P. K. Mallory & Co. Inc., "Composite Material". U.S. Patent 3,573,037 (*Gold Bulletin*, **4** (1971), 79).
86. Deutsche Gold– und Silver–Scheideanstalt, "Dispersion Hardened Materials". German Offen. 1,783,074 (*Gold Bulletin*, **4** (1971), 79).
87. Sastry, S. M. L. and Ramaswami, B., Dispersion and order strengthening of Cu_3Au, *Monogr. Rep. Ser. Inst. Met. London*, **36** (1973), 51–54.
88. Anon., The production of gold foams: wide range of characteristics with high resistance to corrosion, *Gold Bulletin*, **6** (1973), 7.
89. Featherstone, R. W. (Dunlop Holdings Ltd.), "A Method of Making Articles of Porous Metal". British Patent 1,257,796 (1971).
90. Anon., Expanded metal dental reinforcements, *Gold Bulletin*, **8** (1975), 40.
91. Brook, G. B., Gold alloys with shape memory, *Gold Bulletin*, **6** (1973), 8–11.
92. Brook, G. B. and Iles, R. F., "Treatment of Alloys". British Patent 1,346,047 (1970).
93. Brook, G. B. and Iles, R. F., Gold–copper–zinc alloys with shape memory, *Gold Bulletin*, **8** (1975), 16–21.
94. Nakanishi, N., Murakami, Y. and Kachi, S., Pseudoelasticity and elastic anisotropy in the β-phase thermoelastic alloy, *Scr. Metall.* **5** (1971), 433–439.
95. Murakami, Y., Lattice softening, phase stability and elastic anomaly of the ß-gold-copper-zinc alloys, *J. Phys. Soc. Japan*, **33** (1972), 1350–1360 (*C.A.* **77** (1972), 169801n).
96. Murakami, Y. *et al.*, Phase relation and kinetics of the transformations in Au-Cu-Zn ternary alloys, *Jap. J. Appl. Phys.* **6** (1967), 1265–1271.
97. Nakanishi, N. *et al.*, Low temperature stability in ternary gold–zinc based ß-phase alloys, *Jap. J. Appl. Phys.* **7** (1968), 302–303.
98. Nakanishi, N., Murakami, Y. and Kachi, S., Phase stability and elastic anisotropy of the 3/2 electron compounds, *Scr. Metall.* *2* (1968), 673–675.
99. Pops, H. and Massalski, T. B., Some comments on the martensitic transformations in β-brass type alloys, *Acta. Metall.* **15** (1967), 1770–1772.
100. Ellis, P., Gold in photography: evolution from early artistry to modern processing, *Gold Bulletin*, **8** (1975), 7–12.
101. Blay, C. J., Blue toning with gold chloride, *Amat. Photogr.* **96** (1946) (Jan. 16), 39.

102. Nelson, W. C. (Eastman Kodak Company), "Toning Bath for Photographic Prints". U.S. Patent 1,849,245 (1932).
103. Formstecher, F., *Photogr. Ind.* **17** (1922), 378; **18,** 403.
104. Namias, Eder's Handbuch Photogr., Vol. 4, 1928 (via Ref. 100).
105. Crabtree, J. I., Eaton, G. T. and Muehler, L. E., The elimination of hypo from photographic images, *J. Photogr. Soc. Am.* **6** (4) (1940), 6–13.
106. Henn, R. W. and Mack, B. D., A gold protective treatment for microfilm, *Photogr. Sci. Eng.* **9** (1965), 253–261 (cf. Henn, R. W. and Wiest, *Photogr. Sci. Eng.* **7** (1963), 253 ff).
107. James, T. H., Vanselow, W. and Quirk, R. F., Gold and mercury latensification and hypersensitization for direct and physical development, *J. Photogr. Soc. Am.* **14** (1948), 349–353.
108. James, T. H., The site of reaction in direct photographic development. II. Kinetics of development initiated by gold nuclei, *J. Colloid Sci.* **3** (1948), 447–455.
109. James, T. H. and Vanselow, W., Dependence of latensification upon the degree of development of a photographic material, *J. Photogr. Soc. Am.* **15** (1949), 688–693.
110. Hamm, F. A. and Comer, J. J., The electron microscopy of photographic grains. Specimen preparation techniques and applications, *J. Appl. Phys.* **24** (1953), 1495–1513.
111. Sheppard, S. E., Photographic gelatin, *Photogr. J.* **65** (1925), 380–387.
112. Kropff, F., Chemische Sensibilisation, *Photogr. Ind.* **42** (1925), 1145.
113. Carroll, B. H. and Hubbard, D., Sensitization of photographic emulsions by colloidal materials, *J. Nat. Bur. Stand.* **1** (1928), 565–588.
114. Kankelwitz, B., "Verfahren zur Herstellung grau- und gelbschleierfrei arbeitender und haltbarer photographischer Halogensilberentwicklungsemulsionen". German Patent 618,354 (1935).
115. Jenisch, W., Zur Kenntnis der Reifung photographischer Silverhaloidemulsionen, *Z. Wiss. Photogr.* **24** (1926), 248–256.
116. Charriou, A. and Valette, S., Influence de l'argent colloidal et de l'or colloidal sur les émulsions photographiques supersensibles, *Sci. Ind. Photogr.* **2** (6) (1935), 395–396.
117. Schmieschek, U. (Deutsche Versuchsanstalt für Luftfahrt.), "Verfahren zur Herstellung höchstempfindlichen photgraphischer Emulsionen". German Patent 692,828 (1940).
118. Koslowsky, R., Gold in photographischen Emulsionen, *Z. Wiss. Photogr.* **46** (4)–(6) (1951), 65–72.
119. Trivelli, A. P. H. and Smith, W. F. (Kodak Ltd.), "Improvements in Photographic Emulsions". British Patent 636,140 (1950).
120. Leermakers, J. A., Hewitson, E. H. and Yackel, C. (Kodak Ltd.), "Improvements in Photographic Emulsions". British Patent 636,234 (1950).
121. Waller, C., Collins, R. B. and Dodd, E. C., "Improvements in or Relating to Photographic Materials". British Patent 570,393 (1945).
122. Faelens, P. A., Sur le mécanisme de la sensibilisation par les métaux nobles (II), *Sci. Ind. Photogr.* **27** (4) (1956), 121–122.
123. Mueller, F. W. H., Some remarks on gold treatment of photographic silver halide emulsions, *J. Opt. Soc. Am.* **39** (6) (1949), 494–496.
124. Mueller, F. W. H., 30 years of gold sensitization, *Photogr. Sci. Eng.* **10** (6) (1966), 338–343.

125. Mueller, F. W. H., Gold sensitized emulsions and the American photographer, *Photogr. Sci. Tech.* **16B** (1950), 46.
126. Steigmann, A., Sur les mécanismes de la sensibilisation chimique et de la formation de l'image latente en présence de sels d'or, *Sci. Ind. Photogr.* **26** (8) (1955), 289–304.
127. Hautot, A. and Sauvenier, H., Sur la nature des centres de sensibilité, des germes d'image latente et des germes de voile, *Sci. Ind. Photogr.* **28** (2) (1957), 57–65.
128. Faelens, P. A., Present state of the mechanism of gold sensitization, *Photog. Korresp.* **104** (7) (1968), 137–146.
129. Ishitani, A. and Gakula, K., "Photosensitive Material". Jap. Patent 74-17,289 (1974).
130. Myerscough, L., Nuclear properties of gold, *Gold Bulletin*, **6** (1973), 62–68.
131. Hoechst, A. G., "Stabilised Gold Colloid". German Offen. 2,420,531 (1975).
132. Chernyak, A. S. and Shestopalova, L. F., Formation of gold sols in alkaline solutions of amino-acids, *Kolloidh. Zh.* **37** (1975), 202–204.
133. Morlon, S. K. and Weinberg, F., Continuous casting of steel. I. Pool profile, liquid mixing, and cast structure in the continuous casting of mild steel, *J. Iron Steel Inst. London*, **211** (pt. 1) (1973), 13–23.
134. Zeder, H. and Hedström, J., Determination of skin formed in solidification of continuously cast slabs by means of radioactive isotopes, *Radex-Rundsch.* (2) (1971), 407–418.
135. Saito, K. and Tate, M., Solidification and strand guidance of slabs at high casting speeds, *in* Proc. Nat. Open Hearth Basic Oxygen Steel Confer., Vol. 56, 1973, pp. 238–268 (*C.A* **81** (1974), 109249m).
136. Smit, P., The medical uses of gold through the ages, *Proc. Mine Med. Officers Assoc. S. Afr.* **47** (1968), 90–93.
137. Ellery, R. S., On the periphery: gold, *Med. J. Aust.* **41** (1954), 762–763.
138. Block, W. D. and van Goor, K., "Metabolism, Pharmacology and Therapeutic Use of Gold Compounds". Edited by A. C. Curtis, Thomas, Springfield, Illinois, 1959, p. 3.
139. Koch, R., *Deutsch, Med. Wochenschr.* **16** (1890), 756.
140. Landé, K., *Muench. Med. Wochenschr.* **74** (1927), 1132.
141. Anon., Gold therapy in 1975, *Br. Med. J.* (1975), 26th April, 156–157.
142. Empire Rheumatism Council Research Sub-Committee, Gold therapy in rheumatoid arthritis. Final report of a multicentre controlled trial, *Ann. Rheum. Dis.* **20** (1961), 315–334.
143. Fraser, T. N., Gold treatment in rheumatoid arthritis, *Ann. Rheum. Dis.* **4** (1945), 71–75.
144. American Rheumatism Association, Cooperating Clinics Committee, A controlled trial of gold salt therapy in rheumatoid arthritis, *Arthritis Rheum.* **16** (1973), 353–358.
145. Sigler, J. W. *et al.*, Gold salts in the treatment of rheumatoid arthritis. A double-blind study, *Ann. Intern. Med.* **80** (1974), 21–26.
146. Constable, T. J., Crockson, R. A., Crockson, A. P. and McConkey, B., Drug treatment of rheumatoid arthritis, *Lancet* (1975), May 24, 1176–1179.
147. Walz, D. T., Di Martino, M. J. and Sutton, B. M., Design and laboratory evaluation of gold compounds as anti-inflammatory agents, *Med. Chem. Ser. Monogr.* **13** (1974), 209–244.
147a. Sadler, P. J., The biological chemistry of gold: advances in understanding with modern techniques, *Gold Bulletin*, **9** (1976), 110–118.

148. Girdwood, R. H., Death after taking medicaments, *Br. Med. J.* **1** (1974), 501–502.
149. Bond, G. C., The catalytic properties of gold: potential applications in the chemical industry, *Gold Bulletin*, **5** (1973), 11–13.
150. Bond, G. C., "Catalysis by Metals". Academic Press, London and New York, 1962.
151. MacDonald, W. R. and Hayes, K. E., Comparative study of the rapid adsorption of oxygen by silver and gold, *J. Catal.* **18** (1970), 115–118.
152. Yolles, R. S., Wood, B. J. and Wise, H., Hydrogenation of alkenes on gold, *J. Catal.* **21** (1971), 66–69 (*C.A.* **74** (1971), 103546q).
153. Wood, B. J. and Wise, H., Diffusion and heterogeneous reaction. II. Catalytic activity of solids for hydrogen atom recombination, *J. Chem. Phys.* **29** (1958), 1416–1417 (*C.A.* **53** (1959), 7744).
154. Byrne, M. and Kuhn, A. T., Electrocatalytic reduction of ethylene on gold and other substrates, *J. Chem. Soc., Faraday Trans.* **68** (1972), 1898–1907 (*C.A.* **77** (1972), 121411h).
155. Bond, G. C. and Sermon, P. A., Gold catalysts for olefin hydrogenation. Transmutation of catalytic properties, *Gold Bulletin*, **6** (1973), 102–105 (cf. *J. Chem. Soc., Chem. Commun.* (1973), 444–445).
156. Kishimoto, S. and Nishioka, J., Catalytic activity of cold worked and quenched gold for the decomposition of hydrogen peroxide, *J. Phys. Chem.* **76** (1972), 1907–1908 (*C.A.* **77** (1972), 52758n).
157. Schwab, G. M. and Pesmatjoglou, S., Metal electrons and alloy catalysts: the system gold–cadmium, *J. Phys. Colloid Chem.* **52** (1948), 1046 ff.
158. Buchanan, D. A. and Webb, G., Catalysis by group IB metals. Part I. Reaction of buta-1,3-diene with hydrogen and deuterium catalysed by an alumina-supported gold, *J. Chem. Soc., Faraday Trans.* (1975), 134–144.
159. Mobile Oil Corporation, "α-β-Unsaturated Ketones". British Patent 1,152,817 (1969) (*C.A.* **71** (1969), 112504q).
160. Howe, B. K., Hary, F. R. and Carke, D. A. (Laporte Chemicals Ltd.), "Gas Phase Oxidation of Aliphatic and Cyclo-aliphatic 1:2-diols to Carbonyl Compounds". German Offen. 1,923,048 (1969) (*C.A.* **72** (1970), 42798h).
161. Etherington, R. W. and Koei-Liang liauw (Mobile Oil Corp.), "Catalytic Oxidative Dehydrogenation of Alkylpyridines". U.S. Patent 3,553,220 (1971).
162. Hardy, J. G. and Roberts, M. W., Mechanism of the catalytic decomposition of methanol on gold filaments, *J. Chem. Soc., Chem. Commun.* (1971), (10) 494.
163. Blues, E. T., Bryce-Smith, D., Lawston, I. W. and Wall, G. W., Gold (I) ketenide, *J. Chem. Soc., Chem. Commun.* (1974), 513–514.
164. Bryce-Smith, D., Metal ketenides in the catalytic epoxidation of olefins, *Chem. Ind. (London)* (1976), 15th Feb., 154–157 (cf. British Patent 1,409,421).
165. Boudart, M. and Ptak, L. D., Reactions of neopentane on transition metals, *J. Catal.* **16** (1970), 90–96 (*C.A.* **72** (1970), 30942n).
166. Figar, J. and Haidinger, W., A new type of gold catalyst: activity and selectivity regulated by means of hot electrons, *Gold Bulletin* **7** (1974), 100–102 (cf. *Chem. Phys. Lett.* **11** (1971), 545 and **19** (1973), 564).
167. Cwiklinsk, C. and Perichon, J., Anodic oxidation of ethylene and acetylene: the electrocatalytic role of gold electrodes, *Gold Bulletin*, **9** (1976), 20–21.
168. Clarke, J. K. A., Selectivity in catalysis by alloys, *Chem. Rev.* **75** (1975), 291–305.
169. Flank, W. H. and Beachell, H. C., The geometric factor in ethylene oxidation

over gold–silver alloy catalysts, *J. Catal.* **8** (1967), 316–325 (cf. British Patent 1,243,105).

170. van Ham, N. H. A., Nieuwenhuys, B. E. and Sachtler, W. M. H., Oxidation of cumene on silver and silver–gold alloys, *J. Catal.* **20** (1971), 408–411.
171. Farbenfabriken Bayer A.-G., "Catalytic Manufacture of Vinylesters". French Demande 2,001,139 (*C.A.* **76** (1972), 15163y) (cf. British Patent 1,103,125).
172. Gerberich, H. R., Cant, N. W. and Hall, W. K., Catalytic oxidation. I. Oxidation of ethylene over palladium and palladium–gold alloys, *J. Catal.* **16** (1970), 204–219 (cf. U.S. Patent 3,534,093 (1970)) (*C.A.* **73** (1970), 130605u).
173. Armstrong, W. E. (Shell International Research), "Platinum and Gold Catalysts for the Dehydrogenation by Oxidation of Organic Compounds". U.S. Patent 3,156,735 (1962), Belg. Patent 611,379 (1962) (*C.A.* **57** (1962), 14933f).
174. Shell International Research Maatschappy N.V. British Patent 940,710 (1963).
175. Sinfelt, J. H. and Barnett, A. E. (Esso Research & Engineering Co.), "Catalytic Hydroforming of Naphtha". U.S. Patent 3,567,625 (1971), German Offen. 2,003,743 (1970).
176. Merrill, H. E. and Lunt, R. S., "Reforming Naphthas with a Rhenium–Gold Catalyst". U.S. Patent 3,785,960 (1974) (*C.A.* **80** (1974), 135660n).
177. Exxon Research & Engineering Co., "Catalytic Reforming Process". British Patent 1,400,211 (1975).
178. Varta, A. G., "Internal Combustion Engine Exhaust Catalyst System". British Patent 1,401,673 (1975).
179. Allison, E. G. and Bond, G. C., Structure and catalytic properties of palladium–silver and palladium–gold alloys, *Catal. Rev.* **7** (1972), 233–289 (*C.A.* **78** (1973), 88962j).
180. Moss, R. L. and Whalley, L., "Advances in Catalysis", No. 22, 1972, pp. 115–189.
181. Muller, M. C., Compounds with gold–antimony bonds: their potential activity in homogeneous catalysis, *Gold Bulletin*, **7** (1974), 39–40.

Author Index

The numbers in brackets are the reference numbers and those in italic refer to the Reference pages where the references are listed in full.

A

Abbott, W. H., 122 (63), *133*, 170 (154), *177*, 276 (19), *287*
Ackroyd, M. D., 188 (4), *194*
Adamov, M., 103 (88), *110*
Adamson, R. J., 26 (9), 28 (9), 29 (9), *29*
Ahmed, N., 182 (23), 183 (23), *193*
Aichinger, W., 239 (187), 240 (187), 247 (187), *263*
Ainsworth, P. A., 145 (32), *172*
Akabori, S., 126 (94), *134*
Akebi, M., 248 (318), *269*
Akella, J., 126 (96), *134*
Akzo, N. V., 300 (62), *317*
Alexander, J. A., 301 (73), *317*
Aliprandini, G., 213 (39), 247 (291), 249 (329), 252, 253 (340), *257, 260, 269, 270*
Allison, E. G., 313 (179), *322*
American Dental Association, 95, 96(2), 97 (2), *106*, 181 (18), *193*
American Rheumatism Association, 310 (144), *320*
American Society for Testing Materials, 292 (2), *314*
Anderson, E. A., 58 (149), *91*
Anderson, J. C., 122 (68), *133*
Ando, E., 165 (113), 166 (113), *175*
Andreassen, T., 281 (426), *288*
Andreeva, G. F., 209, 210 (29), 211, 221 (29), *256*
Andres, F., 286 (107), *291*
Andrew, R., 191 (64), *194*
Andrews, G. I., 301 (84), *318*
Andrews, J. A., 102 (63), *109*
Andryushchenko, I. A., 39, 85 (226, 227), *86, 87, 94*
Anglo American Corporation of South Africa Ltd., 299 (59), *317*
Annual Reports, Chamber of Mines of South Africa, 6 (3), *24*
Anthony, D. H., 100 (48), 103 (74), *108, 109*
Antler, M., 111 (1), 121 (47), 122 (59, 61), *130, 132, 133*, 184 (29, 30, 31), *193*, 239 (184), 247 (184), *263*
Antony, P., 253 (341), *270*
Anzai, K., 55 (144), *91*
Ardell, A., 55 (141), *91*
Armstrong, W. E., 313 (174), *320*
Asahina, M., 54 (129–132), *90, 91*, 55 (129), 55 (143)
A.S.M., 167 (130), *176*
Atkinson, B., 46 (41), 49 (41), *87*
Atabe, Y., 247 (295), *268*
Atanasyants, A. G., 238 (151), 248 (151), *261*
Aubauer, H. P., 77 (213), *94*, 155 (82), *174*
Augis, J., 122 (57), *133*
Avila, A. J., 213 (45), *257*
Azkan, D., 128, *135*
Azuma, S., 55 (144), *91*

B

Bacquias, G., 232 (127, 128), *233*, 248 (127, 128), *260*, 274 (15), *287*
Badinelli, L. A., 100 (37), *108*
Baker, D., 182 (25), *193*

Baker, K. D., 250, 253 (335), *269*, 284 (67), 285 (95), *289, 290*
Bakish, R., 57, 58 (148), 59 (150, 151), *91, 93*, 68 (181), 69 (181)
Balashova, 250 (336), *269*
Ballard, K. H., 271 (3), *286*
Barb, R. P., 141 (15, 16), *171*
Barham, B. S., 179, *192*
Barlow, H., 277 (28), *287*
Barnett, A. E., 313 (175), *322*
Barry, R. F., 274 (149), *287*
Barshtein, N. P., 39 (36), *87*
Bartlett, E. S., 170 (154), *177*
Barton, J. A., 99 (30), *107*
Barton, P. W., 53 (119), *90*
Bauer, C. L., 224 (106), *260*
Bauer, H., 221 (86), 224, 248 (86), *259*
Baum, L., 95 (10, 11), *106*
Beachell, H. C., 313 (169), *321*
Beasley, J. K., 165 (106), *175*
Bechtold, G., 168 (135), *176*
Bek, E., 234 (131), *260*
Bek, Yu. R., 214, *257*
Belais, D., 44 (43), *87*
Belevantsev, V. I., 199 (17), 200 (17), *256*
Belousova, T. P., 39 (37), *87*
Benham, R. B., 246 (276), 248 (276), *267*
Benner, S. G., 102 (64), *109*
Bergman, M., 99 (29), *107*
Berman, R., 120 (40, 41, 42, 43), *131, 132*
Bergstrom, A. E., 285 (85), *290*
Bernett, M. K., 166 (127), *176*
Berta, J., 103 (77), *109*
Bertorelle, E., 213 (40), *257*
Bibic, N., 103 (88), *110*
Bick, M., 225 (109), 240 (196), 244 (245), 245 (245), 246 (196), 247 (197, 302), 248 (192, 196, 301, 245), *260, 263, 265, 268*
Bihlmaier, K., 235, 237 (137), 238 (137), 244, 247 (137), 248 (137), *261*
Black, A. D., 95 (6), *106*
Blair, A., 121 (149), *132*, 236 (145), 239 (169), 244, 247 (145, 169), 248 (145), *261, 262*
Blay, C. J., 304 (101), *318*
Blazy, A., 71 (185), *93*
Blech, I. A., 122 (71), *133*
Bletry, P. C., 298 (42), *316*
Block, W. D., 309 (138), *320*
Blues, E. T., 312 (163), *321*
BNF Metals Technology Centre, 76 (197), *93*
Boas, W., 79 (220), *94*
Bockris, J. O'M., 205, 206, 216 (63), 219 (63), *256, 258*, 283 (46), *288*
Bogenschütz, A. F., 284 (54), *289*
Bond, G. C., 310 (149, 150), 311 (155), 313 (150), *321, 322*
Bondy, A. R., *87*
Borelius, G., 53 (106), *89*, 119 (35), *131*
Borom, M. P., 102 (55), *108*
Boswell, F. W., 295 (18), *315*
Boudart, M., 312 (165), *321*
Boughton, J. D., *94*, 140 (10), *171*
Bouman, J., 53 (98), *89*
Box, E. R., 48 (73), *88*
Bradford, C. W., 249 (333), *269*
Bramley, L. A., 166 (128), *176*
Brandes, E. A., 54 (128), *90*, 122 (62), *133*, 168 (138), *176*
Brasch, W. R., 247 (146), *261*
Brasunas, A. de S., 99 (34), *107*
Braun, A., 221 (81), 247 (81), *258*
Braun, J. D., 187 (39), *194*
Breitling, R. M., 122 (69), *133*
Brenner, A., 197 (8), 208, 210, 230, 234 (8), 235, 241, 245 (8), *255, 256*, 247 (8), 284 (56), *289*
Brenner, B., 44 (58), *87*, 293 (3), *314*
Brennerman, R., 227 (111), *260*
Brentford, H., 249, *269*
Brepohl, E., 30 (10), *86*
Brewer, D. H., 187 (41), *194*
Brinkworth, B. J., 156 (85, 86), 159, 160, 166 (85), *174*
Brock, J. C. H., 120 (41), *132*
Brook, G. B., 80, 93 (196), *93, 94*, 302 (91, 92, 93), 303, *318*
Brookshire, R. R., 284 (57), *289*
Brooks, M. S., 100 (48), 103 (74), *108, 109*
Brown, A. G., Boveri et Cie, 145
Brown, L., 89 (50), *89*
Brown, M. J., 213, *257*
Brown, R., 213 (48), *257*
Brüche, E., 294 (11, 14), *315*
Brug, J. E., 299 (50), *316*
Brugirard, J., 99 (35), *107*

Bruk, E. S., 240 (205), 247 (205), 248 (205), *264*
Brumfield, R. C., 97, *107*
Bryce-Smith, D., 312 (163, 164), *321*
Buchanan, D. A., 311 (158), *321*
Buchner, G., 65 (170), *92*
Budke, J., 59 (157), 60, 61, *92*
Bullis, W. M., 125 (80), *134*
Burgers, W. G., 53 (98), *89*
Burnett, A. P., 100 (48), 102 (44), 103 (74), *108, 109*, 184 (35), *193*
Burrows, I. R., 214, *257*
Busby, G. E., 285 (101), *291*
Buschle, R., 239 (187), 240 (187), 247 (187), *263*
Buseck, P. R., 125 (88, 90), 126 (93), *134*
Buss, G., 65 (166), *92*
Busser, A., 65 (171), *92*
Bustacchini, G., 188 (45), *194*
Butler, F. P., 240 (216, 217), 246 (216, 217), 247 (216, 217), 248 (216, 217), *264, 268*
Byrne, M., 311 (154), *321*

C

Cadenhead, R. L., 192 (66), *195*
Cady, J. R., 197 (11), *255*, 240 (228), *265*
Cafferty, A. G., 241, 247 (239, 240), *265*
Caley, R. H., 276 (21), *287*
Camp, E. K., 238 (162), *260, 262*
Campana, C. R., 240 (209), 247 (209), 248 (209), *264*
Campbell, R., 55 (139), *91*
Cant, N. W., 313 (172), *322*
Cantwell, K. R., 96 (13), *106*
Carke, D. A., 312 (160), *321*
Carlson, R. O., 125 (79), *134*
Carpenter, R., 53 (102), *89*
Carpenter, R. W., 53 (100), 53 (105), *89*
Carroll, B. H., 306 (113), *319*
Carter, F. E., 34 (21), 40 (21), 44 (54), *86, 87*, 51 (21), 54 (21)
Catalano, E., 180 (12), 181 (12), *192*
Cathrein, R., 223 (104), 232 (104), 233, 245 (126), 246 (104, 126), 248 (104, 126), *259, 260*
Cathro, K. J., 244 (246, 247), *265*
Caul, H. J., 97 (21, 22), *107*
Centre Stephanois de Recherche Mecanique Hydromecanique, 300 (69), *317*
Chamer, E. S., 35 (28), 44 (28), 49, *86*
Chandrasekhar, 166 (120), *176*
Charbeneau, G. T., 96 (15), *106*
Charlot, G., 203, 221 (22), *256*
Charrion, A., 306 (116), *319*
Chaston, J. C., 34 (22), 51 (22), 63 (159a), *86, 92*, 137 (2), 139 (2), 148, 149, *170, 173*
Chaurasia, H. K., 123 (75), 125 (92), *133, 134*
Cheh, H. Y., 213 (46), 219 (54), 220, *257*
Chemical Society, London, *255*
Chemnitius, F., 271 (1), *286*
Chen, E. S., 300 (67), *317*
Cherniak, M. G., 189 (53), *194*
Chernyak, A. S., 308 (132), *320*
Christie, J. S., 77 (210), *94*
C.I.E., 78 (216), *94*
Clarke, J. F., 191 (65), *195*
Clarke, J. K. A., 313 (168), *321*
Clasing, M., 301, *318*
Clipstone, C. J., 55 (136), *91*
Coad, B. C., 144 (21), *171*
Cohen, J. B., 53 (121), 53 (122, 124), *90*
Cohen, M., 180 (9), *192*
Cohen, R. L., 296 (23), *315*
Colbert, W. J., 190 (58), *194*
Colbus, J., 140 (8), *171*
Cole, S. S., Jnr., 102 (54), *108*
Coleman, R. L., 97 (18), *107*
Collat, J. W., 284 (75, 76), *290*
Collins, C. B., 125 (79), *134*
Collins, R. B., 306 (121), *319*
Collumeau, A., 203, 221 (22), *256*
Comer, J. J., 305 (110), *319*
Comptoir Lyon-Alemand Louyout et Cie, 44 (62), *88*, 148 (63), *173*
Coniglio, A., 247 (305), *268*
Constable, T. J., 310 (146), *320*
Corbett, J. M., 295 (18), *315*
Cornely, R., 123 (74), *133*
Cousins, C. A., 28 (10), *29*
Cowger, G. T., 102 (69), *109*
Crabtree, J. I., 305 (105), *319*
Craft, W. H., 284 (72, 73), *289, 290*
Craig, S. E., 219 (75), *258*
Creydt, M., 181 (14), *193*
Crockson, A. P., 310 (146), *320*

Crockson, R. A., 310 (146), *320*
Crowell, W. S., 48 (74), 51 (91), *88, 89,* 97, *107*
Csuthy, B., 285 (87), *290*
Cuff, K. F., 165 (108), *175*
Culbert, D. P. A., 240 (222), 246 (222), 247 (222), 248 (222), *264*
Cusack, N. E., 113 (8), *130*
Custer, R., 103 (87), *110*
Cwiklinsk, C., 312 (167), *321*

D

Daiga, V., 283 (49), *316*
Daiga, V. R., 299 (49, 53), *316*
Dalton, I. M., 240 (197), *263*
Danemark, M., 245 (126), 246 (104, 126), 248 (126), *259, 260*
Danemark, M. A., 212 (31), *256*
Darby, E. C., 239 (186), 247 (186), *263*
Darling, A. S., 47, *88,* 113, 115, 116 (4), 117 (24), 127, *130, 131,* 148, 149, *173*
Darling, H. E., 116 (17), 126 (17), *131*
Davey, N., 277 (28), *287*
Davies, T. A., 239 (175), 247 (175), *262*
Davis, G. L., 124 (78), *134*
De Bruin, A. J., 189 (55), *194*
De Groot, C., 54 (125), *90*
DeGussa, 74 (195), *93*, 95, *106*, 118 (31), 148 (60, 61), *173*, 212, 247 (287, 288), *267*
Dehoff, R. T., 123 (73), *133*
Dehoust, G., 245 (256), 247 (256), *266*
De Keijzer, A., 54 (125), *90*
Delahay, P., 205 (24), *256*
Delahoy, A., 159 (87), *174*
Delgado, E. F., 285 (100), *291*
Denzer, D., 77 (203), *93*
Depierre, J., 246 (267), 247 (267), *266*
Derobert et Cie, 223 (99), 247 (99), *259*
Derouwaux, P., 49 (80), *88*
Despic, A. R., 216 (64), *258*
Desthomas, G., 213 (39), 247 (291), *257, 267*
Dettke, M., 213, 222, 240 (199), 240 (206), 241 (199, 206), 245 (199), 246 (199), 247 (199, 43, 87, 297, 206), 248 (43), 206, 199, *259, 263, 264, 268*
Deutsche Gold - und Silber Scheideanstalt, 247 (37), *257,* 301 (86), *318*
Dickert, J. J., 170 (151, 152), *177*
Dickinson, 230 (3), 231 (3)
Dickson, P. F., 151 (70a), *173*
Dickson, G., 97 (22), 99, *107*
Diehl, R. F., 187 (40), *194*
Dietz, G., 238 (162), *260, 262*
Dietz, J. W., 299 (57), *317*
Dietzel, A., 102 (52, 53), *108*
Di Martino, M. J., 310 (147), *320*
DXIRAT, F., 240 (211), 247 (211), *264*
Disam, A., 242, 245, 246, 247 (251, 241), 248 (241, 251), *265, 266*
Distler, W. B., 221 (80), 240 (80), 247 (80), *258*
Dodd, E. C., 306 (121), *319*
Doduco, K. G., 298 (44), *316*
Dole, M., 247 (206), *268*
Domnikov, L.. 246 (259), 247 (259), *266*
Donnelly, R. G., 88 (60), 141 (13), 146 (13), *171*
Donnelly, R. G., 88 (60), *88*, 141 (13), 146 (13), *171*
Douglas, G. S., 102 (62), *109*
Dowie, D., 8, *24*
Dowson, A., 51 (92), *89*
Dreaper, W. P., 148 (47), *172*
Duckett, R., 188 (44), *194*
Duckworth, W. H., 102 (66), 103 (66), *109*
Duerrwachter, E., 77 (203), *93*
Dufour, G. H., 48 (72), *88*
Duhl, D. N., 180, *192*
Dunano, A., *269*
E. I. du Pont de Nemours and Co., 145, 146 (40), *172*, 277 (27), *287*, 299 (51), *316*
Dusing, W., 188 (48) (49), *194*
Duva, R., 220, 223, 225 (117), 234, 238 (157), 240 (157), 241 (117), 243 (115), 245 (117), 247 (284, 157, 117), 248 (117, 157, 284), 252 (346), *258, 259, 262, 267, 270, 272*

E

Eaton, G. T., 305, *319*
Eash, J. T., 54 (126), *90,* 97 (23, 24), *107*

Eash, T. J., 51 (91), 54 (126), *89, 90*
Edmunds, G., 58 (149), *91*
Edson, G., 226 (108), *260*
Edson, G. I., 247, 248 (304), *268*
Egan, T. F., 121 (46), *132*
Eichhorst, I., 247 (301, 305), 248 (301), *268*
Eichner, K., 103 (75), *109*
Eick, J. D., 97 (21, 22), 99, *107*
Eicke, H., 116 (20), *131*
Eilfield, F., 147 (43), *172*
Eisenmann, E. T., 239 (173), 247 (173), *262*
Elkington, W. E., 103 (82), *110*
Ellery, R. S., 309 (137), *320*
Ellington, T. S., 182 (26), *193*
Ellis, P., 304, 305 (100), *318*
Ellis, R. J., 239 (178, 179), 241, 244, 248 (179), *263*
Ellsworth, H. D., 244 (248), *266*
Embury, J., 53 (111), *89*
Empire Rheumatism Council Research Sub-Committee, 310 (142), *320*
Engel, A., 34 (25), 45 (51), 51, 63 (25), *86, 87*
Engelhard Industries, 301 (81), *318*
Engelhard Minerals and Chemicals Corp., 301 (81), *318*
English Electric Co. Ltd., 144 (27), *171*
Ennos, A. E., 165 (101), *175*, 190 (62), *194*
Erb, R. A., 166 (123, 124, 125, 126), 167 (126), *176*
Erenburg, A. M., 199 (17), 200 (17), 249 (332), *256, 269*
Erhardt, R. A., 223, *259*
Ernst Winter and Sohn, 300 (71), *317*
Etherington, R. W., 312 (161), *321*
Evans, G. E., 68 (180), 69 (180), *93*
Evans, H. E., 170 (156), *177*
Evers, A., 199 (15), *256*
Exxon Research and Engineering Co., 313 (177), *322*
Ezawa, T., 284 (58, 59), *289*

F

Faelens, P. A., 306 (122, 128), *319, 320*
Fahrenwald, F. A., 48 (67–69), *88*
Fairbanks, N. P., 141 (16), *171*
Fairhurst, C. W., 73 (193), *93*, 100 (46, 47, 49), 103 (73), 103 (84), *108, 109, 110*
Falkenburg, G., 44 (66), *88*, 148 (65), *173*
Fan, J. C. C., 301 (75), *317*
Farben Fabriken Bayer A.G., 313 (171), *322*
Faust, C. L., 197, 238 (11), *255*
Featherstone, R. W., 302 (89), *318*
Federal Coin and Currency Inc., 25 (8), 27 (8), *29*
Fedot'ev, N. P., 67 (175), *92*, 209 (29), 210 (29), 211 (29), 221 (79), 221 (29), 221 (82), 231 (82), 237 (147, 150), 238 (155), 240 (226, 227, 231, 234), 241, 245, 246 (223), 247 (238, 257, 308, 150, 226, 227), 248 (231, 226, 155, 257), *256, 258, 261, 262, 265, 266, 268*
Fehrer, F. C., 51 (90), 54 (90), *89*
Fells, P. D., 3, 4, 7, 13, 21, 23, 24 (2), *24*
Fesolowich, A., 53 (118), *90*
Fichter, R., 181 (14), *193*
Field, S., 218 (70), 241, 247 (235), *258, 265*
Figar, J., 312 (166), *321*
Figuli, E. S., 212, 229 (33), *256*
Filatov, A. G., 300 (70), *317*
Finch, G. P., 179 (4), *192*
Finch, R. G., 144 (22), *171*, 276 (20), *287*, 298 (40), *316*
Fine, M., 87 (50), *88*
Finkelstein, N. P., 198 (14), 199 (14, 15, 16, 20), 200 (16), 200, 201 (20), 202, *256*
Finstad, T. G., 281 (426), *288*
Fischer, H., 254 (351), *270*
Fischer, J., 68 (183), *93*, 247 (278), 254 (350), *267, 270*
Fisher, R., 53 (111), *89*
Fitch, H. M., 271 (4–7), 272 (8, 9), *286, 287*
Flank, W. H., 313 (169), *321*
Flatley, T. W., 170 (156), *177*
Flegontov, Yu. N., 145 (36), *172*
Fletcher, A., 244 (242), 246 (242), *265*
Flom, D. G., 170 (153), *177*
Flühmann, W., 122 (60), *133*

Forbes, J., 30 (1), 47 (1), *85*
Flühmann, W., 122 (60), *133*
Forbes, J. S., 30 (1), 47 (1), *85*
Forbes, L., 125 (81, 83, 84, 85, 86), *134*
Ford, P. J., 113 (6, 7), *130*
Formstecher, F., 304 (103), *319*
Forster, K. R., 298 (43), *316*
Fortovova, L. S., 240 (195), 245 (195), 246 (195), 247 (282, 195), 248 (195), *263, 267*
Foster, A. J., 247 (285), 248 (285), *267*
Foulke, D. G., 217 (66), 220, 225 (117), 238 (161), 240 (190), 240 (213), 241 (213), 244, 245 (117), 246, 247 (283, 284, 213, 190, 117), 248 (213, 117, 190), *258, 262, 263, 266, 267,* 284, 285 (90), *290*
Fowler, P., 165 (110), *175*
Frank, F. C., 214, *258*
Fraser, T. N., 310 (143), *320*
Frebel, M., 53 (113, 114, 115), 54 (113, 114, 115), *89, 90*
Freberg, C., 197, 238 (11), *255*
Frens, G., 296 (21, 22), *315*
Frey, W. P., 223 (105), 232, 247 (105), 250 (105), *260*
Freyberger, P., 103 (86), *110*
Friedberg, R., 25 (1), *29*
Fritsche, L., 165 (115), *175*
Fuchs, K. H., 240 (199), 241 (199), 245 (199), 246 (199), 247 (199), 248 (199), *263*
Fukano, Y., 53 (110), *89*
Fulmer Research Institute Ltd., 168 (139), *176*, 285 (84), *290*, 295 (19), 296 (19), *315*
Fulrath, R. M., 102 (59, 60, 61), *108, 109*, 189 (51, 52), *194*
Funk, W., 147 (42), 149, *172*
Fusayama, T., 285 (96), *290*
Fuschillo, N., 123 (74), *133*, 159 (87), *174*, 301 (76), *317, 318*
Fuys, R. A., 100 (49), *108*

G

Gadet, M. C., 254, *270*
Gainsbury, P., 74 (194), 76 (194), *93*
Gakula, K., 306 (129), *320*
Gallagher, C. J., 125 (79), *134*
Gamer, N. T., 44 (61), *88*
Gardam, D. E., *92*
Gardam, G. E., 64 (160), 78 (208), 79 (208), 77 (208, 209), *94*, 218, 231, *258, 260*
Gardil, R., *269*
Gardiner, J. A., 284 (75, 76), *290*
Gardner, L. A., 44 (56), *87*
Gaunt, P., 55 (136), 55 (138), 55 (140), *91*
Gavrilova, J. P., 250 (336, 337), *269*
Geary, A. L., 283 (47), *288*
Geckle, R. A., 72 (189), *93*
Gedansky, L. M., 202, 207 (21), *256*
Gee, G., 30 (7), 49 (7), 84, 76 (7), *85*
Gerberich, H. R., 313 (172), *322*
Gerlach, W., 53 (108), *89*
Gettleman, L., 55 (142), *91*, 99 (34), *107*
Gibala, R., 53 (109), *89*
Gibson, C. S., 272 (10), *287*
Gillham, E. J., 165 (100), *175,* 190 (61), *194*
Gilliland, R. G., 44 (60), *88*
Gimpl, M. L., 301 (76), *317, 318*
Gioria, J. M., 249 (329), 252, 253 (329), 253 (340), *269, 270*
Girard, E. H., 191 (65), *195*
Girdwood, R. H., 310 (148), *321*
Girling, D. S., 169, *177*
Gleekman, L. W., 68 (180), 69 (180), *93*
Glynn, C., 3, 4 (2), 7, 13, 21, 23, 24, *24*
Gmelin's Handbuch der Anorganischen Chemie, 32, 33 (12), 36 (12), 37 (12), 67 (176), 68 (176), 69 (176), 71 (176), *86, 92*
Goedecke, W., 148 (50), *172*
Godfrey, T. B., 116 (16), *131*
Goldie, W., 122 (65), *133*, 197, 218 (3), 225, 229, 232 (3), 233 (3), 234 (3), 238 (3), *255*, 284, 285 (97), *289, 290*
Golla, M., 299 (48), *316*
Gomes, G. S., 285 (102), *291*
Gostin, E. L., 284 (53, 60), *289*
Gordon, T. E., 184 (35), *193*
Gore, J. K., 224, 246 (107), *260*
Graf, L., 58, 59 (152–159), 60, 61, 64 (156), *91, 92*
Grafried, E., 148 (44), *172*
Gragg, J. E., Jr., 300 (60), *317*

Graham, A. K., 246 (269), 248 (269), *266*
Graham, M., 52 (95), 54 (95), *89*
Green, T., 24 (5–7), *24*
Greener, E. H., 103 (82), *106, 110*
Greenspan, L., 235, 238 (163), 240 (210), 243 (140), 246, 247 (163, 140), *260, 261, 262, 264,* 248 (210, 163), 252 (140)
Greenwood, J. C., 169 (146), *177*
Grigor'ev, A. T., 141 (17), 142, *171*
Grevstad, P. E., 168 (144), *177*
Grigorjew, A. T., 148 (54, 55), *173*
Griepink, E., 128 (97), *134*
Grilikhes, S. Ya., 67 (175), *92*
Grimwade, M. F., 76, *93*
Grin, Yu D., 247 (310), *268*
Groenewald, T., 200, 203 (19), *256,* 283 (44), 284 *288*
Grossman, H., 221 (86), 252 (86), *259*
Groszek, A. J., 301 (84), *318*
Groth, R., 162 (95), 163 (95), 164 (95), *174*
Grove, Jr., C. S., 68 (180), 69 (180), *93*
Grover, R. K., 76, *93*
Grube, G., 241, 242, 247 (236), *265*
Guidess, J., 285 (87), *290*
Gyorgy, E. M., 129, 130 (102), *135*

H

Hagon, P. J., 144 (23), *171*
Hagiuda, Y., 284 (83), *290*
Haidinger, W., 312, *321*
Hall, W. K., 313 (172), *322*
Halpern, T., 113 (9), *130*
Haltner, A. J., 170 (153), *177*
Hamm, F. A., 305 (110), *319*
Hamm, K., 89 (50), *89*
Hampy, R. E., 293 (8), *315*
Hancock, R. D., 198 (14), 199 (15), 199 (14) (20), *256,* 200, 201 (20), 202
Handy and Harman, 72 (190), *93*
Hanks, G. S., 140 (6), *170*
Hänsel, 212, 247 (38) *257*
Hansen, M., 32, 33 (13), 36 (13), *86,* 137 (3), 139 (3), 142, 143, 146 (3), 147 (3), *170*
Hanson, R. C., 125 (90), *134*
Harcourt, J. K., *106*
Hardy, H. K., 84 (50), *88*
Hardy, J. G., 312 (162), *321*
Hariada, A., 299 (54), *316*
Harigatani, H., 54 (129), 55 (129), *90*
Harigaya, H., 54 (132), 55 (132), 55 (143), *91*
Hariya, H., 54 (130), 55 (131), *90*
Harman, G. G., 293 (5), *314*
Harmsen, N., 117 (28), 122 (64), *131, 133*
Harmsen, U., (34) *172*
Harper, C. A., 122 (66), *133,* 279 (38), *288*
Harr, R. E., 219 (75), 241, 247 (239, 240), *258, 265*
Harris, L., 165 (106, 107, 108, 109, 110, 111), 166 (118), *175, 176*
Harris, R. P., 25 (1), *29*
Harris, S. J., 239 (186), 247 (186), *263*
Harrison, J. A., 213 (52), 214, *257, 258*
Harrison, J. D., 99 (34), *107*
Harrison, W. N., 102 (64), *109*
Hary, F. R., 312 (160), *321*
Hashimoto, H., 130 (104), *135*
Hatswell, J. S., 85 (228), *94*
Hattori, S., 182 (27), *193*
Hautot, A., 306 (127), *320*
Hayasaka, T., 182 (27), *193*
Hayashi, J., 298 (39), *316*
Hayashi, K., 284 (66), *289*
Hayes, K. E., 310 (151), *321*
Heal, T. J., 84 (50), *88*
Healy, J. H., 102 (63), *109*
Healy, T., 55 (137), *91*
Hedström, J., *320*
Heidelberg, E. X., 299 (50), *316*
Heilmann, G., 212, 247 (289, 35, 36), *257, 267*
Heiman, S., 246 (269), 248 (269), *266*
Heinzel, H., 145 (35), *172,* 181, 187, 188 (44), *193, 194*
Henn, R. W., 305, *319*
Hennig, F., 125 (82), *134*
Henrich, V. E., 301 (75), *317*
Henry, J., 219 (75), *258*
Hensel, F. R., 244 (243), *265*
Hentzschel, H. P. K., 285 (89), *290*
Hepfer, I. C., 285 (86), *290*
Hepler, L. G., 202, 207 (21), *256*

Heraeus, G.m.b.H., 298 (41), *316*
Heraeus, W. C., G.m.b.H., 148 (52, 53), *172, 173*
Hermance, H. W., 121 (46), *132*
Herschlag, V. E., 247 (292), *267*
Hevesy, G., 180 (10), *192*
Hewitson, E. H., 306 (120), *319*
Heymann, K., 285 (88), *290*
Hill, K. A., *317*
Hillier, J., 295 (17), *315*
Hintermann, H. E., 221 (81), 247 (81), *258*
Hirano, K., *192*
Hirsch, P. B., 294 (15), *315*
Hirschhorn, L., 102 (43), *108*
Hischmann, M., 212, 222, *256, 259*
Hitch, T. T., 277 (31), *288*
Hodgson, R. W., *260*
Hodson, J. T., 96 (17), *107*
Hoechst, A. G., 308 (131), *320*
Hof, J., 277 (32a), *288*
Hoffman, L. C., 277 (24, 25, 26), *287*
Hofman, C., 49 (80), *88*
Hofman, E. E., 140 (7), *171*
Hofmann, W., 181, *193*
Hollenback, G. M., 72 (192), *93*, 96 (12), *106*
Holmes, C. L., 277 (22), *287*
Holmes, P. J., 279 (36a), *288*
Holmlund, L., 99 (29), *107*
Holt, L., 239 (176), 239 (177, 178, 179), 241 (177, 179), 244, 247 (176, 177, 178, 179), 248 (177, 179), *262, 263*
Holtzman, A., 296 (24, 25, 26), *315*
Holzmann, H., 148 (46), *172*
Homma, R., 246 (261), 248 (261), *261*
Hoppe, D. J., 25, *29*
Horn, F., 221 (86), 252 (86), *259*
Horn, G. L., *132*
Hornbogen, E., 54 (133), *90*
Houlston, J. F., 122 (56), *133*
Hovan, M., 55 (141), *91*
Howard, M., 102 (67), *109*
Howe, B. K., 312 (160), *321*
Hubbard, D., 306 (113), *319*
Hudson, P. R. W., 169 (147), *177*
Huettner, D. J., 219, 220, 239 (183), 247 (183), *258, 263*
Hughes, E. J., 53 (119), *90*
Huizinga, A., 125 (92), *134*
Hultgren, R., 33 (17), 34 (17), 51 (17), *86*
Hume-Rothery, W., 47, *88*
Hummel, R. E., 122 (69), 123 (73), *133*
Hunt, L. B., 196, *255*, 296 (29), 297 (29), *316*
Huntley, D. J., 120 (40, 41), *132*

I

Ienco, M. G., 46, *87*
Iles, R. F., 93 (196), *93* 302 (92, 93), 303, 304 (91), *318*
Iles, R. G., 93 (196), *93*
Il'in, V. A., 240 (195), 245 (195), 246 (195), 247 (282, 195), 248 (195), *263, 267*
Imai, F., 301 (74), *317*
Imai, H., 240 (223), 248 (223), *264*
Ingersoll, C. E., 100 (40), *108*
Ingraham, R., 95 (8), *106*
Ingri, N., 99 (29), *107*
International Business Machines and Motorola Inc., 144, *171*
International Business Machines Corpn., 144, 145, *171, 172*
International Standard Electric Corpn., 144 (24), *171*
Ishiguro, I., 232 (130), 247 (130, 299), *260, 268*
Ishitani, A., 306 (129), *320*
Ito, H., 284 (58, 59), *289*
Ito, K., 165, *175*, 284 (83), *290*
Iwashita, T., 237 (154), 248 (154, 321), *261, 269*

J

Jackson, R. S., 41 (40), *87*
Jacobs, W., 300 (68), *317*
Jacquet, P. A., 67 (173), 68 (173), 69 (173), *92*
James, T. H., 305 (107, 108, 109), *319*
Jarrett, T. C., 43 (48), *87*
Jenisch, W., 306 (115), *319*
Johannson, C. H., 119 (35), *131*, 148, *173*
Johnson, A., 53 (116), 53 (119), *90*
Johnson, A. A., 53 (117), 53 (120), *90*

Johnson, D. C., 217, *258*
Johnson, D. R., (40), *172*
Johnson, F. M. G., 168 (137), *176*
Johnson Matthey and Company, 34 (27), 51 (27), 85, *86*, 116 (23), *131*, 141 (12), 148, *171, 173*
Johnson Matthey Metals Ltd., 273 (11), *287*
Johnson, P. R., 249 (325), *269*
Johnston, J. F., 103 (77a), *109*
Jones, M. C., 151 (70a), *173*
Jones, D., 240 (197), 247 (197), *263*
Jones, G. R., 167 (134), *176*
Jones, R., 182 (25), *193*
Jones, W. D., 179 (8), *192*
Jordan, D. R., 77 (202), *93*
Jostan, J. L., 284 (54), *289*

K

Kachi, S., 302 (94, 98), *318*
Kakuno, N., 126 (94), *134*
Kanai, M., 247 (300), *268*
Kankelwitz, B., 306 (114), *319*
Kanzawa, Y., 99 (28), *107*
Kard, P. G., 163 (97), *175*
Kardos, O., 217 (66), 218 (66, 71), *258*
Kasai, K., 54 (132), 54 (134, 135), 55 (143), *90, 91*
Katz, M., 240 (198), 247 (198), *263*
Kautz, K., 102 (56, 57, 58), *108*
Kawai, S., 232, 247 (130), *260*
Kawanishi, I., 54 (129), 55 (129), *90*
Kawanishi, K., 54 (130, 131), *90*
Kazakov, V. P., 199 (17), 200 (17) (18), 249 (331), *256, 269*
Keesom, W. H., 119 (35), *131*
Kehrer, H., 117 (26), *131*
Kehrer, H. P., 169 (148), *177*
Kelly A., 50 (85), 50 (88), 52 (85), *88*, 294 (15), *315*
Kennedy, G. C., 126, *134*
Kenyon, R. L., 25 (1), *29*
Kersten, H. J., 248 (250, 322), *266, 269*
Khan, H. R., 125, *134*, 239 (356), 240 (356), 247 (356), 248 (356), *270*
Kim, C. K., 239 (180), 247 (180), *263*
Kimball, O., 53 (122), *90*
Kimball, O. F., 53 (124), *90*
Kimoto, K., 165 (117), *175*
King, B. W., 102 (66), 103 (66), *109*
Kirby, R. S., 140 (6), *170*
Kishimoto, S., 311 (156), *321*
Kitamura, M., 168 (142), *176*
Klatte, H., 59 (156), *92*
Klein, B. J., 122 (70), *133*
Klein, G., 102 (67), *109*
Klingenberg, H. V., 301 (82), *318*
Kloiber, K., 33 (16), 51 (16), *86*
Klomp, J. T., 189 (57), *194*
Knap, F. J., 103 (85), *110*
Knittel, T., 120 (39), *132*
Knödler, A., 238 (158), 239 (158), 239 (167), 239 (170), 239 (185), *262, 263*, 247 (158, 167, 170, 185)
Knoedler, A., 214 (60), *258*
Knosp, H., 151, *172*
Knowlson, P. M., 184 (34), *193*
Kobler, R. J., 300 (68), *317*
Koch, O. F. A., 244 (246, 247), *265*
Koch, R., 309, *320*
Koch, T. U. S., 44 (45), *87*
Koeditz, H., 212 (38), 247 (38), *257*
Koei-Liang, 312 (161), *321*
Kohl, W. H., 134 (77), *134*, 180 (13), *193*
Kojo, H., 240 (224), 248 (224), *264*
Kokotailo, G. T., 170 (150), *177*
Komarov, V. P., 271 (2), *286*
Komura, K., 247 (294), 248 (294), *268*
Konig, W., 252, *270*
Konovalova, M. V., 249 (331), *269*
Kopchikin, D. S., 248 (31), *269*
Kopp, J., 120 (43, 44, 45), *132*
Korbelak, A., 231, *260*
Korovin, N. V., 223, 245 (101), 248 (101), *259*
Koser, J. R., 95 (8), *106*
Koslowsky, R., 306 (118), *319*
Köster, W., 77 (211), *94*, 113 (9), 117 (26, 27), 130, *130, 131, 135*, 152, 154, 155, 158, *174*
Kouchi, S., 165 (102), *175*
Kovaleva, R. G., 240 (205), 247 (205), 248 (205), *264*
Kozak, S. F., 73 (193), *93*
Kralik, G., 53 (104), *89*
Kraljevic, M., 103 (88), *110*
Krasikov, B. S., 247 (307, 310), *268*
Krause, H., 65 (164), *92*

Kretzmann, R., 152, *174*
Krichmar, S. I., 216 (65), *258*
Kroll, H., 240 (202), 247 (202), 248 (202), *264*
Kropff, F., 306 (112), *319*
Kruglova, E. G., 247 (308), *268*
Krumbhaar, W., 168 (136), *176*
Kuhn, A. T., 311 (154), *321*
Kume, M., 298 (38), *316*
Kuprina, V. V., 143 (20), *20*
Kuranova, N. V., 240 (22), 248 (220), *264*
Kurnakow, N., 51 (89), 55 (94), *89*
Kuroda, K., 232, 247 (130, 299), *260, 268*
Kusch, W., 212 (38), 247 (38), *257*
Kushner, J. B., 68 (177), *92*, 254 (349), *270*

L

Lacal, R., 248 (320), *269*
Lacroix, R., 150 (69), *173*
Lalevic, B., 159 (87), *174*
Lainer, V. I., 246 (263), 248 (263), *266*
Lambert, J. L., 161 (92, 93), *174*
Landé, K., 309, *320*
Landsberg, A., 299 (58), *317*
Langley, R. C., 159 (88, 90), 160 (90), 161 (91), 165 (88, 90), *174*, 190 (60), *194*, 272 (9), *287*
Lapshin, A. I., 200 (18), *256*
Lapshina, A. E., 221 (85), 240, 248 (230), *259, 265*
Larson, C., 300 (72), *317*
Lautenschlager, E. P., 95 (3), 103 (80, 81, 82), 106 (91), *106, 109, 110*
Lavine, M. H., 103 (87), *110*
Lawston, I. W., 312 (163), *321*
Lawton, S. L., 169 (150), *177*
Leak, C., 55 (137), *91*
Lea-Ronal Inc., 247 (298), *268*
Lee, F. F. M., 121 (55), *133*
Lee, W. K., 187, *193*
Leermakers, J. A., 306 (120), *319*
Leinfelder, K. F., 51 (93), 52 (96), *89*, 99 (25), 100 (42, 46, 47), 103 (84), *107, 108, 110*
Lendvay, J., 213 (47), 239 (356), 239 (167), 240 (356), 247 (167, 356), 248 (356), *257, 262, 270*
Leone, E. F., 103 (73), *109*
Lerner, L. B., 248 (319), *269*
Lesh, N. G., 281 (42), *288*
Leuser, J., 34 (18), 36 (35), 36 (18), 39 (35), 39 (18), 40 (35), 51 (18), 54 (18), *86, 87*
Levin, P. A., 271 (2), *286*
Levy, D. J., 285, 285 (99, 100, 102, 103, 104), *290, 291*
Lewis, P., 170 (155), *177*
Liao, S. Y., 164 (99), *175*
Lichtenberger, H., 101 (50), *108*
Liebkecht, O., 44 (55), *87*
Lifshits, V. A., 43 (42), *87*
Linde, J. O., 116, 117 (25), *130, 131*, 148, *173*
Lindell, M. E., 219 (26), 226 (26), *256*
Lingenberg, 301 (82), *318*
Lippmann, H., 116 (18), 126 (18), *131*
Loasby, R. G., 277 (28), 279 (36a), *287, 288*
Lochet, J. A., 225 (109), 240 (196), 244 (245), 246 (196), 247 (302), 248 (196, 302, 245), *260, 263, 265, 268*
Lockheed Aircraft Corp., 285 (106), *291*
Lockheed Missiles and Space Co., 285 (105), *291*
Loeb, A. L., 165 (107), *175*
Loebich, O., 37 (30), 40 (30), 46 (30), 57 (30), 62, 76 (30), *86*, 146 (41), 150, 152, 153, 154, 155, 156, 157, 158, *172, 173*, 190 (63), *194*, 223, *259*
Lokshtanova, O. G., 221 (83), 240, 248 (231), *259, 265*
Loram, J. W., 113 (6, 7), *130*
Losi, S., 250, 252, 253 (329), *269*
Losi, S. A., 251 (338), *269*
Lowenheim, F. A., 197 (9), 218 (9), 224 (9), 226 (9), 238 (9), *255*
Lübke, A., 294 (10), *315*
Luce, B. M., 284 (61), *289*
Ludwig, R., 222, 240 (199), 241 (199), 245 (199), 246 (199), 247 (199, 87, 297), 248 (199), *259, 263, 268*
Luebke, H. J., 240 (204), 247 (204), 248 (204), *264*
Lund, M. R., 95 (10), *106*

Lunt, R. S., 313 (176), *322*
Lüthy, H., 99 (26, 27), *107*
Lutz, K., 299 (48), *316*
Lyon, D. M., 102 (69), *109*
Lyons, N. E., 96 (12), *106*
Lythgoe, S., 125 (89), *134*

M

MacArthur, D. M., 214, *257*
McConkey, B., 310 (146), *320*
McCormack, J. F., 284 (65, 77), *289, 290*
McCormick, J. E., 180 (11), 183 (11), 184 (11), 185 (11), *192*
MacDonald, D. K. C., 120 (37), *132*
MacDonald, W. R., 310 (151), *321*
McGinnies, R. T., 166 (118), *176*
McKay, J. A., 151 (79a), *174*
McLean, J. W., 100 (38), *108*
McMullin, J. G., 33 (14), 36 (14), 51 (14), *86*
McMunn, 277 (23), *287*
McNerney, J. J., 125 (88, 90, 91), 126 (93), *134*
Machu, W., 65 (165), *92*
Mack, B. D., 305, *319*
Mahan, J., 96 (15), *106*
Maja, M., 213, 215 (53), *257*
Mallory, P. K., and Co. Inc., 301 (85), *318*
Malm, D. L., 239 (181, 182), 247 (181, 182), *263*
Marakhtanova, Z. H., 247 (286), *267*
Marchon, M. J. C., 203, 221 (22), *256*
Marinaro, A. T., 240 (208), 246 (208), 248 (208), *264*
Marker, B. C., 106 (91), *110*
Marley, J. E., 300 (68), *317*
Marquis, H., 168 (140), *176*
Martin, J., 86 (50), *88*
Martin, K. U., 247 (297), *268*
Maruda, K., *316*
Masing, G., 33 (16), 51 (16), *86*
Massalski, T. B., 302 (99), *318*
Mason, D. R., 121 (49), *132,* 236 (144, 145), 237 (144), 239 (169), 244, 247 (145, 169), 248 (145), 252, 255 (355), *261, 262*
Massin, M., 77, *94*
Masson, D. B., 53 (120), *90*
Mathiessen, A., 113, *130*
Matijevic, E., 296 (27), *315*
Matsubayashi, H., 284 (55), *289*
Matsumara, S., 238 (154), 248 (154), *261*
Matsunaga, M., 284 (83), *290*
Maurer, R. D., 297 (34), *316*
Melnikov, P. S., 247 (290), *267*
Mel'nikov, P. S., 246 (265), 248 (265), *266*
Melrose, S. P. G., 246 (276), 248 (276), *267*
Mentner, J. W., 294 (15), *315*
Menzel, E., 235, 246 (138), 247 (138), 248 (138), *261*
Menzel, T. J., 247 (280), *267*
Merica, P. D., 50 (83), *88*
Merl, W. A., *132*
Mermillod, J., 223 (97), 225, 232, 233 (97), 237, 238 (149), 240 (149), 247 (97, 149), 248 (97), 248 (149), *259, 261*
Merrill, H. E., 313 (176), *322*
Meshii, M., 53 (122), *90*
Metahi, H., 99 (28), *107*
Meyer, A., 213 (39), 218, 247 (291), 249 (329), 250, 253 (329), 253 (340, 341), *257, 267, 269*
Meyer, A. R., *257, 269, 270*
Meyer, C. L., (34), *172*
Meyer, J. N., 103 (78, 79), *109*
Meyer, W. R., 218 (72), *258*
Michael, I., 168 (135), *176*
Michalopoulos, C., 1 (1), *24*
Milazzo, G., 68 (184), *93*
Miller, Ch. B., 102 (69), *109*
Miller, R. C., 213 (44), *257*
Mills, 230 (3), *255*
Mingoia, Q., 249 (326), *269*
Mintern, R. A., 148, 149, *173*
Missel, L., *175*, 183 (28), *193*
Mitchell, N. W., 238 (162), *260, 262*
Mitoff, S. P., 102 (59), *108*
Miyazaki, S., 240 (214), 247 (214), 248 (214), *264*
Mizuhara, H., 140 (11), *171*
Mizushima, Y., 165 (116), *175*
Moak, D. P., 130 (103), *135*
Mobile Oil Corporation, 312 (159), *321*

Moeller, C. E., 130 (105), *135*
Mohamed, H. A., 181 (16a), *193*
Mohan, A., 255 (355), *270*
Mohler, J. B., 71 (185), *93*
Mohri, H., 248 (318), *269*
Molchan, A. J., 274 (14), *287*
Momyer, W. R., 285, *290*
Monnier, D., 103 (78), *109*
Moodie, A. F., 189 (55), *194*
Moore, B. K., 106 (91), *110*
Moore, T. R., 240 (216, 217), 246 (216, 217), 247 (216, 217), 248 (216, 217), *264*
Morabito, J. M., 121 (50), *132*, 42, *288*
Morgan, W. L., 190 (58), *194*
Mori, H., 246 (266), 248 (266),*266*
Morissey, R. J., 213 (49), 221 (49), *257*
Morlon, S. K., 309 (133), *320*
Moss, R. L., 313 (180), *322*
Mueller, F. W. H., 306 (124, 125), *319, 320*
Mueller, L. E., 305, 306 (125), *319*
Mukaiyana, T., 162 (102), *175*
Mukherjee, K., (117), *90*
Mukoujima, H., 246 (273), *267*
Muldawer, L., 55 (139), *91*
Müller, K., 217 (67), *258*
Muller, M. C., 313 (181), *322*
Mulnet, G., 68 (179), 71 (186), *93*
Munier, G. B., 239 (168), 244, 247 (168), *262*
Murakami, Y., 298 (38), 302 (94, 95, 96, 98), *316, 318*
Myers, L. E., 95 (9), *106*
Myerscough, L., 307, 309 (130), *320*

N

Nagai, H., (37), *316*
Naidus, G. G., 189 (53), *194*
Nakanishi, N., 302 (94, 98), *318*
Nally, J. N., 100 (45), 103 (77, 78, 79), *108, 109*
Namias, 305 (104), *319*
Narcus, H., 284 (81), *290*
Nazarova, E. S., 246 (275), 248 (317), *267, 269*
Nedoluha, A., 165 (112), *175*
Nelson, W. C., 304 (102), *319*
Nenadović, T., 103 (88), *110*
Nesbitt, E. A., 129, 130 (102), *135*
Nesse, T., 141 (14), *171*
Newhall, D. H., 116 (17), 126 (17), *131*
Nicholson, R., 50 (85), 52 (85), *88*
Niehoff, R. T., 197, 238 (11), *255*
Nielsen, J. P., 99 (31), 100, 101 (50), 102 (71), *107, 108, 109*
Nieuwenhuys, B. E., 313 (170), *321*
Niney, C., 150 (69), *173*
Nippon Denso Co. Ltd., 167 (131), *176*
Nishikawa, M., 100 (39), 102 (39), *108*
Nishino, M., 295 (20), *315*
Nishioka, J., 311 (156), *321*
Nobel, F. I., 225 (110), 234 (134), 238 (159), 240 (134), 240 (189), 240 (200), 240 (207), 240 (221), 241 (189), 245 (134), 247 (159, 207, 146, 189, 221), 248 (221, 189, 197, 207, 200, 134), *260, 261, 262, 263, 264*
Noguchi, T., 248 (318), *269*
Norton, J. T., 33 (14), 36 (14), 51 (14), *86*
Nowack, L., 34 (19), 48, 51 (19), 54 (19), 76 (198), *86, 88, 93*, 148 (49, 51), *172*
Nuttall, J. L., 294 (15a), *315*
Nutting, J., 294 (15a), *315*

O

O'Brien, W. J., 51 (93), *89*, 99 (25), 100 (47, 49), *107, 108, 109, 110*, 102 (70), 103 (84)
Obrutsheva, A., 180 (10), *192*
O'Connor, G. P., 49 (79), *88*
Oda, T., 284 (66), *289*
Oddo, B., 249 (326), *269*
Ohkubo, K., 246 (273), *267*
Ohring, M., 122 (72), *133*
Ohsuga, A., 238 (154), 248 (154), 248 (321), *261, 269*
Oike and Co. Ltd., 294 (13), *315*
Okinaka, Y., 284 (50, 68, 69, 70, 71, 72, 73), *289, 290*
Okuno, G., 285 (94), *290*
Oldham, P. A., 65 (169), *92*
Olivier, A., 235 (141), 243 (141), *261*
Olsen, T., 281 (42b), *288*
Onota, Y., 240 (212), 248 (212), *264*
Ordonez, J., 53 (117), *90*

Ornellas, D. L., 180 (12), 181 (12), *192*
Ostrow, B. D., 222, 234 (134), 238 (159), 240 (134), 240 (189), 240 (200), 240 (207), 240 (221), 241 (189), 245 (134), 247 (90, 159, 207, 189, 221), 248 (221, 189, 207, 200, 134), *261, 262, 263, 264*
Otter, M., 151, 152, 153 (74), *174*
Oxy Metal Finishing Corpn., 252 (342), *270*

P

Page, R. T., 197, 202, 222 (4), 224, 250, 252, *255*
Pahlke, S., 238 (148), *261*
Parker, E. A., 223, 233 (94), 234 (132), 238 (160), 239 (160), 239 (164), 240 (225), 241 (225), 244, 245 (132), 247 (164, 225), 248 (225, 132, 164), *259, 261, 262, 265*
Parker, W. C., 125 (85, 86), *134*
Pasciak, A., 239 (165), *262*
Pask, J. A., 102 (55, 59, 60, 61), *108, 109*, 189 (51, 52), *194*
Paunovic, M., 283 (48), *288*
Pearce, W. M., 58 (149), *91*
Pearlstein, F., 284 (82), *290*
Pearson, W. B., 120 (36, 37), *132*
Peiffer, H. R., 300 (68), *317*
Pepperhoff, W., 77 (212), *94*, 154 (81), *174*
Perichon, J., 312 (167), *321*
Perrault, R., 49 (78), *88*
Peshchevitskii, B. I., 199 (17), 200 (17) (18), 248 (332), *256, 269*
Peschko, R. J., 48 (71), *88*
Pesmatjoglou, S., 311 (157), *321*
Pestie, J. P., 281 (42a), *288*
Pfaffenbarger, G. C., 97, *107*
Pfeiffer, W., 214 (60), *258*
Pfestorf, G., 151, *173*
Pfund, A. H., 165, *175*
Philip, R., 151, 152, *174*
Philippi & Co., K. G., 240 (191), 240 (218), 247 (191), 248 (218), *263, 264*
Philips, L. S., 182 (24), *193*
Phillips, R. E., 95, *106*
Phillips, R. W., 71 (187), 73 (187), *93*, 102 (68), *109*, 181 (17), *193*
Philofsky, E., 144 (28), *171*, 293 (6), *314*
Philpott, J. E., 167 (132), *176*
Piccard, J., 44 (57), *87*
Piffle, L., 145 (37), *172*
Pinasco, M. R., 46 (52), *87*
Pinkerton, H. L., 242 (269), 248 (269), *266*
Pitts, J. W., 102 (64), *109*
P.M.D. Chemicals Ltd., 240 (201), 246 (201), 247 (201), 248 (201), *264*
Pokras, D. S., 227 (114), *260*, 284 (63), *289*
Polke, R., 189 (54), *194*
Poll, G. H., 121 (48), *132*, 230 (120), *260*
Polti, J. L., 301 (83), *285*
Poniatowski, M., 301 (79), *318*
Popakul, J., 301 (83), *318*
Popov, K. I., 216 (64), *258*
Pops, H., (99), *318*
Poskanzer, A. M., 296 (27), *315*
Pouradier, J., 254 (352), *270*
Powell, A. R., 48 (72), *88*
Powell, R. L., 120 (38), *132*
Powers, J. A., 238 (160), 239 (160), 239 (164), 240 (225), 241 (225), 247 (164, 225), 248 (225, 164), *262, 265*
Predel, B., 53 (113, 114, 115), 54 (113, 114, 115), *89, 90*
Preston, J. S., 165 (100), *175*, 190 (61), *194*
Protzmann, K., (44), *172*
Pruemmer, R., 168 (135), *176*
Ptak, L. D., 312 (165), *321*
Pummerer, R., 148 (45), *172*
Pyatyshev, V. I., 146, *172*

Q

Quarrel, A. G., 179 (4), *192*
Quirk, R. F., 305 (107), *319*

R

Radnoth, M. Sz. V., 103 (75), 103 (80, 81, 83), *109, 110*

Rajchenbaum, N. B., 285 (92), *290*
Raleigh, J. P., 252 (346), *270*
Ramaswami, B., 300 (61), 302 (87), *317, 384*
Ramcke, K., 245 (256), 247 (256), *266*
Ramsay, T. H., 292 (1), *314*
Rantell, A., 296 (24, 25, 26), *315*
Rao, M. U., 274 (14), *287*
Rapson, W. S., 197 (6), *255*, 283 (44), *288*
Rasberry, S. D., 97 (21), *107*
Raub, C., 293 (4), *314*
Raub, C. J., 146 (41), *172*, 213 (47), 239 (185), 247 (185), *257, 263*
Raub, Ch. J., 125, *134*, 239 (356), 239 (167), 240 (356), 247 (167), 247 (356), 248 (356), *262, 270*
Raub, E., 34 (24, 25, 26), 45 (24, 51), 48 (66), 49 (76), 51 (24, 25, 26), 63 (25), *86, 87, 88*, 148, *173*, 197 (12), 207, 208, 210, 214 (60), 217 (67), 222, 223, 234 (135), 235, 237 (137), 237 (148), 238 (137, 152), 239 (170, 185), 241, 242, 244, 245 (251), 246, 247 (27, 237, 251, 256, 153, 241, 137, 170, 185, 91), 248 (137, 152, 153, 241, 251, 255), *255, 256, 258, 259, 261, 262, 263, 265, 266*
Rave, H. P., 117 (27), *131*
Ravnö, A. B., 283 (49), *288*
Razumney, G. A., 216, 219 (63), *258*
Rayne, J. A., 151 (79a), *174*
Raynor, G. V., 30 (11), *86*
Read, S. M., 166 (128), *176*
Rechenberg, 65 (167), *92*
Reddy, A. K. N., 205, 206, *256*
Redomger, V., 247 (311), 248 (311), *268*
Reents, W., 147 (43), *172*
Rehrig, D. L., 213 (50, 51), *257*
Reichelt, W., 162 (95), 163 (95), 164 (95), *174*, 281 (41), *288*
Reichert, M., 68 (182), *93*
Reid, F., 77 (207), *94*
Reid, F. H., 122 (65), *133*, 197 (3), 225, 229, 232 (3), 233 (3), 234 (3), 238 (3), *255*, 285 (97), *290*
Reinacher, G., 147 (42), 149, *172*
Reinheimer, H. A., 239 (174), 245 (253), 247 (174), 248 (253), *262, 266*
Renaud, J. P., 221 (81), 247 (81), *258*
Retajeczyk, T. F., 284 (72), *289*
Reynolds, K. R., 247 (304), 248 (304), *268*
Rhines, F., 168 (141), *176*
Rich, D. W., 284 (78), *290*
Richard, M., 116 (18), 126 (18), *131*
Richards, E. Tn., 76 (201), 77 (201), *93*
Richmond, J. C., 102 (64), *109*
Richter, K. G. P., 48 (70), *88*
Richter, W. A., 96 (13), *106*
Riddiford, A. C., 207 (25), *256*
Riedel, H., 103 (75), *109*
Riedel, W., 247 (297), *268*
Rinker, E. C., 187 (40), *194*, 223, 234 (133), 238 (157), 240 (157), 247 (133), 247 (283, 157), *259, 261, 262, 267*, 248 (157)
Ritz, J. W., 191 (65), *195*
Rivlin, V. G., 80 (223), *94*
Rivlin, V. G., 144 (28a), 150, *172*
Rivory, J., 77 (214), *94*, 155, *174*
Roberts, C. B., 299 (55), *317*
Roberts, E. I., 79 (217), *94*
Roberts, M. W., 312 (162), *321*
Roberts-Austen, W. C., 25 (1), *29*, 179 (5, 6), *192*
Robertson, W. D., 57 (147, 148), 58 (148), 59 (150, 151), 68 (181), *91, 93*, 69 (181)
Robin, S., 151, 152 (73), *173*
Robinson, P. J., 125 (89), *134*
Rochat, R., 223 (98), 247 (98), 248 (98), *259*
Rochel, E., 146 (41), *172*
Rodies, J., 80 (223a), *94*, 150, 156 (70b), *173*
Rose, T. K., 76 (199), *93*
Rosenberg, R., 122 (71), *133*
Rosier, L. L., 125 (83), *134*
Ross, R. G., 113 (8), *130*
Rothenbacher, W., 117 (26), *131*
Rowe, C. N., 170 (151, 152), *177*
Rowland, T., 113 (8), *130*
Rudenko, V. K., 34 (23), 51 (23), *86*
Rule, R. W., 96 (16), *106*
Rundman, M., 53 (122), *90*
Rushton, J. R., 284 (68), *289*
Russell, R. J., 121 (53), *132*, 274 (14, 17, 18), 275 (18), 276 (17), *287*
Ruthardt, K., 148 (44), *172*

Ryge, G., 73 (193), 100 (47), 102 (70), 103 (84, 85), *93, 107, 108, 109, 110*

S

Sachtler, W. M. H., 313 (170), *322*
Sadler, P. J., 310 (147a), *320*
Saeger, K. E., 80, 77 (215), *94*, 150, 156 (70b), *173*, 187, 188 (43), *194*
Safarzynski, S., 248 (328), 254, *269, 270*
Sah, C. T., 125 (83), *134*
Saifullin, R. S., 246 (265), 248 (265), *266*
Saito, K., 309 (135), *320*
Sakata, A., 126 (94), *134*
Sanwald, R. C., 219, 220, 239 (183), 247 (183), *258, 263*
Sard, R., 213, 219, 220, *257*, 284 (68, 72, 73, 74), *289, 290*
Sasaki, T., 247 (345), *270*
Sastry, S. M., 300 (61), *317*
Sastry, S. M. L., 302 (87), *318*
Sato, A., 55 (144), *91*
Saubestre, E. B., 284 (80), *290*
Sautter, F., 210, 222, 223 (30), *256*
Sautter, F. K., 300 (63, 64, 65, 66, 67), *317*
Sauvenier, H., 306 (127), *319*
Sawers, J. R., 146 (40), *172*
Saxer, W., 122 (60), *133*
Sayers, P., 277)30), *287*
Sced, I. R., 100 (38), *108*
Schaber, A., 165 (115), *175*
Scheil, E., 115 (11), *130*
Schering, A. G., 246 (277), 247 (277, 296), 248 (277), *267, 268,* 285 (93), *290*
Scherzer, J., 246 (274), 248 (274), *267*
Schickel, M., 247 (301), 248 (301), *268*
Schiff, K. L., 122 (64), *133*
Schlumberger, H., 25 (1), *29*
Schmellenmeier, H., 218 (73), *258*
Schmid, E., 247 (279), *267*
Schmid, H., 99 (33), *107*, 148 (64), *173*
Schmidt, D. W., 117 (29, 30), *131*
Schmidt, W., 117 (29), *131*
Schmieschek, U., 306 (117), *319*
Schnabl, R., 122 (64), *133*
Schneble, F. W., 284 (65), *289*
Schneider, E. B., 219 (26), 226 (26), *256*
Schneider, H., 286 (108), *291*
Schneider, J. F., 115 (12), *130*
Schnur, K., 247 (279), *267*
Scholze, G., 240 (203), 247 (203), 248 (203), *264*
Schröder, H., 163 (96), *175*
Schubert, R., 122 (57), *133*
Schüller, H. J., 181 (19), *193*
Schulz, L. G., 151, 152, *174*
Schulze, A., 116 (20, 21), *131*
Schulze, K. J., 294 (11, 14), *315*
Schumpelt, K., 225 (116), 243, *260*
Schwab, G. M., 311 (157), *321*
Schwartz, M. M., 140 (5), 141 (5), *170*
Schwarze, W., 247 (278), *267*
Sedgwick, R. D., 125 (89), *134*
Seegmiller, R., 224, 246 (107), *260*
Sekine, E., 54 (127), 55, *90, 91*
Sell, G., 65 (162, 163), 72 (188), *92, 93*
Selman, G. L., 148, *173*
Sel-Rex Corporation, 240 (192), 245 (192), 245 (244), 246 (244, 192, 264), 247 (192, 244, 192), 248 (315, 192, 244, 262), 252 (347), *263, 265, 266, 268, 270*, 285 (91), *290*
Serizawa, S., 245 (252), 248 (252), *266*
Sermon, P. A., 311 (155), *321*
Servais, W. J., 100 (42), *108*
Shashkov, O. D., 34 (23), 51 (23), *86*
Shell International Research Maatschappy, N.V., 313 (174), *322*
Shell, J., 102 (71), *109*
Shell, J. S., 96 (12), *106*
Sheppard, S. E., 306 (111), *319*
Shestopalova, L. F., 308 (132), *320*
Shifo, S., 248 (321), *269*
Shimizu, A., 130 (104), *135*
Shirogami, T., 168 (143), *177*
Shishakov, N. A., 221 (84), 231 (84), 234 (84), 240 (232), *259, 265*
Short, O. A., 298 (45, 46, 47), 299 (52, 56), *316, 317*
Shoushanian, H. H., 213 (49), 221 (49), 250, 252, 253 (334), *269*
Siebert, G., GmbH, 148 (48), *172*
Siegel, B. M., 166 (118), *176*
Sieverts, A., 168 (136), *176*
Sigler, J. W., 310 (145), *320*
Silcox, J., 55 (140), *91*
Silvertsen, J., 53 (107), *89*

Silman, H., 217 (68), *258*
Silver, H. G., 239 (171, 172), *262*, 247 (171, 172)
Silver, M., 102 (67), *109*
Simonian, A., 243 (115), 245 (126), 246 (126), 248 (126), *260*, 285 (90), *290*
Simonian, A. Y., 225 (126), *260*
Sinfelt, J. H., 313 (175), *322*
Sintsov, V. N., 166 (121), *175*
Sistare, G. H., 35 (28), 44 (28), 46 (28), 49, *86*
Sivertsen, J., 53 (112), *89*
Sivil, C. S., 115 (12), *130*
Skinner, E. W., 72 (192), *93*
Slaughter, G. M., 44 (60), *88*, 141 (13), 146 (13), *171*
Sloboda, M. H., 85 (228), *94*, 136, 137 (2), 139, 140 (10), 165 (1), *170, 171*
Slusark, W., 159 (87), *174*
Smagunova, N. A., 240 (188), 247 (281), 247 (293, 188), 250, *263, 267, 268, 269*
Smart, D. C., 295 (18), *315*
Smit, P., 309 (136), 310 (136), *320*
Smith, B. R., 277 (33–35), *288*
Smith, D. L., 52 (97), *89*, 97 (21), 100 (48), *107, 108, 109*, 103 (74), 184 (35), *193*
Smith, E. A., 30 (5), 39 (5), *85*
Smith, P. T., 244 (242), 246 (242), 249 (327), 252 (343), 253 (327), *265, 269, 270*
Smith, W. F., 306 (119), *319*
Smorodina, T. P., 221 (83), *259*
Socha, J., 239 (170), 247 (170), 248 (328), 254 (353), *262, 269, 270*
Société Continentale Parker, 246 (258), 247 (258), 248 (258), *266*
Solomon, A. J., 184 (31), *193*
Solov'eva, Z. A., 221 (84, 85), 231 (84), 234 (84), 240, 248 (229, 230, 232), *265*
Solow, B., 116 (19), *131*
Sommer, G., 102 (54), *108*
Souder, W., 72 (191), *93*
Soundy, G. W., 285 (101), *291*
Spanner, J., 34 (18), 36 (18, 29a), 39 (18), 51 (18), 54 (18), *86*
Sparks, L. L., 120 (38), *132*
Specht, H., 115 (11), *130*
Speidel, H., (44), *172*
Spender, M. R., 148 (66, 67), *173*
Sperner, F., 101 (51), 103, *108*, 113, 114, 117 (28, 3), *130, 131*
Spiro, M., 283 (43), *288*
Spranger, H. J., 77 (213), *94*, 155 (82), *174*
Spreter, V., 223 (96), 225, 232, 233 (97), *259*, 247 (97), 248 (97)
Spring, W., 179 (7), *192*
Stabe, H., 151 (77), *174*
Stagno, E., 46 (52, 53), *87*
Stahl, R., 77 (211, 213), *94*, 152, 155, 158, *174*
Stanyer, J., 239 (176), 239 (178, 179), 241, 244, 247 (176, 177, 178, 179), *262, 263*
Starchenko, I. P., 43 (42), *87*
Steigmann, A., 306 (126), *320*
Steinmann, S., 122 (60), *133*
Steinemann, S., 221 (81), 223 (81), 247 (81), *258*
Steltman, L., 85 (225), *94*
Stern, M., 283 (47), *288*
Sterner-Rainer, L., 30 (4), 33 (4), 33 (15), 36 (4), 39 (4), 44 (4), 48 (4), 51 (15), 62 (15), *85, 86*
Stevenson, P. C., 295 (17), *315*
Stookey, S. D., 297 (32, 33, 36), *316*
Stops, D. W., 128 (99), *135*
Strong, J., (59), *194*
Suito, E., 295 (16, 20), *315*
Sun, P. H., 122 (72), *133*
Sundahl, R., 53 (112), *89*
Sutar, W., 141 (15, 16), *171*
Svitak, J. J., 182 (23), 183 (23), *193*
Sullens, T. L., 284 (63), *289*
Sutton, B. M., 310 (147), *320*
Swan, S. D., 284 (53, 60), *289*
Swartz, M. L., 102 (68), *109*
Sweeney, W. T., 97 (19), *107*
Syutkina, V. I., 34 (23), 51 (23), *86*
Szkudlapski, A. H., 227 (112), *260*

T

Tai, K. L., 122 (72), *133*
Takeuchi, Y. 117 (27), *131*
Tammann, G., 55 (146), 56 (146), 59 (146), *91, 94*, 79 (218, 219)

Tanabe, Y., 284 (55), *289*
Taormina, S. C., 240 (208), 246 (208), 248 (208) *264*
Tardif, H. P., 168 (140) *176*
Tarnapol, L., 33 (17), 51 (17), *86*
Tasch, A. F., 125 (83), *134*
Tate, M., 309 (135), *320*
Taxay, D., 25 (1), *29*
Taylor, D. F., 51 (93), *89,* 99 (25), *107*
Taylor, N. O., 97 (19), *107*
Tazaki, A., 299 (54), *316*
Technic Inc., 240 (193, 194), 246 (270, 271), 247 (194), 248 (193, 270, 271), *263, 266*
Technical Materials Inc., 273 (12), *287*
Tegart, W. J. McG., 67 (174), 68 (174), *92*
Tembe, G., 274 (13), *287, 132*
Templeton, I. M., 120 (36, 37), *132*
Thelan, E., 166 (123), *176*
Theobald, 294 (9), *315*
Thews, E. R., 230 (122), *260*
Thiede, H., 293 (4), *314*
Thiede, M., 293 (4), *314*
Thiele, H., 296 (28), *315*
Thoma, E., 234 (131), *260*
Thomas, J. H., 281 (42), *288*
Thomas, J. L., 116 (15, 22), *131*
Thompson, J., 213 (52), 214, *257, 258*
Thomson, D. W., 225 (110), 247 (146), *260, 261*
Thornton, H. R., 248 (316), *269*
Thwaites, C. J., 185 (37), *172, 193*
Tidswell, N. E., 231 (124), *260*
Tiedema, T., 53 (99), *89*
Tiedema, T. J., 53 (98), *89*
Tissot, P., 99 (26, 27), *107*
Todt, G., 240 (206, 241 (206), 247 (206), 248 (206), *264*
Tokunaga, S., 298 (37), *316*
Tole, A. B., 125 (83), *134*
Toole, F. J., 168 (137), *176*
Traud, W., 283 (45), *288*
Tripp, H. P., 102 (66), 103 (66), *109*
Trivelli, A. P. H., 306 (119), *319*
Trueblood, R. K., 284 (64), *289*
Trueman, W. H., 95 (7), *106*
Tsuchiya, T., 100 (39), 102 (39), *108*
Tuccillo, J. J., 100, 101 (50), *107, 108*
Tugwell, G. L., 161 (89), *174*
Turkevich, J., 295 (17), *315*
Turnbull, J. C., 188 (46), *194*
Turner, P., 219 (75), *258*
Tylecote, R. F., 178 (1, 2), 181 (20), 191 (1), *192, 193,*

U

Ulrich, W., 130 (101), *135*
Upton, P. B., 285 (101), *291*
Uyeda, N., 295 (16, 20) *315*
Vahl, J., 103 (75), *109*
Valette, S., 306 (116), *319*
Van Goor, K., 309 (138), *320*
Van Ham, N. H. A., 313 (170), *322*
Van Houten, G. R., 102 (65), *109*
Van Der Toorn, L., 53 (98), 53 (101), *89*
Van Heerden, J., 30 (2), *85*
Van Osch, G. W. S., 128 (97), *134*
Vanselow, W., 305 (107, 109), *319*
Varker, R. O., 166 (123), *176*
Varta, A. G., 313 (178), *322*
Vasile, M. J., 239 (180, 181, 182), 247 (180, 181, 182), *263*
Vasileva, G. S., 246 (264), 248 (314), 248 (264), *266, 268*
Velichko, Y. A., 246 (263), 248 (263), *266*
Vickery, R. C., 100 (37), *108*
Villot, 301 (82), *318*
Vines, R., 50 (82), 51 (82), 53 (82), 55 (82), *88*
Vines, R. F., 118 (32), *131*
Vinogradov, S. N., 236 (142, 143), 246 (272), 248 (142, 272), *261, 266*
Vogt, C., 113 (5), *130*
Volk, F., 223 (95), 225, *259*
Volkman, G., 53 (103), *89*
Volpe, M. L., 102 (61), *109,* 189 (52) *194*
Volyanyuk, G. A., 221 (82), 231 (82), 247 (257), *258, 266*
Von Krusenstjern, A., 197 (7), *255*
Von Siemens, W., 252 (348), *270*
Voss, W. A. G., 123 (75), 125 (92), *133, 134*
Vozdvizhenskii, G. S., 246 (265), 248 (265), *266*
Vrobel, L., 219 (76), 240 (219), 241 (219), *258, 264,* 245 (219), 247 (219), 248 (219)

Vyacheslavov, P. M., 209 (29), 210 (29), 211 (29), 221 (79, 29, 82, 83), 231 (82), 239 (166), 240 (231, 233), 246 (260), 247 (260, 257, 166, 308), *256, 258, 259, 262, 265, 266, 268*

W

Wachtel, E., 115 (11), *130*
Waghorne, R. M., 144 (28a), *172*
Wagner, E., 36 (35), 36 (29a), 39 (35), 40 (35), *86, 87,* 99 (32), 102 (32), 103 (76), *107, 109*
Wagner, C., 283 (45), *288*
Wagner, W. J., 117 (29, 30), *131*
Wahlbeck, H. G. E., 35 (29), *86*
Wakamatsu, A., 126 (94), *134*
Wall, G. W., 312 (163), *321*
Waller, C., 306 (121), *319*
Walls, W. F., 167 (133), *176*
Walter, P., 34 (26), 49 (76), 51 (26), *86, 88*
Walton, R. F., 284 (62, 63), *289*
Walz, D. T., 310 (147), *320*
Waltz, M. C., 245 (254), 248 (254), *266*
Warble, C. E., 189 (55), *194*
Washburn, J., 181 (16a), *193*
Watson, P., 239 (175), 247 (175), *262*
Weaver, R. V., 298 (47), *316*
Webb, G., 311 (158), *321*
Weil, R., 187 (40), *194*
Wein, S., 284 (79), *290*
Weinberg, F., 309 (133), *320*
Weinrich, A. R., 190 (58), *194*
Weisberg, A. M., 196, 213 (49), 221 (49), 240 (202), 246 (274), 247 (202, 303), 248 (202, 274) *255, 257, 264, 267, 268*
Weise, J., 53 (103), *89*
Weiss, D. E., 155, *174*
Wert, C., 53 (107), *89*
West, R. W., 296 (23), *315*
Weyl, W. A., 297 (31), *316*
Whall, T. E., 113 (7), *130*
Whalley, L., 313 (180), *322*
Wheeler, W. J., 298 (43), *316*
Wictorin, C. G., 148 (57), *173*
Wiehl, H. P. 237 (148), 239 (185), 247 (185), *261, 263*
Dr. Th. Wieland Scheideanstalt, 100 (41), 102 (41), *108*
Wiesner, H. J., 221 (80, 223 (105), 232, 240 (80), 247, 250 (80, 105), *258, 260*
Wildes, R., 106 (91), *110*
Wilkins, D. G., 166 (128), *176*
Wilkinson, P., 121 (49), *132,* 236 (145), 244 (145), 247 (145), *261,* 248 (145)
Wilkinson, P. G., 166 (119), *174, 176*
Willcox, P., 240 (228), *265*
Williams, B. E., 165 (100), *175,* 190 (61), *194,*
Williams, G. I., 144 (28a), *172*
Williams, R. V., 44 (59), *87*
Wilman, H., 179 (4), *192*
Wilson, C., 79 (219), *94*
Wilson, R., 234 (136), 239 (136), 248 (136), *261*
Wilson, T. C., 58 (149), *91*
Wingot, J., (59), *194*
Winkler, J., 213, 248 (41, 42), *257*
Wirsing, C. E., 181 (16), *193*
Wise, E., 82 (50), 90 (50), 51 (82), 53 (82), 55 (82), *88*
Wise, E. M., 30 (3), 33 (3), 34 (20), 34 (3), 35 (3), 36 (3), 39 (3), 44 (49), 45 (51), 51 (20), 51 (91), 53 (20), 54 (20), 54 (126), 57 (3), 68 (178), 69 (178), 71 (178), *85, 86, 87, 89, 90, 93,* 95, 96, 97, *106, 107,* 112, 118 (32), *130, 131,* 140 (9), 142, 145, *171,* 181, *193,* 197 (10) *255,* 281 (39), *288,* 294 (12), *315*
Wise, H., 311 (152, 153), *321*
Witheridge, R. E., 301 (84), *318*
Wittmer, L. L. 125 (85), *134*
Wolanski, Z. R., 248 (316), *269*
Woldt, G., 285 (88), *290*
Wolf, F., 165 (115), *175*
Wolowodiuk, C., 284 (69, 72, 73), *289, 290*
Wood, B. J., 311 (152, 153), *321*
Wood, R. H. 49 (81), *88*
Woolrich, J. S., 248 (324), *269*
Worthing, A. G., 154 (80), *174*
Woycheshin, F. F., 102 (69), *109*
Wullhorst, B., 223, 247 (92), *259*

X

Xhonga, F., 96 (14), *106*

Y

Yackel, C., 306 (120), *319*
Yamada, H., 248 (323), *269*
Yamaguchi, H., 247 (295), *268*
Yamaguchi, S., 100 (39), 102 (39), *108*
Yamakazi, T., 238 (156), 248 (156), *262*
Yamamoto, S., 240 (212), 240 (214), 247 (214), 248 (212, 214), *264*
Yamamura, K., 233, 235 (139), 240 (215), 247 (215, 125), 248 (215), *260, 261, 264*
Yamauchi, C., 285 (94), *290*
Yamomoto, T., 246 (261), 248 (261), *266*
Yampolsky, A. M., 212 (32), *256*
Yasuda, K., 99 (28), *107*
Yates, E. L., 79 (221), *94*
Yeargan, J. R., 125 (84, 85)
Yolles, R. S., 311 (152), *321*
Young, C. B. F., 247 (292), *267*
Young, R. S., (77), *88*
Young, W. H., 285 (87), *290*
Yudina, A. K., 240 (188), 247 (188), 250 (337), *263, 269*

Z

Zaeschmar, G., 165 (112), *175*
Zaitsev, V. N., 248 (313), *268*
Zak, T., 248 (328), 254 (353), *269, 270*
Zander, J. M., 102 (62), *109*
Zeblisky, R. J., 284 (65), *289*
Zeder, H., 309 (134), *320*
Zeien, R. H., 277 (32), *288*
Zelinskii, A. G., 214 (58), *257*
Zimmerman, D. D., 145 (31), *172*
Zimmerman, R., 227 (111), *260*
Zimmermann, K. F., 140 (8), *171*
Zisman, W. A., 166 (127), *176*
Zsigmondy, R., 296 (30), *316*
Zuikova, V. S., 246 (263), 248 (263), *266*
Zuman, P., 296 (27), *315*
Zuntini, F., 253 (340), *270*
Zuntini, F. L., 248 (329), 250, 251 (338), 252, 253 (329), *269*
Zvolner, H., 247 (309), *268*
Zwingmann, G., 48 (75), *88*, 143 (19), *171*

Subject Index

A

Absorption of radiation by gold and gold alloys, *see* Optical properties of gold and gold alloys
Activation analysis for determination of gold, 309
Aerosol plating of gold, *see* Electroless aerosol plating of gold
Age hardening of gold alloys
  alloys containing Al, Co, Cr, Fe, Ge, In, Sn and Ti, 39, 54–55
  Au-Ag-Cu alloys
    effects of Al, Ni and Zn on, 39–41
    effects of composition on, 37–41
    susceptibility to corrosion and stress corrosion cracking, 55–65
  Au-Cu alloys, 33–35, 51, 136–140
  Au-Ni alloys, 44–45, 53, 140
  Au-Pt alloys, 53, 147–150
  casting gold alloys for bonding to dental porcelain, 99–104
  casting gold alloys for dentistry, 96–99
  hardening processes, 34–35, 50–55
  nickel white golds, 41–47
  spinneret alloys, 147–150
Alkaline gold (I) cyanide electrolytes, 224–225
  composition of typical, 226
  conducting, buffering and complexing salts in, 224
  direct effects of excess cyanide in, 224
  secondary effects of cyanide decomposition products in, 224
  See also
    Electrodeposition of gold
    Electrodeposition of gold alloys
Alloying behaviour of gold, fundamental aspects, 30
Alloys of gold, *see*
  Au alloys
  Chain making
  Electrodeposition of gold alloys
  Ferromagnetic gold alloys
  Glassy gold alloys
  Gold alloy coatings, sputtered
  Thick film gold alloy conductors
  Thin film gold alloy conductors
Alumina
  intermediate layer in coating glass with gold, 190
  substrates for thick film processes, 276
Aluminium
  addition to carat gold casting alloys, 77
  addition to dental gold casting alloys, 100
  age hardening gold alloys containing copper and aluminium, 55
  effects on Au-Cu ordering, 39
  hot bonding to gold, 293
  wire, *see* Wire, aluminium
Alums, addition to gold plating baths, 225, 243–244
Annealing
  effect of grain refiners on grain growth in, 35
  effect of work hardening on grain growth in, 36–37
  effect on susceptibility to stress corrosion cracking, 57
  "orange peel" effect, 36–37
Anti-inflammatory agents, *see* Medical use of gold
Antimony, grain refining and brightening effects in electrodeposition of gold, 219, 245–246, 250
Arsenic, grain refining and brightening effects in electrodeposition of gold,

219, 244, 250
Au alloys (after Au, the symbols for the other components of the alloys listed below are arranged in alphabetical order)
Au-Ag alloys
electrodeposition, 221, 241–242, 247
for thermal fuses, 167
for thick film conductors, 272
in electrical contacts, 121–122
in industrial brazing, 139–140
optical properties, 79, 154, 157–158
phase diagram, 32
stress corrosion cracking of, 58–61
superficial corrosion of, 56
Au-Ag-Cd alloys
electrodeposition, 247
use as green carat golds, 37
Au-Ag-Cd-u alloys
electrodeposition, 247
use as green carat golds, 37
Au-Ag-Cd-Cu-Zn alloys, in industrial brazing, 139
Au-Ag-Cd-In alloys
in electrical contacts, 122, 276
resistance to tarnish, 122, 276
Au-Ag-Cu alloys
addition of Cd and Zn to, 35, 39–41
addition of Co, In, Ni, Pd and Pt to, 34–35, 39
age hardening, *see* Age hardening of gold alloys
annealing, 37
effects of composition on
age hardening, 38–39
colour, 37
corrosion, 55–64
melting points, 82
work hardening, 36–39
effects of ordering on phase segregation in, 34
electrodeposition, 213, 247
grain growth during annealing, 37
grain refinement, 35
heat treatment, 34, 38, 39
in electrical contacts, 122
in industrial brazing, 139
in jewellery, *see* Carat golds
liquidus, solidus and solid-solid transformations, 34–36
liquidus surface, 32–34
ordering in, 51
phase diagrams, 32, 33, 36
staining of skin and clothes by, 64–65
stress corrosion cracking of, *see* Stress corrosion cracking
work hardening of, 36–39, 79
Au-Ag-Cu-Ga-Zn alloys, as dental and carat gold solders, 85
Au-Ag-Cu-Ni alloys, electrodeposition, 247
Au-Ag-Cu-Ni-Zn alloys, *see* White golds
Au-Ag-Cu-Pd-Zn alloys
in electrical contacts, 122
See also Casting gold alloys for dentistry
Au-Ag-Cu-Pt alloys, as dental casting alloys, 50
Au-Ag-Cu-Pt-Zn alloys, in electrical contacts, 122
Au-Ag-Cu-Zn alloys
as casting alloys, 124
as substitution brasses, 40
stress corrosion cracking of, 62–63
See also Carat golds
Au-Ag-Ge alloys, as solders, 143
Au-Ag-Ge-Si alloys, as solders, 143
Au-Ag-In alloys, electrodeposition, 247
Au-Ag-Ni alloys
electrodeposition, 247
in electrical contacts, 122
Au-Ag-Ni-Pd alloys, electrodeposition, 247
Au-Ag-Pd-Pt alloys, *see* Casting gold alloys for bonding to dental porcelain
Au-Ag-Pt alloys
as solders for white gold, 84
in electrical contacts, 122
Au-Ag-Sb alloys, electrodeposition, 247
Au-Ag-Si alloys, as solders, 143
Au-Al alloys, *see*
Purple plague
Violet gold
Au-Bi alloys, electrodeposition, 245–246, 247
Au-Bi-In alloys, as solders, 45
Au-Bi-Sn alloys, as solders, 45
Au-Cd alloys
electrodeposition, 242–243, 247

shape memory of, 302
Au-Cd-Cu alloys
electrodeposition, 212, 222, 237, 247
use in electrical contacts, 237
Au-Cd-Cu-Ni alloys, electrodeposition, 232, 247
Au-Cd-Ni alloys, electrodeposition, 247
Au-Cd-Sb alloys, electrodeposition, 232, 247
Au-Cd-Sn alloys, electrodeposition, 247
Au-Co alloys
age hardening, 55
as dental casting alloys, 99
electrodeposition, 221, 236–240, 247
for resistors, 116
in connectors, 121
in low temperature thermocouples, 119
Au-Co-Fe alloys, magnetic properties, 130
Au-Co-In alloys, electrodeposition, 247
Au-Co-In-Ni alloys, electrodeposition, 247
Au-Co-Ni alloys, electrodeposition, 247
Au-Co-Sb alloys, electrodeposition, 247
Au-Cr alloys
electrodeposition, 247
for film resistors, 116
for wire resistors, 116, 130
optical properties of sputtered films of, 159
Au-Cr-Co-Zr alloys, in electrical contacts, 122
Au-Cr-Ni alloys, in industrial brazing, 140
Au-Cr-Zn alloys, electrodeposition, 247
Au-Cu alloys
buckling and cracking of, due to ordering
electrodeposition, 209–212, 221–222, 231–233, 247
in electrical contacts, 121–122
in gold coins, 26, 28
in industrial brazing, 136–137
optical properties, 154–155
ordering, 34, 51
phase diagram, 32
rate of ordering and its control, 34
stress corrosion cracking of
by aqueous reagents, 57–61
by mercury, 64
Au-Cu-Fe-Ni-Pd (+B+Si) alloys, in industrial brazing, 141
Au-Cu-In-Ni alloys, in industrial brazing, 140
Au-Cu-Ni alloys
age hardening, 41–43
Au-Cu ordering in, 41–43
effects of Zn on, 41–43
electrodeposition, 213, 247
in industrial brazing, 140–141
metallographic studies, 46
phase diagram, 45
stress corrosion cracking, 62–63
Au-Cu-Ni-Cr (+B) alloys, in industrial brazing, 140
Au-Cu-Ni-Zn alloys
electrodeposition, 248
metallographic studies, 46
See also White golds
Au-Cu-Pd alloys
in industrial brazing, 142
electrodeposition, 248
Au-Cu-Sb alloys, electrodeposition, 248
Au-Cu-Sn alloys, electrodeposition, 248
Au-Cu-Zn alloys
hot working, 76
shape memory in, 302–303
Au-Fe alloys
age hardening, 54
electrodeposition, 221, 240–241, 248
for low temperature thermocouples, 119–120
for resistors, 115
See also Violet gold
Au-Fe-Ni alloys, magnetic properties, 130
Au-Fe-Pd alloys, for resistors, 115
Au-Fe-Pt alloys, for permanent magnets, 130
Au-Ge alloys, in bonding of Au to Ge, 142–145
Au-Ge-Sb alloys, as solders, 146
Au-Ge-Sn alloys, as solders, 145
Au-In alloys
as solders, 145
electrodeposition, 243–244
evaporated films of, in microelectronics, 281
Au-In-Pt alloys, electrodeposition, 248
Au-Ir alloys, electrodeposition, 246, 248

Au-Mn alloys
electrodeposition, 248
for low temperature resistance thermometry, 118
Kondo effect in, 118
Au-Mo-Ni alloys, in industrial brazing, 140
Au-Mo-Ni-Zn alloys, electrodeposition, 248
Au-Mo-Pd alloys, for potentiometer wire, 116–117
Au-Nb alloys, in industrial brazing, 146
Au-Ni alloys
age hardening, 53–54
casting, heat treatment and metallographic studies, 46
electrodeposition, 221, 237–240, 248
in electrical contacts, 121
in industrial brazing, 140
optical properties, 156
phase diagram, 45
precipitation in, 45, 53
See also Mechanically clad gold coatings
Au-Ni-Pd alloys
electrodeposition, 248
in industrial brazing, 141–142
Au-Ni-Ta alloys, in industrial brazing, 140–141
Au-Ni-Zn alloys
electrodeposition, 248
metallographic studies, 46
See also White golds
Au-Os alloys, electrodeposition, 246, 248
Au-Pb alloys
electrodeposition, 244–255, 248
See also Contamination of gold and its alloys
Au-Pb-Sn alloys
electrodeposition, 248
in soft soldering of gold, 185–188
Au-Pd alloys
as diffusion barriers, 168
conductors, 278–279
electrodeposition, 246, 248
for film resistors, *see* Thick film gold alloy pastes
for high temperature thermocouples, 118–119
for spinnerets, 148
for standard resistances, 113–116
for thermal fuses, 167
in fuel cells, 168
in industrial brazing, 141
in spark plug electrodes, 166
optical properties, 156
ordering in, 113
phase diagram, 141–144
temperature coefficients of resistance, 113
See also White golds
Au-Pd-Pt alloys
in industrial brazing, 142
in electrical contacts, 122
Au-Pd-Rh alloys, electrodeposition, 248
Au-Pd-V alloys
in potentiometers, 116–117
properties, 117
Au-Pd-Zn alloys, electrodeposition, 248
Au-Pt alloys
conductors, 278–279
electrodeposition, 246, 248
for containers, 167
for crucibles, 167
for film resistors, *see* Thick film gold alloy pastes
for spinnerets, 147–150
phase diagram, 149
precipitation hardening of, 53
review of, 47
Au-Pt-Ir alloys, for spinnerets, 148
Au-Pt-Re alloys, for spinnerets, 148
Au-Pt-Rh alloys
for crucibles, 150
for spinnerets, 149–150
phase diagram, 47
precipitation hardening of, 148
Au-Rh alloys, electrodeposition, 246, 248
Au-Sb alloys, electrodeposition, 245–246, 248
Au-Sb-Sn alloys, electrodeposition, 248
Au-Si alloys
in bonding of Au to Si, 142–144
properties of, 144
Au-Sn alloys
as solders, 145
electrodeposition, 244–245, 248
phase diagram, 146
Au-Sn-Zn alloys, electrodeposition, 248
Au-Ta alloys, in industrial brazing, 146

Au-Ti alloys, age hardening, 52
Au-U alloys, electrodeposition, 247, 248
Au-V alloys, for standard resistances, 117
Au-Zn alloys
  electrodeposition, 242–243, 248
  optical properties, 155
Autocatalytic plating of metals
  electroless plating of gold, 283–284
  reaction mechanisms, 282–283
Autocatalytic plating of plastics
  activation by gold, 284–285
  process, 282
Autoclaves, gold coating of by explosive bonding, 168

B

Bars, gold
  for investment purposes, 28–29
  "good delivery", *see* Good delivery bars
Beam lead integrated circuits, precision plating of, 229
Beam leads, *see* Thick film gold conductors
Beryllium, effect on gold wire, 293
Beryllium oxide insulators, application of heat conductive gold coatings to, 167–168, 281
Biological chemistry of gold, review on, 310
Bismuth oxide
  in coating of glass with gold, 164, 190–191
  in liquid golds, 191, 272
  in thick film conductor pastes, *see* Thick film gold pastes
Black bright gold alloy electrodeposit, 238
Bleaching of gold, *see* White golds
Blue gold alloys, *see* Violet gold
Bonding of gold to ceramics
  factors involved in, 101–103, 191–192
  in liquid gold processes, 191
  in production of evaporated and sputtered gold coatings, 192
  in thick film processes involving pre-metallisation of the surface, 192
  in thick film processes using glass frit, 191
  in thick film processes using reaction bonding, 191
  See also Casting gold alloys for bonding to dental ceramics
Bonding of gold to glass
  frit bonding of gold to glass in thick film conductor pastes, 276–278
  gilding of glass, gold mosaics, 188
  gold films on glass by vacuum processes, 190–191
  mechanism of bonding, 188–190
Bonding of gold to metals
  cold bonding, 181–182
  diffusion bonding, 179–181
  explosive bonding, 184
  factors affecting, 178–179
  friction bonding, 183–184
  fusion welding, 184
  gold coatings evaporated on to metals, 279–280
  gold coatings sputtered on to metals, 279–280
  hot bonding, 182–183
  sliding contacts, wear by cold bonding, 183–184
  soldering, 185–188
  solders for bonding to gold coatings, 185–188
  thermocompressive bonding, 182–183, 292–293
  ultrasonic bonding, 182–183
  use of epoxy adhesives, 192, 279
Boron
  in carat gold alloys, 31
  in casting gold alloys, 35
Brazes, *see*
  Bonding of gold to metals
  Carat gold solders
  Dental gold allloy solders
  Industrial gold brazing alloys
Brazing alloy pastes, *see* Carat gold solders
Brightening agents in gold plating baths, *see* Growth of gold electrodeposits
British gold coins, *see* Coins, gold
Bullion
  flow to and from Europe, 23–24
  purity, 28–29

See also
Good delivery bars
Refining of crude bullion
"Burned" electrodeposits, *see* Current density-potential relationships in electrodeposition
Bursting discs, gold, 167

C

Cadmium
addition to Au-Ag-Cu alloys, 35
in 18 carat green golds, 32, 37
in electrodeposited green gold alloy coatings, 233–234
in solders for soldering to plated gold surfaces, 188
toxicity of fumes from, 82
Cadmium oxide, in thick film gold conductor pastes, 277
Calcium, in gold powder for dentistry, 96
Caratage, definition, 30
Carat gold alloys
22 and 20 carat Au-Ag-Cu alloys, 35
18 carat Au-Ag-Cu alloys, 36–37
16 and 14 carat Au-Ag-Cu alloys, 38–39
10, 9 and 8 carat Au-Ag-Cu alloys, 40–41
composition of commercially available, literature on, 30–32
See also
Age hardening of gold alloys
Au-Ag-Cu alloys
Carat gold solders
Casting of gold and gold alloys
Chemical etching of gold alloys
Colouring of carat golds by surface enrichment
Colour matching of carat golds
Colours of carat golds
Contamination of gold and its alloys
Corrosion of gold alloys
Electropolishing of gold alloys
Hot working of gold alloys
Stress corrosion cracking of gold alloys
White carat golds
Carat gold jewellery, amounts of gold used in, 9, 13–16
Carat gold solders
brazing pastes, 81
cadmium in, 82
caratage and colour requirements, 80
compositions and melting ranges of typical, 81–83
current practice, 80–81
metallurgical aspects, 83–85
See also Solders
Casting of gold and gold alloys
by investment, *see* Investment casting
continuous, *see* Continuous casting
into ingots, 71
of dental gold alloys, 71–74
of white golds, *see* White golds
Casting gold alloys for bonding to dental porcelains
age hardening of, 100–101
bonding processes, 102–104
composition of, 100, 102
See also Bonding of gold to ceramics
Casting gold alloys for dentistry
age hardening of, 97–99
compositions of, 98
corrosion of, 99
grain refining of, 97, 99
specifications for, 97, 98
See also Casting gold alloys for bonding to dental porcelains
Catalysis by gold
heterogeneous catalysis
catalytic hydrogenation, 310–311
epoxidation, 311–312
isomerisation of hydrocarbons, 312
oxidation, 312
oxidative dehydrogenation, 312
homogeneous catalysis by soluble complexes, 313–314
Catalysis by gold alloys, 312–313
Centenario, Mexican gold, *see* Coins, gold
Ceramic ware, decoration of with gold, *see*
Gold powders
Liquid golds
Cerium oxide, as disperse phase in gold composites, 122, 301
Chain making, gold alloys for, 46, 292
Chemical etching of gold alloys, 71

Chemical polishing of gold alloys, 71
Chervonetz, Russian gold, *see* Coins, gold
Chromium, age hardening effects of in Au alloys, 54
Chrysotherapy, *see* Medical use of gold
Clad alloys, *see* Mechanically clad gold coatings
Clips, gold alloys for, 41
Cobalt
  age hardening effects in bulk gold alloys, 54–55
  age hardening effects in electrodeposited Au-Co coatings, 240
  effect on Au-Cu ordering, 34
  grain refining action in Au-Ag-Cu alloys, 35
  in carat gold alloys, 31, 35, 39
  in electrodeposited decorative yellow gold alloy coatings, 234–235
  in white golds, 49
Coherent precipitation, *see* Precipitation hardening
Coins, gold
  alloys used in, 26
  dimensions, 25–26
  fake, 19–20
  fineness, 26–27
  gold content, 26–27
  Latin Monetary Union, 26
  official, amounts of gold used in, 20–21
  production, 26–27
  reference works on, 25
  weights, 26–27
Cold bonding, *see*
  Bonding of gold to metals
  Friction
Cold welding, *see*
  Bonding of gold to metals
  Friction
Colloidal gold
  activation of autocatalytic plating of plastics by, 285
  production and properties of gold sols, 295–297
  See also
    Glass
    Purple of Cassius
    Radioactive gold sols
Colour characterisation by physical methods, *see* Colour matching of carat golds
Colouring of carat golds by surface enrichment
  to achieve enriched gold colours, 65
  resulting from surface oxidation and pickling of nickel white golds, 47, 65
Colour matching of carat golds, 77–80
Colours of carat golds, effects of
  cold working, 79
  polishing of cast alloys, 79–80
  surface enrichment in gold, 47, 65, 80
  See also
    Green golds
    Optical properties of gold and gold alloys
    White golds
Colour standards
  for carat golds, 77
  for watchcase plating, 77–78
  See also Colour matching of carat golds
Communist bloc
  gold production, *see* Production of newly mined gold
  net gold sales, *see* Supply of gold
Committeé Internationale de l'Éclairage (C.I.E.), *see* Colour matching of carat golds
Complexes of gold
  relative stabilities, qualitative aspects, 197–199
  relative stabilities, quantitative aspects, 199–201
    determination of stability constants, 199
    estimation of stability constants, 199–200
    stability constants of a selection of complexes, 200
  stereochemistry of Gold(I) and Gold (III) complexes, 197
  sulphite, amine-sulphite and other mixed complexes, 248–250
  thiocyanate complexes, *see* Photography
  thiosulphate complexes, 252, 254
  thiourea complexes, *see* Thiourea

Composites, *see*
  Gold composite coatings
  Gold composites
Condensers, gold coated, 166–167
Conductive metal powder connectors, *see* Electrical connectors
Conductors, *see*
  Film conductors
  Wire conductors
Connectors, *see* Electrical connectors
Consolidated Gold Fields, Limited, reports on gold, 4
Contacts, *see* Electrical contacts
Containers, gold and gold alloy, for corrosive materials, 167–168
  See also
    Autoclaves
    Crucibles
Contamination of gold and gold alloys
  with lead, 76
  with oxides, 76–77
  with phosphorus, 76
  with silicon, 77
  with sulphur, 46–47, 77
Continuous casting of gold alloys, 74–75
Continuous casting of steel, tracer studies of, *see* Radioactive gold
Continuous precipitation, *see* Precipitation hardening
Cookers, domestic, gold coated reflectors in, 161
Copper
  diffusion through gold coatings on copper, 121
  effects on age hardening of Au-Ag-Cu alloys, 38
  effects on colour of Au-Ag-Cu alloys, 37
  effects on melting behaviour of Au-Ag-Cu alloys, 81–82
  in electrodeposited decorative pink gold coatings, 231–233
  in gold coins, *see* Coins, gold
  in nickel white golds, *see* White golds
Copper oxides, in thick film gold conductor pastes, 277
Corrosion of dental gold alloys, *see* Casting gold alloys for dentistry
Corrosion of gold alloys
  effects of gold content, 55–57
  effects of phase separation, 45, 55
  intergranular corrosion, 57–64
  See also
    Casting gold alloys for dentistry
    Corrosion resistant gold alloys
    Staining of skin and clothes by gold alloys
    Stress corrosion cracking of gold alloys
Corrosion resistant gold alloys, *see*
  Bursting discs
  Containers
  Corrosion of gold alloys
  Crucibles
  Glass
  Spinnerets
Cracking of gold alloys, *see*
  Age hardening of gold alloys
  Annealing of gold and gold alloys
  Contamination of gold and gold alloys
  Ordering of Au-Cu alloys
  Stress corrosion cracking of gold alloys
  Work hardening of gold alloys
Creep resistance
  of gold, effect of other metals on, *see* Recrystallisation temperature of gold
  of gold composites, *see* Gold composites
Crowns, Austrian, gold, *see* Coins, gold
Crucibles
  Au-Pt alloy for, 167
  Au-Pt-Rh alloy for, 150
Cryogenic engineering, use of gold films in, 165
Cryogenic resistance thermometry, *see* Resistance thermometry
Current density-potential relationships in electrodeposition
  factors affecting limiting current density
    agitation, 207
    complexing agents and their concentrations, 207
    diffusion, 206–207
    metal concentration in electrolyte, 207
    temperature, 207
    ultrasonic fields, 207

for deposition of gold alloys from cyanide baths, 207–212
for deposition of gold from cyanide baths, 207–212
limiting current densities, 206–207
typical polarisation curve, 206
Current density-potential relationships in electropolishing, 66–67
Current efficiencies in electrodeposition
effects of co-deposition of hydrogen, 204–205
See also
Hydrogen co-deposition in gold plating
Hydrogen overvoltage
Current modulation in plating of gold and gold alloys
effects on structures of gold and gold alloy electrodeposits, 213
modulation without reversal, pulse plating, 213
periodic reversal of current, 212
Cyanide baths, *see* Cyanide electrolytes
Cyanide electrolytes
historical development, 222–223
reviews of, 196–197
See also
Alkaline gold(I) cyanide electrolytes
Decorative gold alloy-electrodeposits
Electrodeposition of gold
Electrodeposition of gold alloys from cyanide electrolytes
Neutral and acid gold(I) cyanide electrolytes

D

Decorative gold alloy electrodeposits
alloying metals used, 230
plating of green golds, 233–234
plating of pink, rose and red golds, 231–233
plating of white golds, 235–236
plating of yellow golds, 234–235
pretreatment of basis metals, 230
specifications and legal requirements, 230
See also Black bright gold alloy electrodeposit
Demands for gold
fabrication, *see* Fabrication of gold
in relation to supplies, *see* Production of newly mined gold
official purchases, 1–4
private purchases, 1–4, 21–22
speculation and investment, 1–4, 21–22
Dendritic electrodeposits, *see* Growth of gold electrodeposits
Dental gold alloy solders, 104–106
See also Solders
Dental gold alloy wires, *see* Wire, gold or gold alloy
Dental gold casting alloys, *see*
Casting gold alloys for dentistry
Casting gold alloys for bonding to dental porcelains
Dental gold powders, 95–96
Dentistry, amounts of gold used in, 18–19
Dewar flasks, gold films on, 165
DF yellow 9 carat gold, 41
Diamonds, gold alloy brazes for bonding to, 146
Diffusion barriers, gold and gold alloy coatings as, 168
Diffusion bonding, *see* Bonding of gold to metals
Diffusion controlled electrodeposition, *see* Current density-potential relationships in electrodeposition
Diffusion effects
in electrodeposited gold coatings, 121
in mechanically clad gold coatings, 276
self diffusion in gold, 179–181
See also
Bonding of gold to ceramics
Bonding of gold to metals
Diffusion barriers
Electrical connectors
Diodes, gold bonded, 144–145
Discontinuous precipitation, *see* Precipitation hardening
Dispersion hardening, 50–51
See also
Age hardening
Gold composites
Distribution centres for marketing gold,

23–24
Distribution routes for marketing gold, 23–24
Domestic cookers, *see* Cookers, domestic
Doping, 124–125, 145
Doublé:
White gold alloys for, 46
See also Mechanically clad gold coatings
Drawing
gold alloys for, 41
See also Hot working
Dropwise condensation of steam, *see* Condensers, gold coated
Ducats, Austrian gold, *see* Coins, gold

E

Eagles, American gold, *see* Coins, gold
Electrical conductivity of gold films, 122–123
Electrical connectors
conductive metal powder connectors, 298
gold coatings for, 120–122
sliding damage to, 122, 183–184
Electrical contacts
amounts of gold used in, 111
electrodeposited gold alloys for, 120–122
gold alloys used for, *see* Au alloys
mechanically clad gold alloys for, 120–122, 273–276
polymer formation on, 121
tarnishing of, 121, 276
See also
Electrical connectors
Electrical contacts, make and break
Electrical contacts, sliding
Electrical contacts, make and break
$Au-CeO_2$ composites for, 301
Au-WC composites for, 300
See also
Au alloys
Electrical contacts
Electrical contacts, sliding
gold composites with carbides of W and Ti, 300
gold composites with graphite, 300
gold composites with sulphides of Mo, Nb and W, 300
lubrication of, 122
sliding damage to, 122, 183–184
See also
Bonding of gold to metals
Electrical contacts
Electrical insulators, gold coatings to assist dissipation of heat from, 168–169
Electrical resistivity of gold and gold alloys
effects of other metals on resistivity of gold, 113
selected data, 112
See also Wire resistors
Electrochemical displacement, gold coatings formed by, 281–282
Electrochemical levelling (microthrowing power), *see* Growth of gold electrodeposits
Electrocrystallisation, *see* Growth of gold electrodeposits
Electrodeposition of gold
historical development, 196–197, 222–223, 248
See also
Complexes of gold
Current density potential relationships in electrodeposition
Current efficiencies in electrodeposition
Current modulation in plating of gold and gold alloys
Cyanide electrolytes
Electrodeposition of gold from cyanide baths
Electrodeposition of gold from gold (III) halide baths
Electrodeposition of gold from sulphite baths
Electrodeposition of gold from thiocyanate baths
Electrodeposition of gold from thiomalate baths
Electrodeposition of gold from thiosulphate baths
Electroforming
Growth of gold electrodeposits
Electrodeposition of gold alloys from

cyanide baths
electrochemical aspects, 208–212
references to plating of individual alloys, *see* Au alloys
structure of gold alloy electrodeposits, 221–222
See also
Current modulation in plating of gold and gold alloys
Decorative gold alloy electrodeposits
Electrodeposition of gold alloys from dialkyldithiocarbamate baths, 246
Electrodeposition of gold alloys from sulphite baths, 250–252; *See also* Decorative gold alloy electrodeposition
Electrodeposition of gold alloys from thiomalate baths, 252
Electrodeposition of gold from cyanide baths, *see*
Alkaline gold(I) cyanide electrolytes
Current modulation in plating of gold and gold alloys
Cyanide electrolytes
Gold(III) cyanide electrolytes
Growth of gold electrodeposits
Neutral and acid gold(I) cyanide electrolytes
Electrodeposition of gold from gold(III) halide complexes, 214, 252
Electrodeposition of gold from gold(I) thiocyanate baths, 214
Electrodeposition of gold from gold(I) thiomalate baths, 252
Electrodeposition of gold from sulphite baths, 250–252
Electrodeposition of gold from thiosulphate baths, 252–253
Electrodes, ion sensitive, *see* Gold powder
Electroetching of gold alloys, 66–71
Electroforming of gold, 254–255
Electroless aerosol plating of gold, 285–286
Electroless gold coatings, *see*
Autocatalytic plating of metals
Electrochemical displacement
Electroless aerosol gold plating
Electrolytic refining of gold, *see* Refining of crude bullion
Electromigration, *see* Film conductors
Electronics, amounts of gold used in, 13, 17
Electroplating of gold and gold alloys, *see*
Electrodeposition of gold
Electrodeposition of gold alloys
Electropolishing of gold and gold alloys, 66–71
Embossing with gold leaf, *see* Gold leaf
Emissivity of gold and its alloys, *see*
Optical properties of gold and gold alloys
Solar energy collection
Emulsion sensitisation, *see* Photography, use of gold in
Epitaxial growth, *see* Growth of gold electrodeposits
Epoxidation with a gold catalyst, *see* Catalysis by gold
Epoxy adhesives, *see*
Bonding of gold to metals
Gold-epoxy conductors
Etching of gold alloys, *see*
Chemical etching
Electroetching of gold alloys
Evaporated gold coatings, *see* Gold coatings, evaporated
Expanded gold gauze, *see* Gauze, gold, expanded
Explosive bonding, *see*
Autoclaves
Bonding of gold to metals

F

Fabrication of gold
amounts of gold used in, 9–21
See also
Carat jewellery
Dentistry
Electronics
Industrial and decorative applications
Medals, medallions and fake coins
Official gold coins
Fake gold coins, *see* Coins, gold
Ferromagnetic gold alloys
for permanent magnets, 130
See also

Switches, magnetic gold alloy cores for
White golds
Film conductors, gold
electrical properties of ultra-thin gold films, 123
electromigration in, 122–123
gold film conductors in electronic circuitry, 120–121, 274–275
See also
Humidistat, electronic
Microwave tubes
Oscillatory circuits
Solar cells
Thick film processes
Thin film processes
Vacuum tubes
Film resistors
Au-Cr alloys, 116
ultra-thin gold films as resistors in strain gauges, 128
See also Resistance thermometry
Fineness, definition, 30
Fire gilded gold coatings, 286
Flip chip bonding, 183, 293
Florins, Austrian, gold, *see* Coins, gold
Foil, gold, *see* Gold foil
Fountain pen nibs, stress corrosion cracking of 14 carat, 62
Free world, supply of gold, *see* Supply of gold
French red gold, 34
Friction
between gold surfaces, 121–122, 183–184
See also
Bonding of gold to metals
Gold composite coatings
Gold composites
Fuel cells, use of gold in, 168

G

Gallium
doping of gold with, for bonding to Ge chips, 144–145
effects of addition to gold cyanide plating baths, 243–244
in carat gold and dental solders, 85
Gauze, gold, expanded, 302
Geometrical levelling in electrodeposition of metals, *see* Growth of gold electrodeposits
Germanium
alloys of with gold, *see* Au alloys
bonding of gold to, in semi-conductor devices, 144–145
properties of Au-Ge eutectic, 144
See also Glassy gold alloys
Gilding, *see*
Bonding of gold to glass
Decorative gold alloy electrodeposits
Fire gilding
Glass
behaviour of gold and gold alloys in contact with, 188–190
electrically heated gold coatings on, 164–165
gilding of, 188
gold coated, for buildings, 161–164
gold containing glasses
for infra-red lamps, 298
for photographic filters, 297–298
ruby glasses, 297
with reversible colour change in light, 298
See also Bonding of gold to glass
Glass frit, *see* Thick film solid pastes
Glassy gold alloys
Au-Ge, 144
Au-Si, 144
Gold alloy coatings, sputtered, 281
Gold alloy resistors, *see*
Film resistors
Wire resistors
Gold alloys, *see* Alloys of gold
Gold beating, *see* Gold leaf
Gold blacks, 165–166
Gold bonded diodes, 144–145
Gold catalysts, *see*
Catalysis by gold
Catalysis by gold alloys
Gold chain, *see* Chain making
Gold(III) chloride, *see* Electrodeposition of gold from gold(III) halide complexes
Gold coatings, evaporated, 279–280
See also
Bonding of gold to ceramics

Bonding of gold to glass
Glass, gold coated for buildings
Thin film processes
Gold coatings, sputtered, 280
See also
Bonding of gold to ceramics
Bonding of gold to glass
Gold alloy coatings, sputtered
Thin film processes
Gold coins, *see* Coins, gold
Gold complexes, *see* Complexes of gold
Gold composite coatings
Production by electrodeposition, 300
Production by vacuum techniques, 301
See also
Electrical contacts, make and break
Electrical contacts, sliding
Gold composites
Gold composites
For jewellery, 301
For metal friction couples, 301
Production techniques, 300–301
See also
Gold composite coatings
Gold-diamond composites
Gold containing glasses, *see* Glass
Gold(I) cyanide electrolytes, *see* Cyanide electrolytes
Gold(III) cyanide electrolytes, 214, 222
Gold determination by activation analysis, *see* Radioactive gold
Gold(I) dialkyldithiocarbamates, *see* Electrodeposition of gold alloys from dialkyldithiocarbamate baths
Gold-diamond composites as heat sinks for electronic devices, 169
Gold doublé, *see* Mechanically clad gold coatings
Gold-epoxy conductors, 279
Gold filled materials, *see* Mechanically clad gold coatings
Gold foil
In dentistry, 95–96, 181
Platinum centred, 96
See also Gold leaf
Gold gauze, expanded, *see* Gauze, gold, expanded
Gold grains, *see* Radioactive gold
Gold interference layer systems, *see* Glass, gold coated, for buildings
Gold jewellery
Consumption of gold in fabrication of, 9, 14–16
See also
Carat golds
Carat gold solders
Decorative gold alloy electrodeposits
Gold composites, for jewellery
Gold(I) ketenide, as a catalyst, 312
Gold leaf
Alloys used in, 294
Production by beating, 294
Production by sputtering or evaporation, 294
Structure, 294–295
Use in printing and embossing, 294
Gold, liquid, *see* Liquid gold coatings
Gold mercaptides, *see* Liquid gold coatings
Gold mosaics, *see* Bonding of gold to glass
Gold, photographic uses, *see* Photography
Gold powders
Production of in different forms, 298–299
Properties of, as sold by one company, 299
Use in decoration of ceramic ware, 298
See also
Carat gold solders
Conductive metal powder connectors
Dental gold powders
Gold blacks
Gold composites
Gold epoxy conductors
Ion sensitive electrode
Postage stamps
Thick film gold pastes
Gold production, *see* Production of newly mined gold
Gold skeletal foams, 302
Gold sols, *see*
Autocatalytic plating of plastics
Colloidal gold
Radioactive gold
Gold standard, 25
Gold sulphite complexes, *see* Complexes

of gold
Gold terpene mercaptides, *see* Liquid gold coatings
Gold therapy, *see* Medical use of gold
Gold thiocyanate complexes, for image stabilisation, *see* Photography
Gold(I) thioglucose, 309
Gold(I) thiomalate, in treatment of rheumatoid arthritis, 309
See also Electrodeposition of gold from gold (I) thiomalate baths
Gold(I) thiosulphate electrolytes, 252–254
Gold thiourea complexes
  for image stabilisation, 305
  See also Electropolishing
Gold toning, *see* Photography
Gold, toxicity of compounds, *see* Medical use of gold
Gold, wetting of, *see* Condensers, gold coated
Gold wire, *see*
  Wire, gold or gold alloy
  Wire, gold plated, Kovar
  Wire resistors
Gold delivery bars
  composition, typical, from Rand Refinery, 28–29
  production from mine bullion, 28–29
  specifications for, *see* London Gold Market
Grain growth, *see* Annealing
Grain refining, *see*
  Annealing
  Casting gold alloys for dentistry
  Iridium
  Rhenium
  Yttrium
Grain refiners in gold plating baths, 219
See also
  Grain size of gold electrodeposits
  Growth of gold electrodeposits
Grain size of gold electrodeposits
  effect of pulsed current upon, 213
  See also Grain refiners in gold plating baths
Graphite, brazes for bonding to, 141, 146
Green gold alloys
  Au-Ag-Cd alloys, 37
  Au-Ag-Cd-Cu alloys, 37
  Au-Ag-Cu alloys, 37
Green gold coatings, *see* Decorative gold alloy electrodeposits
Green gold solders, *see*
  Carat gold solders
  Green gold alloys
Grid
  in humidistat, 129
  wire, in vacuum tubes, 123–124
Growth of gold electrodeposits
  basic considerations, 214–215
  effects of cyanide electrolysis products on, 219–220, 224–225
  factors which decrease surface irregularities, 217–219
  factors which increase surface irregularities, 215–217
  modes of growth of pure gold electrodeposits, 219–221, 224–225

H

Hallmarking of gold jewellery, 30
Hardening, *see*
  Age hardening of gold alloys
  Dispersion hardening
  Ordering
  Precipitation hardening
  Solid solution hardening
  Work hardening
Hardness, *see* Hardening
Hardness and wear resistance
  of gold composite coatings, 300–301
  of gold composites, 300–302
  of mechanically clad golds, 274–275
Heat and radiant energy reflectors, gold coatings for, *see*
  Cryogenic engineering
  Domestic cookers
  Glass, gold coated, for buildings
  Infrared heating and drying
  Lasers
  Missiles
  Optical properties of gold and gold alloys
  Solar cells, thermoelectric
  Solar energy collectors
  Space exploration

Heat dissipation, use of gold to assist, *see*
Electrical insulators
Gold-diamond composites
Heat reflective glass, *see*
Gold coatings, sputtered
Heat and radiant energy collectors, gold coatings for
High pressure gauges, gold alloys for, 126–127
High velocity gold plating, 229–230
Hoarding, of gold, 21–24
Hot bonding, *see* Bonding of gold to metals
Hot shortening, *see* Cracking of gold alloys
Hot working of gold alloys, 75–76
Humidistat, electronic, gold film conductors in, 129
Hydrazine sulphate, hardening effects on gold electrodeposits, 225
Hydrogen codeposition in gold plating
effect upon deposits, 204–205, 208
effect upon pH around cathode, 205
See also
Current efficiencies
Hydrogen overvoltage
Hydrogen overvoltage, *see*
Current efficiencies
Hydrogen codeposition in gold plating
Hydrogenation with gold catalysts, *see* Catalysis by gold

I

Image stabilisation with gold, *see* Photography
Incoherent precipitation, *see* Precipitation hardening
Indium
in age hardening gold alloys, 55, 100
in 18 carat gold alloys, 39
in carat gold solders, 85
in industrial gold brazing alloys, 145
in white golds, 41, 49
See also Casting gold alloys for dentistry
Industrial gold brazing alloys, 136–147
See also Au alloys (for references to individual alloys)
Infrared detectors, gold in transistors for, 125
Infrared heating and drying, heat reflectors for, 161
Infrared lamps
gold coated glass for, *see* Infrared heating and drying
gold containing glass for, *see* Glass
Inlays, clad metal, *see* Mechanically clad gold coatings
Instrument slip rings, *see* Electrical contacts, sliding
Insulation of buildings, *see* Glass, gold coated for buildings
International Monetary Fund, sales of gold, 4
Investment casting of gold alloys, 71–74
Investment in gold, *see*
Demands for gold
Supply of gold
Ion sensitive electrode, 128
Iridium
grain refining action of, in carat golds, 35, 48
grain refining action of, in dental gold alloys, 99–101
see Casting gold alloys for dentistry
in spinneret alloys, 148
in white golds, 48
Iron
blue appearance of certain alloys with gold, 49
effect on ordering in Au-Cu alloys, 34
in age hardening gold alloys, 54–55
in dental casting alloys, 100–102
in white gold, 41, 48
Isomerisation of hydrocarbons by gold catalysts, *see* Catalysts by gold

J

Jewellery, *see* Gold jewellery

K

Kinetics and mechanisms of electrodeposition of gold, *see* Electrodeposition of gold

Kondo effect, 118
Kovar, *see* Wire, gold plated, Kovar
Kruger rand, South African, *see* Coins, gold

L

Lasers, high power density, gold coated reflectors for, 161
Lasers, use in welding, 184
Latensification, use of gold in, *see* Photography
Latent image intensification (Latensification), *see* Photography
Latin Monetary Union, *see* Coins, gold
Lead
  as a contaminant of gold and its alloys, 76
  electrodeposition with gold, 244–245, 248
Levelling agents, *see* Growth of gold electrodeposits
Levelling in electrodeposition of metals, *see* Growth of gold electrodeposits
Limiting current densities, *see* Current density-potential relationships in electrodeposition of metals
Liquid gold coatings
  bonding to substrate, 272–273
  development of liquid golds, 271
    formulation of liquid golds, 272–273
  synthetic gold mercaptide types, 272
  terpene gold mercaptide types, 271
  See also
    Bonding of gold to ceramics
    Bonding of gold to glass
    Solar cells, thermoelectric
Logic circuit, 125
London Gold market, specifications for "good delivery bars", 28
Lubrication
  gold as solid film lubricant, 170
  gold in lubricating oil additives, 169–170
  sliding contacts, 120–122

M

Macrothrowing power in gold plating, *see* Growth of gold electrodeposits
Magnetic cores, *see* Ferromagnetic gold alloys
Magnetic gold alloys, *see*
  Ferromagnetic gold alloys
  Switches, magnetic cores for
  White golds
Malleability of gold, review article, 294–295
Manganese
  electrodeposition with gold, 247–248
  in white gold alloys, 41, 49
  See also Thermometry, low temperature resistance
Marketing of gold, *see*
  Demand for gold
  Distribution routes in
  London Gold Market
  Supply of gold
Mat gold, in dentistry, 95–96
Mechanically clad gold coatings
  manufacture, 273–274
  properties, 274–276
  specifications, 273
  use for decorative purposes, 121, 274
  use in electrical contacts and connectors, 276
Medals, medallions and fake coins, amounts of gold used in, 19–20
Medical use of gold, 309–310
  See also Radioactive gold
Mercury
  detection using gold film, 125–126
  removal from waste water using gold, 126
  stress corrosion cracking by, 64
  use to promote diffusion bonding of gold to gold, 180–181
  See also Fire gilded gold coatings
Mercury gilded gold coatings, *see* Fire gilded gold coatings
Metal friction couples, gold composites for, 301–302
Microforms, image stabilisation of, 305
Microthrowing power (electrochemical levelling), *see* Growth of gold electrodeposits

Microwave strip lines, *see* Thick film gold alloy conductors
Microwave tubes, gold coatings as conductors in, 123
Miller chlorination process, *see* Refining of crude bullion
Missiles, gold films on, to attenuate RF radiation, 163–164
Molymanganese process, 192
Mosaic art, gold in, *see* Bonding of gold to glass
Mosfets, *see* Semiconductor devices
Moving coil suspensions, Au-Ag-Cu alloy wires for, 126
Multilayer circuitry, *see* Thick film gold alloy conductors

N

Neutral and acid gold(I) cyanide electrolytes
  application to plating of integrated circuits, 229–230
  composition of typical acid baths, 228
  composition of typical neutral baths, 227
  See also
    Electrodeposition of gold
    Electrodeposition of gold alloys
    High velocity gold plating
    Selective gold plating
Newly-mined gold, *see* Production, newly mined gold
Nickel
  as a diffusion barrier between gold coatings and metallic substrates, 121
  effects on age hardening of 14 carat alloys, 39
  effects on ordering of Au-Cu alloys, 34
  electrodeposition with gold, 231, 237–240, 248
  grain refining action in Au-Ag-Cu alloys, 35
  in decorative pink gold electrodeposits, 231–232
  in decorative yellow gold electrodeposits, 234–235
  in decorative white gold electrodeposits, 235–236
  in white golds and white gold solders, *see* White golds
  precipitation hardening of alloys with gold, 53–54
Nickel white golds, *see* White golds
Noble metal white golds, *see* White golds
Non-communist world, *see* Free world

O

Optical properties of gold and gold alloys
  Au-Ag, Au-Cu and other Au alloys, 154–156
  mechanisms underlying variations in, 156
  opaque gold films and massive gold, 150–151
  semi-transparent gold films, 152–154
  See also
    Colours of carat golds
    Heat and radiant energy reflectors, gold coatings for
Orange peel effect, *see* Annealing
Ordering
  Au-Ag-Cu alloys, 51
  Au-Cu alloys, 34, 51
  Au-Pd alloys, 113
  See also
    Age hardening
    Casting gold alloys for dentistry
    Hardening
Oscillatory circuits, stabilisation by gold films, 128
Oxidation using gold catalysts, *see* Catalysis by gold
Oxidative dehydrogenations using gold catalysts, *see* Catalysis by gold
Oxide films on gold, 179–180
Oxides as contaminants of carat gold alloys, 76–77

P

Palladium
  as activator for autocatalytic plating of plastics, 284–285
  use in 18 carat golds, 39, 47–49

See also
Casting gold alloys for dentistry
Noble metal white golds
Spark plug electrodes
Spinnerets, gold alloys for
Thermocouples, gold alloys for
Thick film gold alloy pastes
Thin film gold alloy conductors
Wire resistors
Particle size
of gold particles in gold sols, 295–296
of gold powders in thick film pastes, 298–299
Particle size distribution
in gold powders, 298–299
in gold sols, 295–296
Periodic reversal of current in gold plating, *see*
Current modulation in plating of gold and gold alloys
High velocity gold plating
Permalloy, 130
Permanent magnets, *see* Ferromagnetic gold alloys
Peso, Mexican 50 peso gold coin, *see* Centenario
Phosphorus, effects on carat gold alloys, 76
Photocopying, *see* Photography
Photographic light filters, *see* Glass
Photography, use of gold in, 304–306
Photoresists
in electroforming, 254–255
in the gold plating of integrated circuits, 229–230
Photosensitivity of gold compounds, *see* Photography
Pickling
of nickel white golds, 47, 65
of other carat golds, 65
Pink gold coatings, *see* Decorative gold alloy electrodeposits
Plastics, electroless plating of, 282, 284–285
Plating of gold and gold alloys, *see*
Electrodeposition of gold
Electrodeposition of gold alloys
Platinum
as an activator for autocatalytic plating of plastics, 284
use in 18 carat gold alloys, 39, 47–48
See also
Age hardening of gold alloys
Casting gold alloys for dentistry
Noble metal white golds
Spinnerets, gold alloys for
Thin film gold alloy conductors
Wire resistors
Platinum metals, in gold bullion, 28
Polarisation curves, *see* Current density-potential relationships in deposition of metal
Polishing of gold alloys, 66–71, 80
See also
Chemical polishing
Electropolishing
Porosity
of cast gold alloys, 73–74, 106
of electrodeposited gold coatings, effects of pulsed currents upon, 213, 221
of mechanically clad gold coatings, 274–275
Postage stamps, relief gilding using gold powders, 298
Potentiometer wires, gold alloys for, *see* Wire resistors
Powdery electrodeposits, *see*
Current density-potential relationships in electrodeposition of metals
Growth of gold electrodeposits
Precipitation hardening
by different alloys, *see* Au alloys
by nucleation and growth, 51–52
by spinodal decomposition, 52–53
See also Age hardening
Pressure
Au-Cr alloys in high pressure gauges, 126–127
high, calibration scale, 126
Pretreatment of basis metals before electroplating with gold, *see* Decorative gold alloy electrodeposits
Primary current distribution in gold plating, *see* Growth of gold electrodeposits
Printing with gold leaf, *see* Gold leaf
Production of newly mined gold
in free (non-communist) world, 2
in South Africa, 3–4, 6

in U.S.S.R., 7–9
Pulsed current plating of gold, *see* Current modulation in plating gold and gold alloys
Purchases of gold, official and private, *see*
Demands for gold
Supply of gold
Purple of Cassius, 296–297
Purple plague, 293
Pyrolytic carbon resistors, application of gold contacts to, 281

Q

Quartz crystals, coating with gold for use in oscillatory circuits, 128

R

Radioactive gold
activation analysis of gold, 309
production and properties of gold radioisotopes, 307
radioactive gold sols, 308–309
therapeutic and diagnostic applications of, 307–309
use in tracer studies, 309
Radioactive gold sols, *see* Radioactive gold
Radio-frequency sputtering, *see* Gold alloy coatings, sputtered
Radioisotopes of gold, *see* Radioactive gold
Radiometers, *see* Gold blacks
Rand, Kruger, South African, *see* Coins, gold
Rare earth metals, effect on recrystallisation temperature of gold, 77, 293
Reaction bonding of gold and gold alloys, *see*
Bonding of gold to ceramics
Bonding of gold to glass
Casting gold alloys for bonding to dental porcelains
Evaporated gold coatings
Liquid golds
Sputtered gold coatings
Recrystallisation temperature of gold, effect of yttrium or rare earth metals upon, 77, 293
Red gold coatings, *see* Decorative gold alloy electrodeposits
Red golds, *see* Carat golds
Red gold solders, *see* Carat gold solders
Reduction potentials
effect of stability constants of complexes on, 197–204
standard reduction potentials, 201–204
standard reduction potentials for the formation of gold from its complexes, 203
Reed relays, *see* Electrical contacts
Refining of crude bullion
by electrodeposition, 29
by Miller chlorination process, 28
Reflection of radiation by gold and its alloys, *see* Optical properties of gold and gold alloys
Reflectors, *see* Heat and radiant energy reflectors, gold coatings for
Resistance, *see* Electrical resistivity of gold and gold alloys
Resistance thermometry
Au-Mn alloys for low temperatures, 118
gold films for gas temperature measurement, 117
gold films for surface temperature measurement, 130
Resistors, *see*
Film resistors
Wire resistors
Rhenium, use in
Carat golds, 35
Casting gold alloys for dentistry, 49
White golds, 99, 102
Rheumatoid arthritis, *see* Medical use of gold
Rhodium
grain refining action, 48, 99, 148
in liquid golds, 272–273
use for coating of white golds, 46
See also
Crucibles
Spinnerets
Rolled gold, *see* Mechanically clad gold coatings

Rolling of gold, *see* Malleability of gold
Rolling of gold alloys, *see*
  Hot working of gold alloys
  Work hardening of gold alloys
Rose gold coatings, *see* Decorative gold alloy electrodeposits
Ruby glass, *see* Glass
Ruthenium
  grain refining action in carat golds, 35, 48, 99, 100–101
  See also Casting gold alloys for dentistry

S

Safety pickle, 65
Sales of gold, official and private, *see*
  Demands for gold
  Supply of gold
Sea sediments, *see* Radioactive gold
Season cracking, *see* Stress corrosion cracking
Secondary current distribution in gold plating, *see* Growth of gold electrodeposits
Selective gold plating, 196, 230
Self-diffusion in gold, 180
Semiconductor devices
  doping of silicon devices for high speed switching devices, 124–125
  gold as a p-type impurity in silicon, 124
  gold in metal-oxide-silicon-field effect transistors (MOSFET's), 125
Shape memory, gold alloys with, 302–304
Shrinkage in cast gold alloys, *see* Investment casting
Silicon
  bonding of gold plated Kovar headers to silicon chips, 185
  effects on nickel white golds, 47, 77
  properties of Au-Si alloys, *see* Au-Si alloys
  use in carat gold casting alloys, 31, 41, 47
  See also
    Casting gold alloys for bonding to dental porcelains
    Semiconductor devices
Silver
  alloys with gold, *see* Au-Ag alloys
  as a brightener in electrodeposited golds, 222
  as an activator for auto catalytic plating of plastics, 284
  electrodeposition with gold, 221, 241–242, 247
  liquid silvers, 272
Skeletal foams, *see* Gold skeletal foams
Sliding contacts, *see* Electrical contacts
Slip rings, *see* Electrical contacts
Smuggling of gold, 24
Sodium aurothiomalate, *see* Medical use of gold
Solar cells, gold coatings as current collectors in, 123
Solar cells, thermoelectric, liquid gold coatings for, 159–161
Solar energy collectors
  gold coatings in, 156–161
  See also Gold blacks
Soldering
  of gold plated surfaces, 185–188
  of thick film gold alloy conductors, 278
  of thick film gold conductors, 188, 278
  See also
    Solders
    Solders, tin-lead
Soldering pastes, *see* Solders
Solders
  for soldering to gold plated surfaces, 185–188
  See also
    Bonding of gold to metals
    Carat gold solders
    Dental gold alloy solders
    Industrial gold brazing alloys
Solders, dental gold alloy, *see* Dental gold alloy solders
Solders, tin-lead, modification of for soldering to gold plated surfaces, 187–188
Solid solution hardening, 50
South Africa
  gold coins, *see* Coins, gold
  gold-producing areas, 6
  gold production, *see* Production of newly mined gold

Kruger rand, 20, 26–27
South African Mint, gold alloys supplied by, 32
Sovereigns, *see* Coins, gold
Soviet, *see* U.S.S.R.
Space exploration, 165
Spark plug electrodes, Au-Pd alloys in, 166
Specifications
for colours of decorative gold coatings, 77–78
for industrial gold brazing alloys, 136, 138–139
for thickness of decorative gold coatings, 230
Spectacle frames, gold filled, *see* Mechanically clad gold coatings
Speculation and investment in gold, *see*
Demands for gold
Supply of gold
Spinnerets, gold alloys for, 147–150
Spinning, gold alloys for, 41
Spinodal decomposition, *see* Precipitation hardening
Splat cooling, *see* Glassy gold alloys
Springs, gold alloys for, 41
Sputtered gold alloy coatings, *see* Gold alloy coatings, sputtered
Sputtered gold coatings, *see* Gold coatings, sputtered
Stability constants of gold complexes, *see* Complexes of gold
Staining of skin and clothes by gold alloys, 64–65
Stamping
gold alloys for, 41
white golds for, 46
See also Hot working
Standard resistances, *see*
Film resistors
Wire resistors
Standard resistors, *see* Wire resistors
Stereochemistry of gold complexes, *see* Complexes of gold
Strain gauges, ultra thin gold film resistors, *see* Film resistors
Stress corrosion cracking of gold alloys
dependence on gold content of alloy, 57
dependence on nature of reagent, 59–61
dependence on stress, 57–62
effect of degree of work hardening on, 57–58
effect of phase separation upon, 62–63
intergranular cracking, 57
mercury and low melting solders as a cause of, 64
nature of, 57–58
precautions against, 57, 62–63
resistance of carat gold alloys to, 62–64
transgranular cracking, 57–58
Structure of gold alloy electrodeposits, *see* Electrodeposition of gold alloys from cyanide baths
Submarine telephone cable repeaters, gold plating in, 169
Substitution brasses, low carat gold alloys as, 40
Sulphite electrolytes, *see*
Electrodeposition of gold alloys from sulphite baths
Electrodeposition of gold from sulphite baths
Sulphur, effects on carat gold alloys, 47, 76–77
Superconductivity of gold alloys, 125
Supply of gold
from communist bloc, 1–9
from official holdings, 1–9
from private holdings, 1–9
from scrap, 1
in relation to demand, 1–9
newly mined, *see* Production of newly mined gold
Surface temperature measurement, *see* Resistance thermometry
Switches, magnetic gold alloy cores for, 129–130

T

Tarnishing of gold alloys, *see*
Corrosion of gold alloys
Electrical contacts
Staining of skin and clothes by gold alloys
Tarnish resistance

Tarnish resistance
  Electrodeposited gold coatings, 120–123
  Mechanically clad gold coatings, 274–276
Temperature, surface, measurement, *see* Resistance thermometry
Temperature measurement, *see*
  Resistance thermometry
  Thermocouples
Terpenes, for production of gold terpene mercaptides, *see* Liquid golds
Thallium
  effects on rates of Au dissolution in alkaline cyanide baths, 244
  effects on throwing power of gold cyanide baths, 243
  grain refining and brightening effects on gold electrodeposits, 225, 244
Thermal fuses for electric furnaces, 167
Thermocompressive welding of gold, *see* Bonding of gold to metals
Thermocouples, gold alloys for
  Au-Co alloys for low temperatures, 119–120
  Au-Fe alloys for low temperatures, 119–120
  Au-Pd alloys for high temperatures, 118–119
Thermometry, low temperature, *see*
  Resistance thermometry
  Thermocouples, gold alloys for
Thermopiles, *see* Gold blacks
Thick film gold alloy conductors, 278–279
Thick film gold conductors, 278
Thick film gold alloy pastes, 276–278
Thick film gold pastes
  frit bonding, 276–277
  nature, 276
  reaction bonding, 101–104, 277
  See also
    Bonding of gold to ceramics
    Gold-epoxy conductors
Thick film processes, treatise on technology of, 279
Thin film gold alloy conductors, *see* Thin film processes
Thin film processes, *see*
  Gold alloy coatings, sputtered
  Gold coatings, evaporated
  Gold coatings, sputtered
Thiosulphate electrolytes, *see* Electrodeposition of gold from thiosulphate baths
Thiourea, in electropolishing baths, *see*
  Electropolishing of gold and gold alloys
  Stability constants for gold(I) complexes
  Image stabilisation with gold in photography
Tin
  alloys with gold, phase diagram, 145–146
  alloys with gold, use as solders, 145
  electrodeposition with gold, 244–245, 248
  formation of intermetallics with gold in soft soldering gold, 186
  in age hardening gold alloys, 55, 101
  in dental gold alloys, 51, 101–102
  in white golds, 49
  oxides of, in liquid golds, 272
Titanium
  as alternative to aluminium for metallisation of silicon chips, 293
  in age hardening gold alloys, 54
  in dental gold alloys, *see* Casting gold alloys for bonding to dental porcelains
Toxicity of gold compounds, *see*
  Biological chemistry of gold
  Medical use of gold
Tracer studies using gold, *see* Radioactive gold
Transmission of radiation by gold and its alloys, *see* Optical properties of gold and gold alloys
Transistors, *see* Semiconductor devices
Tutankhamun gold, purple films on, 49
Two-tier market, effect on price of gold, 4

U

Ultrasonic agitation of gold plating baths, 207, 219
Ultrasonic treatment, for removal of investment from castings, 72

Ultrasonic welding of gold, *see* Bonding of gold to metals
United States of America
  gold coins, *see* Coins, gold
  supply and demand for gold, *see*
    Demands for gold
    Supply of gold
U.S.S.R.
  gold producing areas, 7, 9
  gold production in, *see* Production, newly mined gold
  Russian chervonetz, *see* Coins, gold

V

Vacuum tubes
  gold coatings as conductors in, 123–124, 188
  See also Bonding of gold to glass
Vanadium, alloys with gold, *see* Wire resistors
Vanadium oxide
  in liquid golds, 272
  in thick film conductor pastes, 277
Violet gold, 49
  See also
  Purple plague
  Tutankhamun gold

W

Watchcase plating, standard colours in, 77–78
Wedding rings, 35
Welding of gold, *see* Bonding of gold to metals
Whiskers, *see*
  Growth of gold electrodeposits
  Submarine telephone repeaters
White gold coatings, *see* Decorative gold alloy electrodeposits
White golds
  nickel white golds
    annealing of, 47
    casting of, 46
    composition of, 44, 46
    effects of S and Si on, 47, 76–77
    electropolishing of, 71
    heat treatment of, 45
    machining of, 45–46
    magnetic properties of, 47
    metallographic studies of, 46
    pickling of, 65
    recovery from wastes, 47
    rhodium plating of, 46
    surface oxidation of on heating, 47
  noble metal white golds
    composition, 47–48
    properties, 48
  other white gold alloys, 49
White gold solders, *see*
  Carat gold solders
  White golds
Windscreens, electrically heated, *see*
  Bonding of gold to glass
  Glass, electrically heated gold coatings on
Wiper alloys, *see* Electrical contacts
Wire
  potentiometer, *see* Wire resistors
  resistances, *see* Wire resistors
Wire, aluminium, bonding to gold surfaces, 293
Wire conductors, *see* Wire, gold or gold alloy
Wire, gold or gold alloy
  dental gold alloy wire, 104–105
  gold bonding wire in electronics
    additions to increase recrystallisation temperature of, 293
    bonding to aluminium surfaces, 293
    production, 292
    specification for, 292
    See also
      Chain making, gold alloys for
      Wire resistors
Wire, gold plated Kovar
  bonding to copper printed circuitry tracks, 185
  bonding to thick film printed circuitry, 188
  in vacuum devices, *see* Vacuum tubes
Wire, grid, in vacuum tubes, 124
Wire resistors
  Au-Co alloys, 116
  Au-Cr alloys, 116
  Au-Fe alloys, 115
  Au-Fe-Pd alloys, 115

Au-Mo-Pd alloys, 116
Au-Pd alloys, 113–115
Au-Pd-V alloys, 116–117
Au-Pt alloys, 113
Au-V alloys, 117
Work hardening of gold alloys
18 carat Au-Ag-Cu alloys, 36–37, 50
14 carat Au-Ag-Cu alloys, 39, 50
effects on colours of gold alloys, 79
effects on grain growth during annealing, 36–37
effects on stress corrosion cracking of Au-Cu alloys, 58
See also Hot working of gold alloys

Y

Yellow gold alloys, *see* Carat golds
Yellow gold coatings, *see* Decorative gold alloy electrodeposits
Yellow gold solders, *see* Carat gold solders
Yttrium
effect on grain growth of a gold alloy, 46, 293
effect on recrystallisation temperature of gold, 77, 293

Z

Zinc
as a deoxidiser in carat golds, 35
effects on Au-Cu ordering, 35
effects on corrosion of carat golds, 40
effects on stress corrosion cracking of carat golds, 62
in 14 carat Au-Ag-Cu-Zn alloys, 40
in 10, 9 and 8 carat Au-Ag-Cu-Zn alloys, 40–41
in dental gold alloys, 97–106
in gold alloys with shape memory, 302–304
in nickel white golds, 41–47
in noble metal white golds, 47–48
loss from gold alloys in vacuum casting, 46
role in carat gold solders, 80–85
Zirconium, control of discontinuous precipitation with, 54